Science and Technology of Rapid Solidification and Processing

NATO ASI Series

Advanced Science Institutes Series

A Series presenting the results of activities sponsored by the NATO Science Committee, which aims at the dissemination of advanced scientific and technological knowledge, with a view to strengthening links between scientific communities.

The Series is published by an international board of publishers in conjunction with the NATO Scientific Affairs Division

A Life Sciences **B Physics**	Plenum Publishing Corporation London and New York
C Mathematical and Physical Sciences **D Behavioural and Social Sciences** **E Applied Sciences**	Kluwer Academic Publishers Dordrecht, Boston and London
F Computer and Systems Sciences **G Ecological Sciences** **H Cell Biology** **I Global Environmental Change**	Springer-Verlag Berlin, Heidelberg, New York, London, Paris and Tokyo

PARTNERSHIP SUB-SERIES

1. Disarmament Technologies	Kluwer Academic Publishers
2. Environment	Springer-Verlag
3. High Technology	Kluwer Academic Publishers
4. Science and Technology Policy	Kluwer Academic Publishers
5. Computer Networking	Kluwer Academic Publishers

The Partnership Sub-Series incorporates activities undertaken in collaboration with NATO's Cooperation Partners, the countries of the CIS and Central and Eastern Europe, in Priority Areas of concern to those countries.

NATO-PCO-DATA BASE

The electronic index to the NATO ASI Series provides full bibliographical references (with keywords and/or abstracts) to more than 30000 contributions from international scientists published in all sections of the NATO ASI Series.
Access to the NATO-PCO-DATA BASE is possible in two ways:

– via online FILE 128 (NATO-PCO-DATA BASE) hosted by ESRIN, Via Galileo Galilei, I-00044 Frascati, Italy.

– via CD-ROM "NATO-PCO-DATA BASE" with user-friendly retrieval software in English, French and German (© WTV GmbH and DATAWARE Technologies Inc. 1989).

The CD-ROM can be ordered through any member of the Board of Publishers or through NATO-PCO, Overijse, Belgium.

Series E: Applied Sciences - Vol. 278

Science and Technology of Rapid Solidification and Processing

edited by

Monde A. Otooni
Materials and Aeroballistic Technology Division,
Armament Engineering Directorate,
U.S. Army Armament Research, Development and Engineering Center (ARDEC),
Picatinny Arsenal, New Jersey, U.S.A.

Springer Science+Business Media, B.V.

Proceedings of the NATO Advanced Research Workshop on
Science and Technology of Rapid Solidification and Processing
West Point Military Academy, New York, NY, U.S.A.
June 21–24, 1994

A C.I.P. Catalogue record for this book is available from the Library of Congress.

DOI 10.1007/978-94-011-0223-0

Printed on acid-free paper

TABLE OF CONTENTS

PREFACE

This book represents a collection of papers presented at the NATO Advanced Research Workshop (NATO/ARW) on "Science and Technology of Rapid Solidification and Processing", held at Hotel Thayer, West Point Military Academy, New York, N.Y., during June 21-24, 1994. The workshop was attended by over forty scientists representing several NATO member countries as well as representatives from Japan, China (PRC), Taiwan and India.

The purpose of this NATO/ARW conference was to review the major advances made in most recent years in both the theoretical and experimental areas of rapid solidification technology and processing. In accordance with the NATO/ARW format, the agenda for the conference was so arranged to offer in depth presentation of the latest developments in the subject area as well as to encourage follow-up discussions by the participants. There was seven sessions each opened with a lecture by an invited guest speaker. Sessions 1-4, covered two days of the conference and focused mainly on Processing Technologies of Rapid Solidification and Thermodynamic Properties (Practical Applications). Sessions 4-6 concentrated on Thermodynamics of Metastable Alloys, Relaxation, Diffusion, Magnetic and Electric Properties (Fundamentals). Session 6 was devoted to the Structural Characterization of Supercooled Melts, Ultra Fine Polycrystalline Materials (New Innovations and Techniques).

There were two equally important aspects of this NATO/ARW conference which must be mentioned. Firstly, this is the first NATO/ARW conference on Science and Technology of Rapid Solidification and Processing held in the United States. It was particularly appropriate that this conference was held at the Military Academy, emphasizing as it does the bringing together scientists from the United States and those from other countries. Secondly, for the first time young scientists (Graduate students or Postdoctoral fellows) were invited to participate with the hope that this tradition will continue in the future and will prepare the younger generation of scientists and engineers for a better appreciation of tomorrow's need for scientific cooperation among all nations. The sponsorship of this conference by the NATO Scientific Affairs Division presented just such an opportunity and it is gratefully acknowledged.

I am most thankful to Professors N. Grant (MIT), H. Jones (Sheffield U), and Prof. T. Egami (U. of P) for their support and encouragement in organizing this NATO/ARW conference. An international meeting such as this takes a considerable effort in planning and organization. It is indeed a pleasure for me to acknowledge the help and advice I received from Dr. Surya N. Bulusu for giving me the benefit of his past experience in organizing a NATO sponsored conference. I am very much indebted to my wife, Madame Beatrice Otooni, for her numerous suggestions in developing the program and planning the logistics of the conference. In addition to the main lecturers it was an honor to have Professor Morris Cohen (MIT), Dr Lance Davis (DOD) and Dr R. Damadian (Fonar Cor.) as guest speakers. I express my deep gratitude to these individuals for their excellent presentations.

Equally important was the generous help I received from Dr Xiaoman Duan, Ms Elaine Tirrelle, and Dr I. Ücok (all from MIT) and Mrs I. Clements (U. P). Of course, I will be always grateful to Mrs Nel de Boer of Kluwer Academic Publishers who so diligently kept me informed of all details regarding timely submission of manuscripts. My many thanks to the managing staffs at the Hotel Thayer who made every effort to make smooth running of the conference possible. The meeting was highly successful and enjoyable due to diligent efforts of all participants giving a strong indication of goodwill and trust which comprise the foundation of human spirit among all mankind.

Monde A. Otooni

PROCEEEDING OF THE NATO ADVANCED WORKSHOP ON SCIENCE AND TECHNOLOGY OF RAPID SOLIDIFICATION AND PROCESSING

Held at: Hotel Thayer, U.S. Military Academy
New York, USA
June 21-24, 1994

INTERNATIONAL ADVISORY COMMITTEE

Dr. M. A. Otooni (Director) USA

Prof. N. J. Grant	(MIT)	USA
Prof. H. Jones	(Sheffield)	U.K
Prof. H. Kronmuller	(Max-Planck)	Germany
Prof. T. Egami	(U of P)	USA
Prof. W. Johnson	(Caltec)	USA
Prof. B. Cantor	(Oxford)	U.K
Prof. R. O'Handley	(MIT)	USA
Prof. C. Wagner	(U. Cal)	USA

CONFERENCE COMMITTEE

Prof. N. J. Grant
Dr. M. A. Otooni
Prof. T. Egami
Prof. W. Johnson
Prof. H. Jones
Prof. T. Ando (B. U)
Dr. L. Davis (DOD)
Dr. X. Duan (MIT)
Prof. B. Cantor
Prof. H. Kronmuller

LOCAL PLANNING COMMITTEE

Prof. N. J. Grant
Dr. M. A. Otooni
Prof. T. Egami
Prof. R. O'Handley
Prof. T. Ando

Cosponsor of the NATO/ARW Conference

ALCOA Corporation of America

GRAIN BOUNDARY DEFORMATION AND FRACTURE OF A FINE GRAINED, HIGH PURITY AL-2% MG ALLOY AT 150° C (423K)

H. C. CHANG* AND NICHOLAS J. GRANT**

* *University of Science and Technology Beijing, People's Republic of China*

** *Massachusetts Institute of Technology Cambridge, MA 02139-4307, USA*

ABSTRACT. Intercrystalline deformation and fracture at elevated temperatures are well established phenomena, and are relatively well defined in terms of temperature, stress and grain size. Grain boundary sliding, leading to intercrystalline fracture, is usually considered to be an embrittling behavior, and worsens with decreasing strain rate, decreasing grain size and increasing temperature. Very coarse grained metals and alloys avoid intercrystalline failures even at very high temperatures by undergoing slip and slip band deformation, even up to the melting temperature. Based on the rule that the slip band spacing is inversely proportional to the stress, coarse grained structures can accommodate coarse slip band spacings to very high temperatures. Following this reasoning, a high purity Al-2% Mg alloy, prepared with a grain size finer than about 10 mm, undergoes grain boundary sliding, minor grain boundary migration, and intercrystalline cracking at 150° C (423K). The extent of grain boundary sliding and cracking are a function of the grain size and the stress.

1. Introduction

The importance of the influence of grain size on the mechanical properties and creep behavior of polycrystalline materials cannot be overemphasized. Because crystalline materials are composed of grains and grain boundaries, the deformation and fracture behavior of materials is naturally related to the individual and combined patterns of both structural entities. Grain size, grain shape and grain and grain boundary geometry will affect the mechanical behavior patterns by undergoing different modes and amounts of deformation and fracture over a range of stresses and temperatures. Our understanding of grain size effects and grain boundary behavior has been greatly enhanced by significant experimental and theoretical work in the past.[1-14] Because mechanical behavior depends strongly on temperature and stress, deformation and fracture patterns are usually labeled as high temperature and low temperature types.

For "high temperature behavior" it is routinely accepted that grains boundaries are weaker than grains leading to the concept of an equicohesive temperature[1], a temperature at a specific strain rate, wherein fractures may be crystalline or intercrystalline or both. All metallic alloys will show the existence of a transition deformation and fracture behavior when creep and stress rupture properties are plotted over a significant stress and temperature range.[2,3,6,8,9] Appropriately, microscopic and macroscopic grain boundary

M. A. Otooni (ed.), Science and Technology of Rapid Solidification and Processing, 1–11.

sliding were observed and extensively measured over a wide range of temperatures and strain rates [15-20], leaving no doubt that grain boundary sliding (gbs) was an extremely important "high temperature" deformation process. The influence on tensile, stress rupture, creep strength and ductility were recorded for many metals and alloys and found to be predictably responsive to temperature and stress[1-22]. This led to the development of new creep and creep-rupture theories and was particularly important in understanding the issue of ductility under conditions which lead to intercrystalline cracking.[6,20,22,23] The grain boundaries act as strengthening agents, because they block slip from easy transmission from grain to grain at "low" temperatures, and act as weakening agents because they undergo easy sliding under conditions involving higher temperatures, lower stresses and finer grain sizes.

During granular deformation, the development of fine slip leads to the formation of slip bands which gradually develop inhomogeneously with increasing plastic strain, resulting in a minimum slip band spacing at saturation, responding to the stress as it is influenced by temperature and the amount of strain.[3,4,9,24]

In the early forties, Orowan[24] proposed a theory of the inverse relationship between stress and the slip band spacing; this relationship was confirmed as being independent of temperature.[3,24] Many years later, Servi and Grant's extensive data on high purity Al provided the confirming experimental evidence to prove the validity of Orowan's theory as shown in Fig. 1.[2,3]

In the "low temperature behavior" region, it was found that the effect of grain size on strength generally followed the Hall-Petch relationship. Most theories were developed on the experimental fact that grain boundaries impede crystalline slip transfer from one grain to another; and the models which were proposed were, in essence, of the mantle and core type[26-28]. In contrast to high temperature behavior, only grain boundary blocking effects were considered, and grain boundary sliding was not considered as a factor at low temperatures, at least not explicitly, though grain boundary sliding had been reported at low temperatures.[29,30]

Practically all of our understanding of the hot deformation of polycrystalline pure materials and simple high purity alloys is based on experimental work with relatively coarse grained materials, significantly larger than 100 μm and up to 2 mm or more. This is understandable because coarse grains are much more stable than fine grains, and it is much easier to differentiate the behavior of grains and grain boundaries by using coarse grained specimens. We know that fine grained materials have better mechanical properties at low temperatures, and instinctively strive to produce fine grained materials. Although columnar, oriented, coarse grained and single crystal turbine blades have been extremely successful for very high temperature applications, most materials, particularly large structures intended for high temperature use, because of difficult processing and cost considerations, are generally "fine grained".

Coarse grained materials are less ductile at room and low temperatures, whereas the ductility of fine grained materials is superior and more reproducible at both high and low temperatures.[2] Great interest has been generated in fine and ultra-fine grained materials with the growth of rapid solidification technology and makes it particularly important to predict the mechanical behavior of such materials at both very low and high temperatures.

With this in mind we initiated a program to study the deformation behavior of grain boundaries in very fine grained materials, particularly the behavior with decreasing, low temperatures. We have found in high purity aluminum (99.995%) that it was extremely difficult to produce fully recrystallized, equiaxed grain sizes finer than about 15 to 20 mm.[2] In this study, we present the results of an investigation of the creep deformation of a "fine grained," high purity, solid solution Al-2% Mg alloy at 150° C (423K). The stability of this single phase alloy and its resistance to grain coarsening are greatly superior to that of high purity aluminum for the above test conditions. This presentation deals with creep transitions and slip band spacing; a second paper will discuss deformation and fracture characteristics and ductility.

2. Experimental

Specimen Preparation The mechanical working and annealing sequences were chosen to obtain very fine, equiaxed grain structures. High purity Al-2% Mg was selected based on prior experience with this alloy[8,9] and was supplied by the Alcoa Research Laboratory as chilled cast rods. The chemical analysis of this alloy is shown in Table I.

Table 1 Chemical Analysis of High Purity Aluminum -2% Mg

Al	Mg	Ti*	B*	Si, Fe, Cu, Mn, Cr, Ni, Zn, Be
bal.	2.01%	0.03%	0.005%	each less than 0.00%

*Added for grain refinement and stabilization.

Production of very fine grained specimens required many variables to be closely controlled, such as the amount of prior cold deformation, the temperature of hot upsetting and rolling, the initial grain size before cold working, the amount of room temperature rolling and cross-rolling and pressing, the amount of reduction at liquid nitrogen temperatures, and the final annealing temperatures and times.

After a great deal of experimentation it became evident that to get fairly uniform, fully recrystallized grain sizes finer than about 10 μm, in sheet form (about 2 mm thick), was very difficult. Not unexpectedly, the most important variable was found to be the amount of cold rolling reduction, especially the amount of reduction in the final pass.

A typical grain structure is shown in Fig. 2, which was taken from the enlarged end of test specimen No. 2 after a small amount of strain at 150° C (423K) to help outline the grain structure more clearly. Measurements showed that most grains were in a range of 6-10 μm, with some grains between 10 and 20 μm.

Creep Testing Specimens were cut parallel to the final rolling direction from the rolled strips, were machined, annealed and electropolished. To facilitate observations of grain boundary sliding, shallow scratches were made on one of the polished surfaces of each specimen by directional rubbing with very fine diamond paste. Scratches, parallel anc perpendicular to the length of the specimen, were made on different parts of the specimer surface. Tests were run at 150° C (423K) at a number of stresses. Test data are shown ir Table 2. After creep deformation, observations were made of grain boundary sliding, anc the average slip band spacing was measured on all specimens.

Table 2 Creep Tests of High Purity Al-2% Mg Alloy at 150° C (423° K)

Specimen No.	Stress psi (MPa)	Test Time	Elongation %	Remarks
1	9,540 (66.8)	240 hr.	13	unbroken
2	12,140 (85)	89 hr.	56	broken
3	14,000 (98)	10 hr.	60	unbroken
4	16,000 (112) 19,000 (133)	35 min. plus 25 min.	9.4 plus 39.6	broken

3. Results and Discussion

Slip Band Spacing The details of the deformation characteristics in this work will b described in a follow-up paper. For this section it suffices to note that the coarsest grain among the four tested specimens (Table 2), with dimensions greater than about 10 µn (twice the observed slip band spacing for this alloy at these stresses), developed norma slip bands and the predicted slip band spacings. Average slip band spacings wer measured and stresses were calculated based on the initial loads after taking int consideration the amount of creep strain. These data are plotted as x's on the plot of lo stress versus log average slip band spacing in Fig. 1. Also shown in Fig. 1 are similar dat from previous publications dealing with similar creep deformation studies, confirming th new data.

The reference curve is that based on the Orowan theory[24] (lower left) and i critical in that it predicts the 45° slope for a plot of log stress versus log slip band spacin (at saturation).

The dashed curve is a plot of data from Yamaguchi[25], the first effort to test th Orowan prediction successfully with experimental test data, using commercially pur aluminum (99.2%). Tests covered a stress range of only one order of magnitude. Th long solid curve is a plot of the data of Servi and Grant[2,3] for creep rupture tests of hig purity (99.995%) aluminum, covering a stress range of more than two orders of magnitud and a temperature test range from 93° - 593° C (366-866 K). Solid dots are data fror Gemmell and Grant's work on creep rupture tests of a high purity Al-1.92% Mg alloy a

260° C (533 K)(9), a considerably stronger, more stable alloy than the high purity aluminum used by Servi and Grant.

It can be seen that the data from the present work can be extended in line with the data from the previous referenced studies. As was predicted at the start of these studies, by reducing the grain size as much as possible and running creep rupture tests at progressively lower temperatures, the validity of Orowan's relationship can be extended to considerably higher stresses (from prior levels of 10,000 psi (70 MPa) now to 25,000 psi (175 MPa).

4. Transition Behavior

Fig. 3 shows a log stress versus log rupture time plot for the high purity Al-2% Mg alloy. Curves at 260 to 482° C (533-755° K) present creep rupture data for a relatively coarse grained (0.9 mm) high purity Al-1.92% Mg alloy(8); the crosses (x) are the data of this work for the fine grained Al-2% Mg alloy (largely 6 - 10 μm) at 150° C (423K). Point (A) on the 260° C (533K) curve(8) represents the transition from low temperature type crystalline deformation and fracture at the higher stresses to the initiation of high temperature grain boundary sliding and migration at the lower stresses. At the lower stresses beyond point A, grain boundary sliding takes place along with some intercrystalline cracking; however, the final fractures were largely transcrystalline, in part because the early deformation reduces the cross sectional area and thereby increases the stress and strain rate commensurately, promoting transcrystalline fracture.

In this work, the Al-2% Mg alloy had a very much finer grain size (approximately 6-10 μm, see Fig. 2) and the test temperature of 150° C (423 K) was also sharply lower. In contrast to the data for the coarser grained specimens(8), this work, with the very much finer grain size, showed extensive grain boundary sliding and intercrystalline cracking at all stresses used at this much lower temperature of 150° C (423 K), even at the highest stresses which were used. However, of importance, while there was extensive grain boundary sliding, there were no obvious signs of grain boundary migration. Fig. 4 is a micrograph taken from the lower strain rate test specimen, which shows extensive grain boundary sliding and intercrystalline cracking. The details of this grain boundary behavior will be presented in greater detail in a follow-up paper.

The data points shown as crosses in Fig. 3 represent "high temperature type" intercrystalline deformation and cracking behavior. This work demonstrates, as expected, that the grain size has a profound effect on the transition of crystalline to intercrystalline deformation and fracture. It is also expected that if the grain size can be further reduced, grain boundary activity will occur at much higher stresses and lower temperatures, and the trans-to-intercrystalline transition will extend proportionately to higher stresses and lower temperatures.

5. Conclusions

The following conclusions can be drawn from creep rupture tests of a high purity, fine grained, Al-2% Mg alloy at 150° C.

1. Orowan's stress versus average slip hand spacing relationship can be extendec from a previously noted stress of 10,000 psi (70 MPa) to 25,000 psi (175 MPa), Fig. 1 and will therefore occur at lower temperatures for still finer grain sizes.

2. This transition of granular to intergranular behavior was extended from a stres: of 5,000 psi (35 MPa) and a temperature 260° C to a stress 19,000 psi (133 MPa) and a temperature of 150° C (Fig. 3).

3. It is expected that with still finer grain sizes, for a high purity Al-2% Mg alloy grain boundary sliding and intercrystalline cracking will be observed at still lowe temperatures. This is not to be interpreted as a weakening or embrittling phenomenon fo this class of alloy, as demonstrated in the current study.

Acknowledgements

The authors wish to express their appreciation to Dr. John Benjamin of the Alcoa Researcl Labs for providing the Al-Mg alloy; to Dr. Teiichi Ando and Dr. Monde Otooni for thei helpful discussions; and to Dr. Xiaoman Duan, Dr. Ibrahim Ucok, Dr. Satoru Matsuo anc Dr. Ralph Mason for their valuable assistance.

References

1. D. Hansen and M. A. Wheeler: *Jour. Inst. of Metals* 45 (1931) 229.

2. I. S. Servi and N. J. Grant: *J. of Metals* 3 10, (Oct. 1951) 909.

3. I. S. Servi and N. J. Grant: *J. of Metals* 3 10, (Oct. 1951) 917.

4. D. McLean: *Jour. Inst. of Metals* 80 (1951-52) 507.

5. H. C. Chang and N. J. Grant: *Trans. AIME* 194 (1952) 619.

6. N. J. Grant and A. G. Bucklin: *Trans. ASM* 45 (1953) 151.

7. H. C. Chang and N. J. Grant: *Jour. Inst. of Metals* 82 (1952-53) 229.

8. A. W. Mullendore and N. J. Grant: *Trans. AIME* 200 (1954) 973.

9. G. D. Gemmell and N J. Grant: *J. of Metals* 9, No. 4, (April 1957) 417.

10. V. Ramachandran and E. P. Abrahamson II: *Scripta Met.* 6 (1972) 287.

11. N. Hanson: *Acta Met.* 25 (1977) 863.

12. J. W. Wyrzykowski and M. W. Grabski: *Mat. Sci. and Eng.* 56 (1982) 869.

13. J. T. Al-Haidary, N. J. Petch and E. R. DeLosRios: *Phil Mag*. A47 (1983) 869.

14. A. Lasalmonie and J. L. Strudel: *J. Mat. Sci.* 21 (1986) 1837.

15. R. King, R. Cahn and B. Chalmers: *Nature* 161 (1948) 682.

16. D. McLean: *Jour. Inst. of Metals* 81 (1952-53) 293.

17. H. C. Chang and N. J. Grant: *Trans. AIME* 197 (1953) 305.

18. H. C. Chang and N. J. Grant: *Trans. AIME* 206 (1956) 169.

19. H. Brunner and N. J. Grant: *Trans. AIME* 218 (1960) 122.

20. A. W. Mullendore and N. J. Grant: *Trans. AIME* 227 (1963) 319.

21. G. Rai and N. J. Grant: *Trans. AIME* 14A (1983) 1451.

22. S. J. Hales and T. R. NcNeeley: *Acta Met.* 36 (1988) 1229.

23. W. D. Nix and J. C. Gibeling: *1983 ASM Material Sci. Seminar - Flow and Fracture at Elevated Temperatures,* edited by R. Raj, pp 1-64.

24. E. Orowan: *Nature* 147 (1941) 452.

25. K. Yamaguchi, *Sci. Papers Inst. Phys. and Chem. Res., Tokyo* 8 (1928) 289.

26. M. F. Ashby: *Phil. Mag.* 21 (1970) 339.

27. U. F. Kocks: *Met Trans.* 1 (1970) 1121.

28. A. W. Thompson, M. I. Baskes and W. F. Flanagan: *Acta Met.* 21 (1973) 1019.

29. F. B. Cuff, Jr. and N. J. Grant: *Trans. AIME* 212 (1958) 355.

30. G. Y. Chin, W. F. Hosford, Jr. and W. A. Backofen: *Trans AIME* 230 (1964) 437.

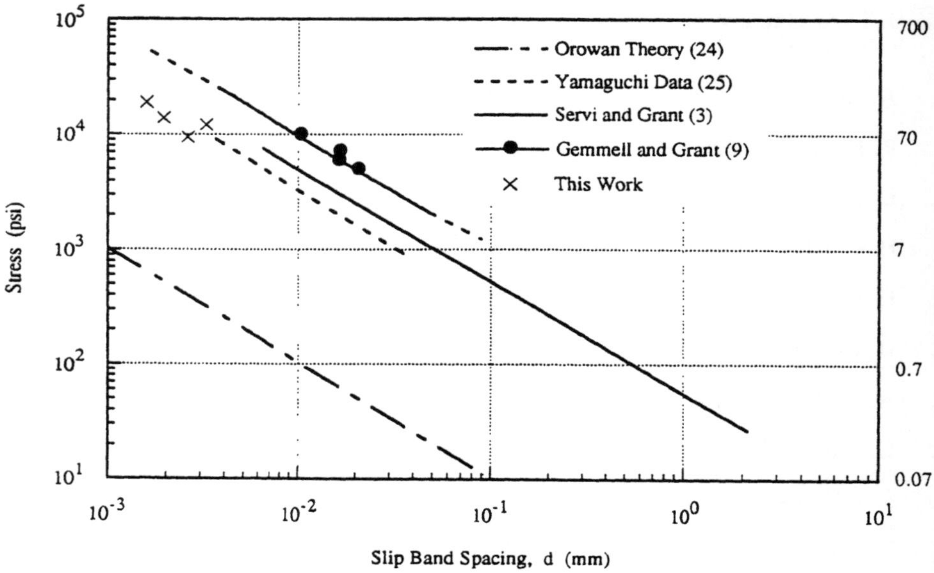

Fig. 1 Log stress vs. log slip band spacing at high strains.
Yamaguchi: pure aluminum; tested 25° to 230°C.
Servi and Grant: high purity aluminum, coarse grained, tested 100° to 590°C.
Gemmell and Grant: Al+2%Mg, 25°C to 250°C.
This work: Al+2%Mg, 6-10 μm grain size, 150°C.

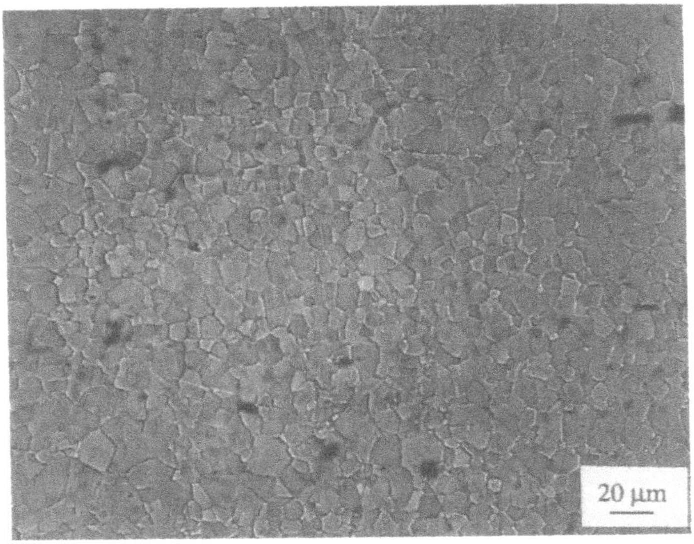

Fig. 2 Microstructure of specimen No. 2, creep deformed at 150°C (423K), showing grain boundary shear and intercrystalline cracking. Average grain size is 6-10 µm. See Table 2.

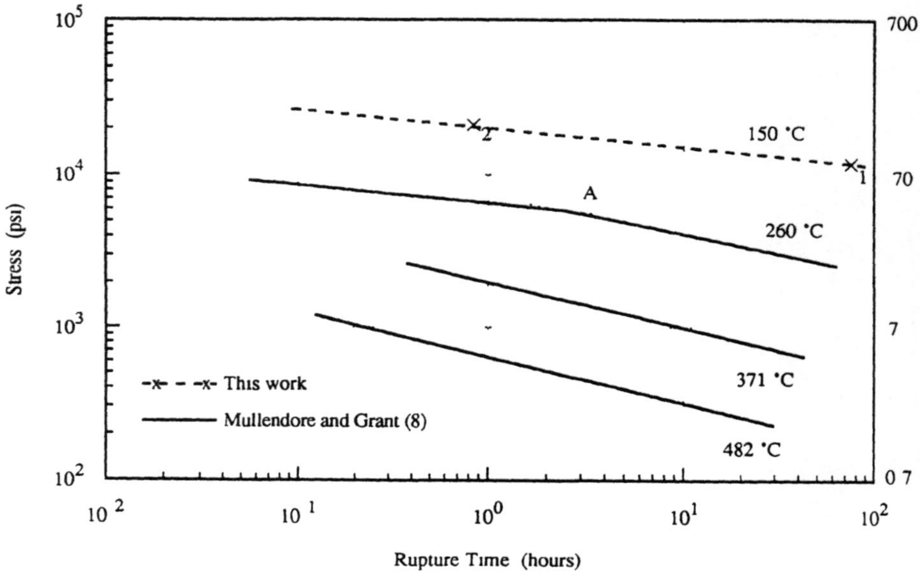

Fig. 3 Log stress vs. log rupture time for high purity Al-2% Mg alloy. Mullendore and Grant's data are for coarse grained (0.9 mm) specimens. Fractures at 260° (right of point A), 371° and 482°C are intercrystalline; to the left of point A fractures are transcrystalline. At 150°C, grain size of 6-10 μm, fractures are intercrystalline.

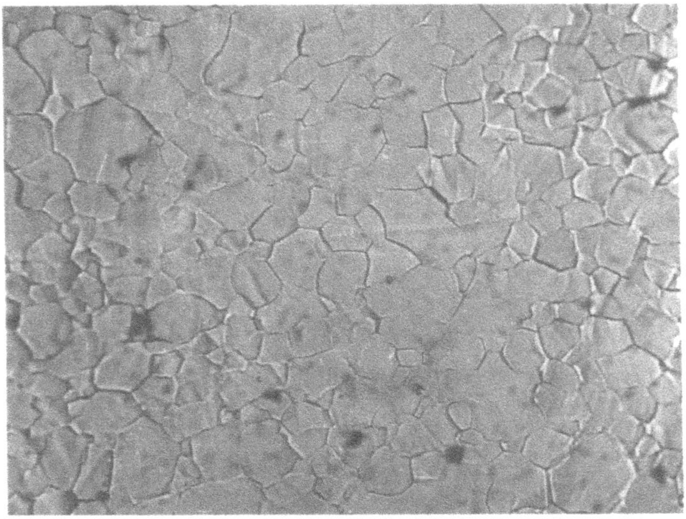

Fig. 4 Micrograph from specimen X_2 (see Fig. 3) after a creep strain of 30% for a high purity Al-2% Mg alloy. Note significant grain boundary sliding, intercrystalline cracking, and slip bands in the coarser grains.

GROWTH PARAMETERS IN FORMATION AND STABILITY OF RAPIDLY SOLIDIFIED MICROSTRUCTURE

H JONES
Department of Engineering Materials
University of Sheffield
Mappin Street
Sheffield, S1 4DU, UK

ABSTRACT. Some recent progress in experimental determination and model prediction of microstructure formation and stability in rapidly solidified materials is surveyed and its significance for engineering applications is discussed.

1. Introduction

The foundations for kinetic modelling of nonequilibrium constitution and microstructure formation in rapidly solidified materials were laid down two decades ago by Turnbull [1] and Uhlmann [2] for the particular case of glass formation. They applied classical nucleation and growth theory to predict the conditions of cooling during solidification under which fraction crystallized during cooling from the liquidus to the glass temperature would be insignificant, so allowing a glass to form. The assumptions of homogeneous nucleation and continuous growth involved were subsequently applied [3] to prediction of grain size of fully crystalline pure metal melt spun ribbons. The application of such approaches to prediction of the much more complex multiconstituent microstructures that are more typical of rapid solidification is a much more recent phenomenon. A number of concurrent developments which have made this possible include:

- extension of the theory of dendritic and eutectic growth to rapid solidification conditions.

- use of containerless processing and related methods that allow nucleation to be triggered at specific levels of high undercooling prior to rapid solidification.

- use of high velocity Bridgman, TIG weld, laser or electron beam surface traverse melting techniques which allow coupling between the advancing solidification front and the imposed velocity of withdrawal or traversing.

M. A. Otooni (ed.), Science and Technology of Rapid Solidification and Processing, 13–24.

These developments have been well documented in a series of recent review articles e.g. [4-7]. The present purpose is to review some recent contributions with which the author has been associated either directly or indirectly. These include determination and interpretation of:

- microstructure selection diagrams in Al and Zn based systems.
- growth temperature, morphology and spacing of constituents versus growth conditions.
- thermal stability of duplex microstructures in Al and Mg based alloys.

2. Determination and Interpretation of Solidification Microstructure Selection Diagrams (SMSDs)

Fig 1a [8] compares experimental fact with predictions for the most Al-rich part of the Al-Fe binary system modifications of which form the basis of several competing alloy chemistries for service at temperatures up to 450°C. Predictions indicate the growth velocity V at which a eutectic and primary phase or two eutectics grow at the same temperature. Such predictions require equations for growth of the competing constituents either based on a theoretical model or determined experimentally. A complete first principles prediction of such an SMSD requires knowledge of [9]:

- the identify of the contending constituents.
- their thermodynamic characteristics, such as formation temperature as a function of alloy concentration C_O including nonequilibrium extensions beyond limiting compositions embodying any effects of solute trapping at high V.
- constitutive equations for growth of the constituents.

Additional requirements when the selection process is not occurring under conditions of steady state growth with all contending constituents already present at the outset, are:

- constitutive equations for nucleation of the constituents.
- the position and time dependence of temperature and/or front velocity in the transforming system and their relationship to external variables such as imposed cooling or heat dissipitation rate.

Partial SMSDs of the type shown in Fig 1a are now available for some twenty systems (see [10] for a listing made in 1990: additional SMSDs have since become available for Zn-Mg [11] (Fig 1b), Zn-Al [12], Al-Cu [13] and Co-Cr_7C_3 [14] with extension of earlier [15] results for Al-Fe to higher Fe contents [16]). Predictions made for the Al-Fe-Si and Al-Cu-Si are in qualitative agreement with preliminary experimental data for Al-Fe-Si [17], differences being at least partly attributable to uncertainties in data from the phase diagram. The predictions for Co-Cr_7C_3, as for dilute Al-Fe alloy concentrations (Fig 1a), are in reasonable accord with the experimental observations.

3. Growth Temperature, Morphology and Spacing of Constituents versus Growth Conditions

Although crucial to the verification of growth models and correct prediction of SMSDs, rather few measurements of growth temperature, or associated solid phase composition C_S*, have been reported. The measurements made for eutectics fit the predicted relation [18].

$$T_{EU} - T_G = K_1 V^{1/2} \qquad (1)$$

where T_{EU} is eutectic temperature and T_G is growth temperature with $0.023 < K_1 < 0.26$ $Ks^{1/2}/\mu m^{1/2}$ for typical systems and $10^{-3} < V < 10$ mm/s [19]. Significant departures from this parabolic relationship are predicted [20,21] for values of V approaching the stability limits for high velocity eutectic growth. Our measurements for the αAl-Al_3Fe and αAl-Al_6Fe eutectics extrapolate to $T_{EU} = 655.1 \pm 0.1$°C and 652.9 ± 0.2°C respectively with $K_1 = 0.41 \pm 0.03$ and 0.20 ± 0.01 $Ks^{1/2}/\mu m^{1/2}$ respectively for Bridgman growth of Al-3 wt%Fe over the range $10 < V < 340$ μm/s [22]. Corresponding results [23,24] for primary Al_3Fe dendrites in Al-3.5 to 14 wt%Fe alloys over the range $10 < V < 1000$ μm/s and temperature gradient G ~ 10 K/mm (Figs 2a,b) fit the constitutive equation [24].

$$T_L - T_G = K_2 V^n C_o^{-m} \qquad (2)$$

where T_L is liquidus temperature, T_G is growth temperature, C_o is alloy concentration with $K_2 = 34 \pm 3$ K $(\mu m/s)^{-n}$ $(wt\%)^m$, $n = 0.36 \pm 0.02$ and $m = 1.23 \pm 0.04$. Measured T_L-T_G agrees with predictions for needle-like dendritic growth within a factor of 2, even though the observed final morphology of Al_3Fe is plate-like for which very much larger values of T_L-T_G are predicted. This apparent contradiction was resolved by a combination of deep etching of longitudinal sections of samples and polishing back from the Al_3Fe growth front [25]. This revealed that the tip of Al_3Fe dendrites is, indeed needle-like, for these conditions, the plate-like final morphology being developed by branching behind the needle-like dendrite tip. Corresponding studies [26] for primary silicon in the commercially very important Al-Si system indicate a predominant morphology of clustered polyhedral particles linked along their edges, with formation temperature relatively independent of V over the range 60 to 700 μm/s and $10 < G < 40$ K/mm for Al-18 wt%Si. This is consistent with control of formation temperature by heterogeneous nucleation with a contact angle near to 30 degrees. A similar result was obtained by Ho and Cantor [27] for formation of primary Si in Al-3 wt%Si at different cooling rates, also attributed to control by heterogeneous nucleation.

Unfortunately no direct measurements of growth or formation temperature have been reported for the higher values of V and G that are characteristic, for example, of laser or electron beam melt traversing. Measurements of cell or dendrite spacing have been reported, however, as a function of V for such extreme conditions [28-30], fitting the relationship:

$$\lambda = K_3 V^{-n} C_o^{-m} \tag{3}$$

with $n = ½$ and, for αAl in Al-Fe alloys with $0.5 < C_o < 8$ wt%Fe, $m = 0.67$ and $K_3 = 54 \pm 4\ \mu m^{3/2}\ s^{-1/2}\ (wt\%)^{2/3}$ alloys. These results are in good accord with predictions of an analytical model [31] for which marginal stability at the tip is assumed to determine the growth condition with the spacing assumed to be related to the tip radius by a scaling factor ~2. A recent numerical model which treats the growth of an array of cells or dendrites and does not require the operating condition at the tip to be specified also makes predictions of λ in excellent agreement with available measurements (Fig 2a) [32]. This new model also enables spacings and growth temperatures of cells and dendrites to be predicted separately for a given condition (Figs 3a,b) and thus predicts transitions from cellular to dendritic and then back to cellular growth with increasing V, as observed experimentally.

Measurements [33] of primary spacing for Al_3Fe dendrites in hypereutectic Al-Fe alloys is consistent with the predicted relationship [34-36]:

$$\lambda_1 = K_4\, G^{-1/2}\, V^{-1/4} \tag{4}$$

with $K_4 = 43 \pm 8\ \mu m^{3/4}\ K^{1/2}\ s^{-1/2}$ for $9.5 < C_o < 20.0$ wt%Fe, $0.01 < V < 5$ mm/s and $G =$ 10K/mm. A marked decrease in $\lambda_1 G^{1/2}$ near the condition for extinction of growth of primary Al_3Fe by growth of αAl-Al_xFe eutectic was attributed to the decrease in $T_G(Al_3Fe) - T_G(EU)$ towards zero as this condition is approached. Analysis [37] shows that equation (4) is widely obeyed within the regime of dendritic growth and it appears also to be obeyed (Fig 4a) for primary silicon in hypereutectic Al-Si alloys for a wide range of conditions [38,39]. These latter results combined with earlier published results also fit the empirical relationship with cooling rate $\dot{T}$ [39]:

$$\lambda_1 = K_5\, \dot{T}^{-1/3} \tag{5}$$

with $K_5 = 250 \pm 30\ \mu m\ (K/s)^{1/3}$ for $10^{-2} < \dot{T} < 10^6$ K/s, where $\dot{T} = GV$, as shown in Fig 4b.

4. Thermal Stability of Duplex Microstructures in RS Al and Mg Based Alloys

Some level of thermal stability is a requirement of all products of RS processing both to meet service performance requirements and to survive or benefit from thermal excursions necessarily involved in product fabrication following the rapid solidification stage of processing. The novel microduplex high performance RS Al and Mg base alloys [40] that depend on a submicron dispersion of aluminide or silicide for their thermal stability are good examples. These dispersoids both strengthen the matrix grains and prevent matrix grain coarsening during processing and in service. A key feature is the stability of the dispersoid to coarsening at elevated temperature. A recent analysis of available data [41] provides a perspective on the exceptional performance of $Al_3(Zr,V)$ and $Al_{12}(Fe,V)_3Si$-related dispersoids in Al-Zr-V and Al-Fe-V-Si alloys. Under circumstances where solute diffusion along grain boundaries or along dislocation pipes make only minor contributions to coarsening, the operative relation for lattice diffusion control is expected to be [42,43]:

$$r^3 - r_o^3 = Kt \tag{6}$$

where r_o is initial mean radius of the dispersoid particles, r the corresponding value after time t at temperature T, and K is a material constant which is expected to depend on temperature T according to a relation of the form:

$$K = (K_o/T) \exp(-Q/RT) \tag{7}$$

where Q is the operative activation enthalpy for coarsening and R is the gas constant. Determination of K from the slope of plots of r^3 versus t allows relative rates of coarsening of different precipitates or dispersoids in a matrix to be compared, for example by plotting log(KT) versus 1/T [44], as in Fig 5 for various precipitates and dispersoids in Al alloy matrices. This shows that even the highest values of KT for $Al_{12}Fe_3Si$ in Al-8Fe-1.5V-1.5Si and $L1_2$-Al_3Zr in Al-0.8Zr are five to six orders of magnitude less than for Si, for example, in the applicable temperature range (375 to 475°C), while the lowest values, for $L1_2$-$Al_3(Zr,V)$ in Al-1.2Zr-1.8V are six to eight orders of magnitude smaller. The values of KT for these phases in these alloys, at 10^{-29} to 10^{-26} m^3 K/s in this temperature range (0.7 to 0.8 T_m), are in fact identical to those for θ" in Al-Cu alloys at much less elevated temperatures (0.4 to 0.5 T_m). The low rates of coarsening of $Al_{12}(Fe,V)_3Si$ in Al-Fe-V-Si alloys at 0.7 T_m are attributable mainly to the low diffusivity and solubility of Fe in the α Al matrix. There are indications that the bcc variant of this phase favoured by higher Fe:V ratios resists coarsening more effectively than the primitive cubic variant favoured otherwise [45]. To what extent this derives from improved lattice or interfacial matching, interfacial segregation or lower effective diffusivity or solubility in the αAl lattice remains to be established.

5. Conclusions

1. The critical condition for displacement of one rapidly solidified microstructure by another can be predicted on the basis of competitive transformation kinetics provided that the magnitudes of the relevant materials parameters are known with sufficient accuracy.

2. Although such measurements are available for only a few cases, they agree reasonably well with model predictions provided that conditions at the transformation front are distinguished from those prevailing in the fully developed microstructure.

3. Stability towards coarsening at elevated temperature is a key requirement of microduplex products of RS processing and a number of questions remain to be answered concerning the remarkable behaviour in this regard of such as $Al_{12}(Fe_1V)_3Si$ and related phases in high performance RS aluminium alloys.

Acknowledgements

The author is grateful to the Brite Euram Directorate, Brussels, the UK Science and Engineering Research Council, DRA Farnborough and the University of Sheffield for contributing financial support to programmes at Sheffield referred to in this paper.

References

1. D Turnbull, Contemp Phys, 1969, 10, 473-488.
2. D R Uhlmann, J Non-Cryst Solids, 1972, 7, 337-348.
3. P G Boswell and G A Chadwick, Scripta Met, 1977, 11, 459-465.
4. W Kurz and R Trivedi, in 'Principles of Solidification and Materials Processing', eds R Trivedi et al, Trans Tech Publ, Aedermansdorf, 1990, Vol 1, pp 67-81.
5. D M Herlach, R F Cochrane, I Egry, H J Fecht and A L Greer, Internat Mater Rev, 1993, 98, 273-347.
6. W Kurz and R Trivedi, in 'Microstructural Development and Control in Materials Processing', eds D R Durham and A Saigal, ASME, 1989, pp 47-55.
7. H Jones, Mater Sci Eng A, 1991, A137, 77-85.
8. P Gilgien, A Zryd and W Kurz, to be published.
9. H Jones, Mater Sci Eng A, 1994, A179/180, 1-8.
10. H Jones, Mater Sci Eng A, 1991, A133, 33-39.
11. H Y Liu and H Jones, Acta Met Mater, 1992, 40, 229-239.
12. H Y Liu and H Jones, Acta Met Mater, 1992, 40, 2003-2009.
13. S C Gill and W Kurz, Acta Met Mater, 1993, 41, 3563-73.
14. X Mao, H Fu, A Frenk and W Kurz, Proc 1st Pacific Rim Conf on Advanced Materials and their Processing, MMMS, 1992, p 193-197.

15. I R Hughes and H Jones, J Mater Sci, 1976, 11, 1781-96.
16. D Liang and H Jones, Mater Sci Eng A, 1993, A173, 109-114.
17. H Jones, P N Hansen, J D Hunt and W Kurz, Final Report on Contract BREU-0065-C, March 1993.
18. K A Jackson and J D Hunt, Trans Met Soc AIME, 1966, 236, 1129-42.
19. W Kurz and D J Fisher, Internat Met Rev, 1979, 24, 177-204.
20. M Zimmermann, M Carrard and W Kurz, Acta Met Mater, 1989, 37, 3305-13.
21. M Pierantoni, M Gremaud, P Magnin, D Stoll and W Kurz, Acta Met Mater 1992, 40, 1637-44.
22. D Liang and H Jones, Z Metallkunde, 1992, 83, 224-226.
23. D Liang and H Jones, Scripta Met Mater, 1991, 25, 2855-59.
24. D Liang and H Jones, Scripta Met Mater, 1993, 28, 7-10.
25. D Liang and H Jones, Mater Sci Eng A, 1993, A173, 109-114.
26. D Liang, Y Bayraktar and H Jones, submitted to Acta Met Mater.
27. C R Ho and B Cantor, Mater Sci Eng A, 1993, A173, 37-40.
28. W J Boettinger, L A Bendersky, S R Coriell, R J Schaefer and F S Bianganiello, J Cryst Growth, 1987, 80, 17-25.
29. J A Juarez-Islas and H Jones, Mater Lett, 1989, 7, 322-4.
30. M Gremaud, M Carrard and W Kurz, Acta Met Mater, 1990, 38, 2587-99.
31. W Kurz, B Giovanola and R Trivedi, Acta Met Mater, 1986, 34, 823-830.
32. J D Hunt and S-Z Lu, Mater Sci Eng A, 1993, A173, 79-83.
33. D Liang, W Jie and H Jones, J Cryst Growth, 1994, 135, 561-4.
34. J D Hunt, in 'Solidification and Casting of Metals', The Metals Society, London, 1979, pp 3-9.
35. W Kurz and J D Fisher, Acta Met, 1981, 29, 11-20.
36. R Trivedi, Met Trans A, 1984, 15A, 977-982.
37. S A Moir and H Jones, Mater Lett, 1991, 12, 142-146.
38. S A Moir and H Jones, J Cryst Growth, 1991, 113, 77-82.
39. Y Bayraktar, D Liang, S A Moir and H Jones, Mater Lett, 1992, 15, 314-316.
40. S K Das and L A Davis, Mater Sci Eng A, 1988, 98, 1-12.
41. L M Rylands, D M J Wilkes, W M Rainforth and H Jones, J Mater Sci, 1994, 29, 1885-1900.
42. I M Lifzhitz and V V Slyozov, J Phys Chem Solids, 1961, 19, 35-56.
43. C Wagner, Z Elektrochem, 1961, 65, 581-591.
44. A J Ardell, in 'Phase Transformations 1987', The Inst of Metals, London, 1988, pp 485-494.
45. D J Skinner, R L Bye, D Raybould and A M Brown, Scripta Met, 1986, 20, 867-872.

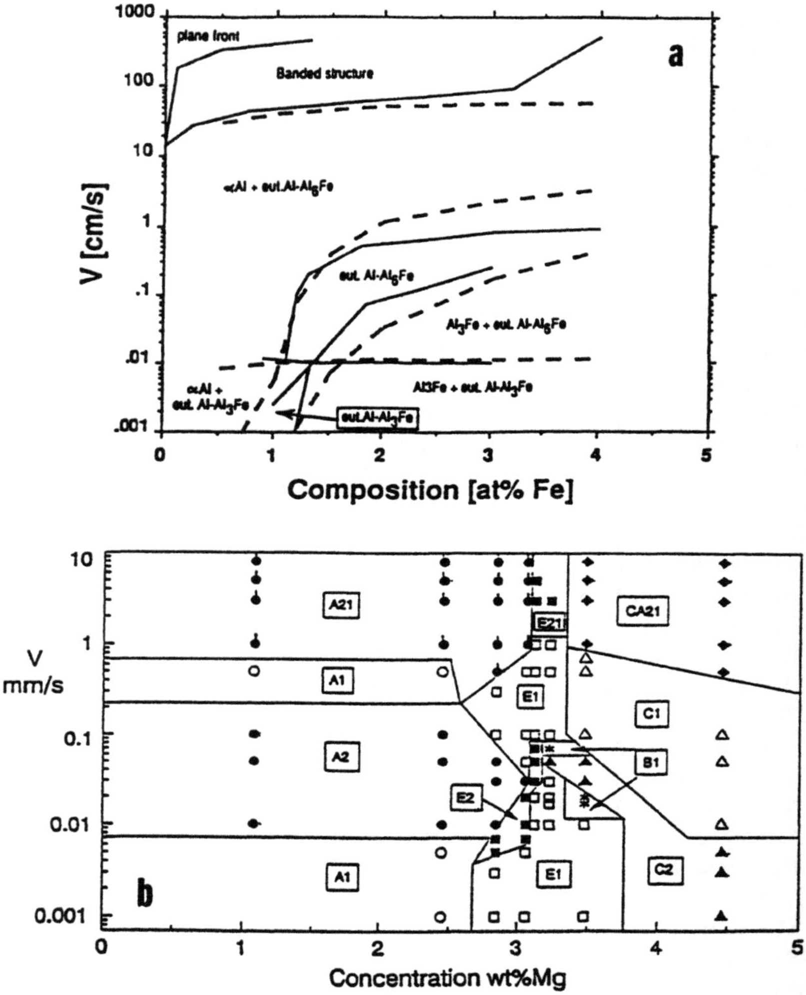

Fig 1: Solidification microstructure selection diagram for dilute (a) Al-rich Al-Fe [8] and (b) Zn-rich Zn-Mg [11] alloys. In (a) full lines are experimentally observed boundaries and broken lines are predictions of competitive growth modelling. Results for V > 100 mm/s are from laser surface melt traversing (G ~2000 K/mm) while those for V < 50 mm/s are for Bridgman growth (G ~ 10 K/mm). In (b) Al = αZn + EU1; A2 = αZn + EU2; A21 = αZn + EU2 in EU1; B1 = β_1 + EU1; C1 = β_2 + EU1; C2 = β_2 + EU2; CA21 = β_2 + αZn + EU2 + EU; E1 = EU1; E2 = EU2; E21 = EU2 in EU1 where $\beta_1 = Mg_2Zn_{11}$, $\beta_2 = MgZn_2$, EU1 = α + β_1 eutectic and EU2 = α + β_2 eutectic. Results are for Bridgman growth (G = 15 K/mm).

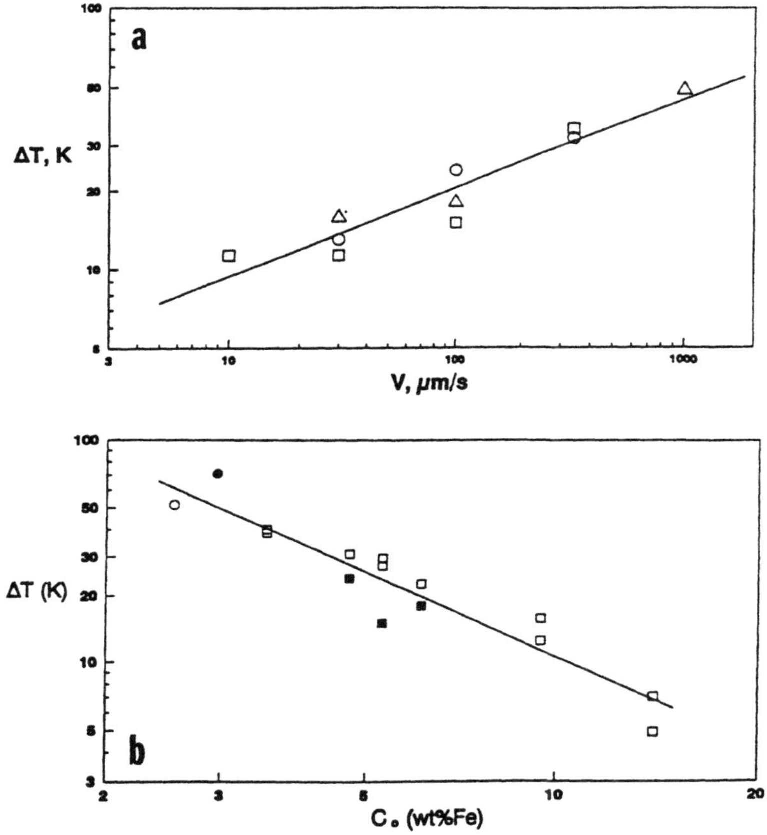

Fig 2: Growth undercooling ΔT for primary Al_3Fe in dilute Al-rich Al-Fe alloys, (a) versus growth velocity V for O 4.7 □5.3 △6.1 wt%Fe [23] and (b) versus alloy concentration C_0 for V = 100 μm/s ■ [23] □ [24] direct measurements O ● from observed limiting condition for growth of primary Al_3Fe in competition with αAl-Al_6Fe eutectic.

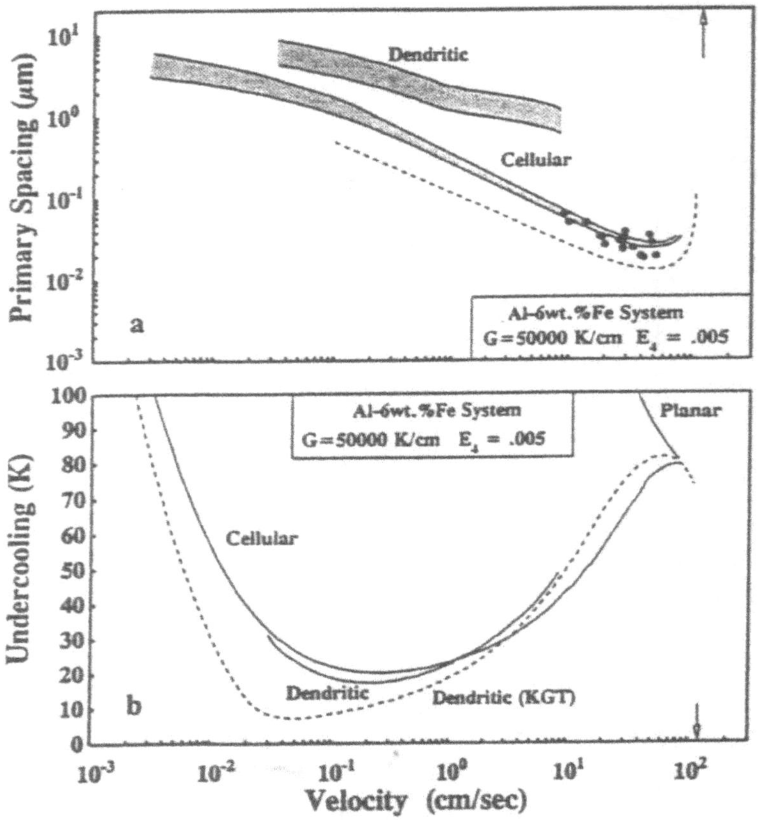

Fig 3: Dependence on growth velocity of (a) primary spacing and (b) growth undercooling for αAl cells and dendrites in Al-6 wt%Fe. Full lines are predictions of the array model, broken lines are predictions of the tip marginal stability model [31] and experimental points are measurements from laser surface melt traversing [30]. Arrows indicate the absolute stability velocity. From [32].

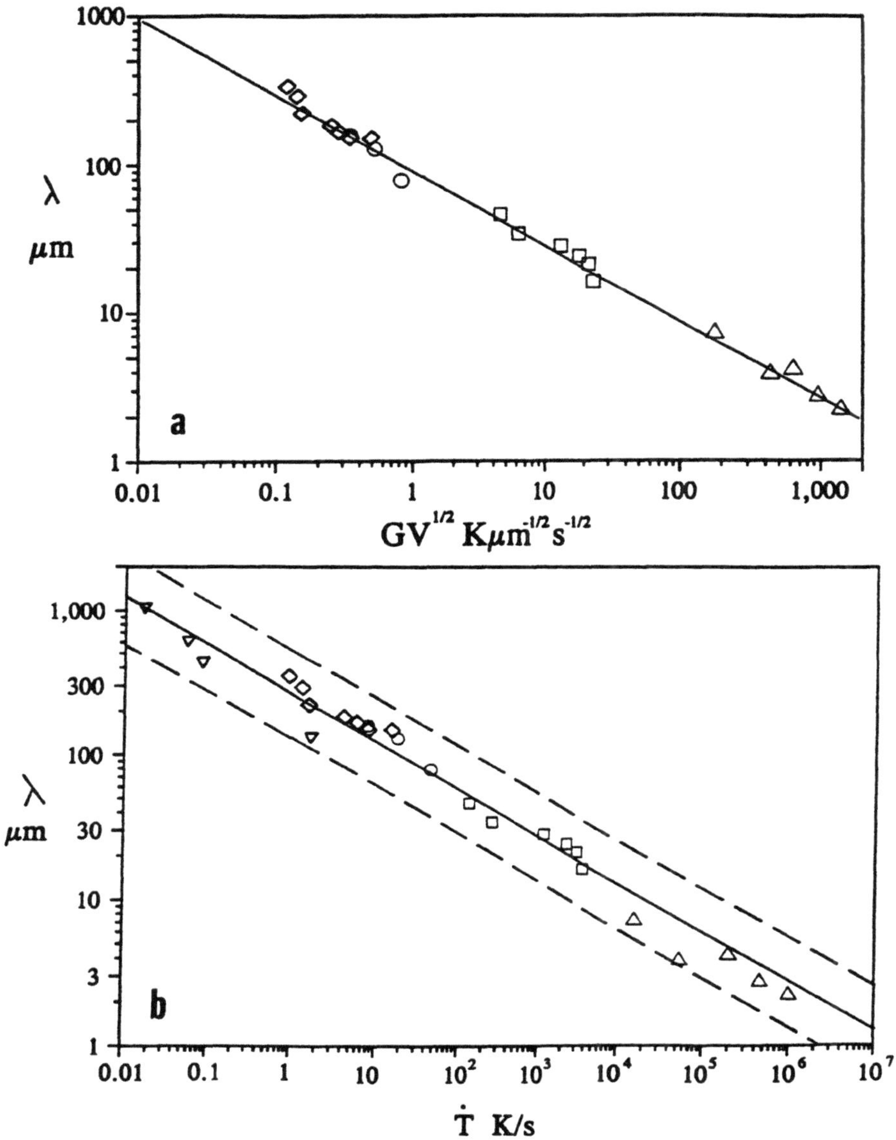

Fig 4: Interparticle spacing λ of primary silicon versus (a) $GV^{1/2}$ and (b) cooling rate $\dot{T}$ during solidification of hypereutectic Al-Si alloys. ▽ continuous cooling ◇ O Bridgman growth □ TIG weld traversing △ laser surface melt traversing. Line shown with factor of 2 scatter band represents $\lambda\dot{T}^{1/3} = 250\ \mu m\ (K/s)^{1/3}$. From [39].

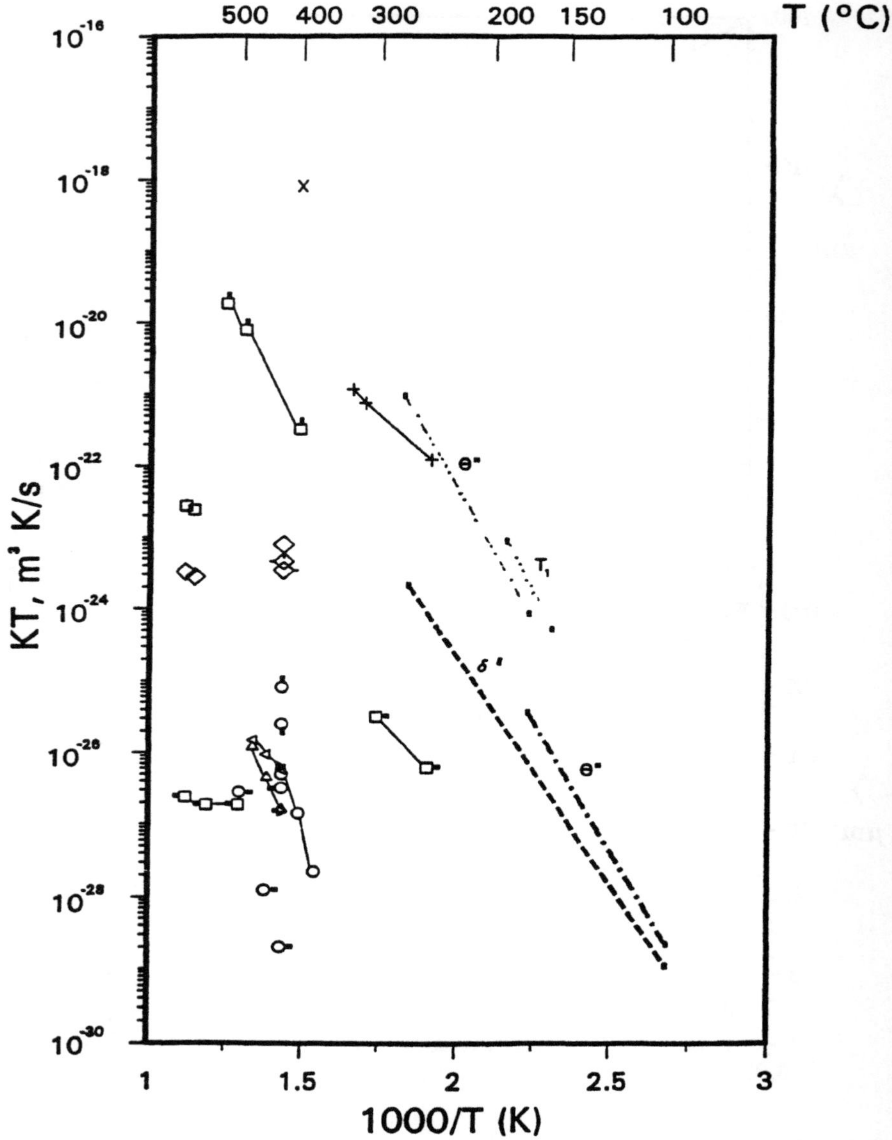

Fig 5: Coarsening parameter KT plotted logarithmically against l/T for precipitates and dispersoids in aluminium alloys.
Key: X Ag_2Al □- Ge ⊔ Si □ Al_3Ni ◇-◇ $Al_{13}Fe_4$ + Al_3Mg_2 -▷ ▷ ◁ △ ▽ $Al_{12}Fe_3Si$ O Ll_2-Al_3Zr -O O- Ll_2-$Al_3(Zr_1V)$ ♁ DO_{23}-Al_3Zr ♀ DO_{23}-$Al_3(Zr_1V)$ -□ Al_6Mn ◇- $Al_{13}(Fe,Mo,V)_4$ ◇ Al_8Fe_4Ce. From [41].

SYNTHESIS AND PROPERTIES OF BULK METALLIC GLASSES

W.L. JOHNSON AND A. PEKER

Keck Laboratory of Engineering, California Institute of Technology, Pasadena, CA, 91125, USA

Abstract

During the past few years, a number of metallic alloys have been reported in the literature which exhibit glass forming ability when cooled from the melt at cooling rates of less than 100 K/s and as low as 1 K/s. This permits formation of large three dimensional specimens of metallic glass. Further, these glasses remain stable against crystallization when heated to temperatures far above T_g, the glass transition temperature. The undercooled melt, or labile, can be studied and characterized over a broad temperature range between T_g and T_l, the latter being the thermodynamic liquidus temperature of the alloy. A brief summary of the constitution, synthesis, glass forming ability, and thermophysical properties is given for one family of these new bulk metallic glasses. The origins of the exceptional glass forming ability are discussed.

1. Introduction

Formation of metallic glasses from the liquid state was first reported by Duwez et. al. [1] in 1960, and was a result of their development of rapid solidification methods capable of quenching liquid metals at ~ 10^6 K/s. Shortly thereafter, Turnbull et. al. [2] proposed that the glass forming ability of these alloys was directly related to deep eutectic features in the phase diagrams of these materials. Using the classical theory of nucleation and growth of crystals in an undercooled alloy melt, Turnbull argued that low lying eutectic features in the liquidus of the alloy permitted one to achieve a high reduced glass transition temperature, $t_g = T_g/T_l$, where T_l is the liquidus temperature of the alloy. This in turn leads to suppression of the rate of homogeneous nucleation of crystals in the undercooled liquid during cooling. A glass is formed when the melt undergoes configurational freezing at T_g. Based on nucleation theory, Turnbull proposed [3] that t_g values greater than 1/2 would favor glass formation during rapid quenching. Further, he

M. A. Otooni (ed.), Science and Technology of Rapid Solidification and Processing, 25–41.

suggested that when t_g equals or exceeds a value of 2/3, the rate of homogeneous nucleation of crystals should be so small as to permit bulk glass formation at much lower cooling rates. Experimental support of this criteria was initially provided by studies of undercooled Pd-Ni-P alloys ($t_g = 0.66$) [4] in which repeated melting in a boron oxide flux was used to dissolve heterogeneous nucleation sites on the melt surface. This resulted in glass formation at cooling rates below 10 K/s and the production of glassy ingots ranging up to ~1 cm in diameter. Other experiments by Lee et. al. [5] on Au-Pb-Sb ($t_g = 0.66$) alloys employing containerless solidification in a drop tower produced mm sized sphere of metallic glass. Recently, Inoue, Masumoto et. al. have studied alloys of La-Al-Ni-Cu [6], Mg-Y-Cu [7], and Zr-Ni-Cu-Al [8,9], having $0.58<t_g<0.66$. These alloys form bulk metallic glass when cast in copper molds at cooling rates ranging from 10-300 K/s. Amorphous alloy rods ranging up to 1.6 cm in diameter were reported for the latter alloy [10]. Most recently, Peker and the author [11] have studied glass formation in Zr-Ti-Ni-Cu-Be quaternary and pentiary alloys with typical t_g values in the 0.66-0.68 range. These alloys are essentially "self fluxing" in that most heterogeneous nucleation sites are naturally dissolved by the liquid alloy. When this is accomplished, glass formation is achieved at cooling rates of less than 1 K/s. Ingots of these metallic glasses weighing 0.2 Kg and having cross sections of several cm dimensions are formed directly on alloying the constituents by plasma arc melting [12]. In fact, a lower cooling rate limitation for synthesizing these bulk metallic glass has not yet been established.

This remainder of this paper presents a brief survey of the glass forming characteristics and properties of these alloys. In section 2, the features of the relevant phase diagrams, the glass forming compositions, and thermodynamic properties of these alloys are outlined. Section 3 discusses the formation of bulk metallic glass, undercooling, and the necessary conditions for suppression of crystallization. Section 4 deals with heterogenous nucleation of crystals and the metastability of the undercooled liquid under various processing conditions. Finally, in section 5, a brief summary of the some of the characteristic properties of the glass is presented.

2. Phase Diagrams, Alloy Constitution, and Low Lying Eutectics in Multicomponent Systems.

To obtain an excellent glass forming alloy, the basic conclusion of Turnbull's analysis of homogeneous nucleation serves as a good starting point. Simply stated, one seeks alloys in which the liquidus surface falls to values for which $T_g/T_l = t_g \geq 2/3$. As a rough approximation, the glass transition temperature of a liquid metallic alloy (where the atomic coordination number is 12-13) should scale with the bonding strength or cohesive energy of the alloy. For a binary alloy $A_{1-x}B_x$, the total bonding enthalpy (or cohesive energy) can be written as:

$$H_{coh} = (1-x)H_A^o + xH_B^o + \Delta H_{mix}$$

where the first and second terms are the weighted contributions of the elemental constituents and the third term is the heat of mixing. Except in rare cases, the latter term is generally smaller than the first two terms. Further, the heat of mixing term is slowly varying with composition. Since T_g is roughly proportional to H_{coh}, it varies slowly with composition in binary and higher order alloys. Experimental results generally support this expectation. On the other hand, the liquidus curves (surfaces) in alloys often show dramatic variation, with composition exhibiting sharp maxima at congruent melting compounds, and deep minima at eutectic features which lie at compositions corresponding to multiphase equilibria in the solid state. As a simple example of these features, the Ti-Cu phase diagram is shown below in Fig. 1. Included

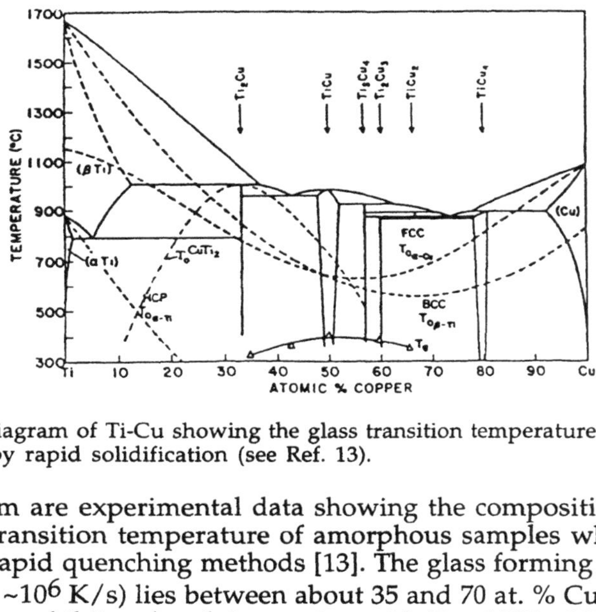

Fig. 1. Phase diagram of Ti-Cu showing the glass transition temperature of amorphous alloys formed by rapid solidification (see Ref. 13).

in the diagram are experimental data showing the composition dependence of the glass transition temperature of amorphous samples which were obtained by rapid quenching methods [13]. The glass forming region (by rapid quenching at ~10^6 K/s) lies between about 35 and 70 at. % Cu where the liquidus curve exhibits a low lying region with $T_l \sim 1200 - 1300$ K. In this region, one has $T_g \sim 600$-670 K. The reduced glass transition temperature, t_g, ranges from 0.47 - 0.53 in this region. The binary Zr-Cu [13], Zr-Ni [15] and Ti-Ni [16] systems show generally similar features with low lying eutectic features over the central portions of the phase diagrams. In all of these binary systems, glass formation is observed by rapid solidification of liquid alloys near low lying eutectics. Perhaps the best glass forming binary alloy among these is $Zr_{38}Cu_{62}$. At this composition, a eutectic feature at 1158 K is found in the phase diagram. The rapidly quenched glass at this composition has T_g = 759 K giving t_g = 0.64, among the highest values found for any binary alloy. Gillessen and Herlach have studied droplets of this alloy solidified during free fall in a drop tower and obtained solid glass spheres ranging up to 0.3 mm in diameter [17]. They estimate the cooling rate to be 8 x 10^3 K/s, a substantially lower rate than achieved in conventional rapid solidification.

The Zr-Be and Ti-Be systems differ from the above binary systems in that crystalline intermetallic phases are confined to the Be-rich portion of the phase diagram. For example, in Zr-Be shown in Fig. 2, the most Zr-rich

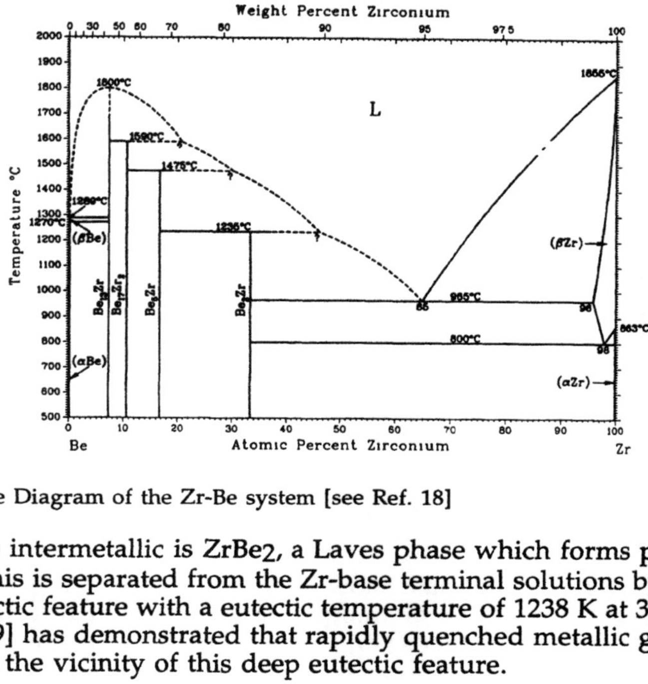

Fig. 2. Phase Diagram of the Zr-Be system [see Ref. 18]

crystalline intermetallic is $ZrBe_2$, a Laves phase which forms peritectically at 1508 K. This is separated from the Zr-base terminal solutions by a broad and deep eutectic feature with a eutectic temperature of 1238 K at 35 at. % Be. Tanner [19] has demonstrated that rapidly quenched metallic glasses are formed in the vicinity of this deep eutectic feature.

The Zr-Ti-Ni-Cu-Be alloys of interest here can best be viewed as pseudo-ternary alloys of the type $(Zr\text{-}Ti)_{1-x-y}(NiCu)_xBe_y$, or more generally as:

$ETM_{1-x-y}\ LTM_x\ Be_y$.

where ETM is an early transition metal (e.g. Ti, Zr, Nb, V, etc.) and LTM is a late transition metal (e.g., Cu, Ni, Co, Fe) [20]. The essential motivation in developing the alloys was to find the deepest eutectic features obtainable in the pseudo-ternary phase diagrams. The absence of Zr-rich (Ti-rich) phases in the Zr-Be and Ti-Be systems provides a key. It suggests, for instance, that Be will have limited solubility in crystalline binary phases such as Zr_2Cu, ZrCu, Zr_2Ni, and ZrNi, since analagous phases are absent in the Zr-Be system. As further support for this conjecture, the metallic radius of Be is substantially smaller than that of Cu and Ni (R_{Be}= 0.112 nm, while R_{Cu}=0.128 nm, and R_{Ni}=0.124 nm), and much smaller than that of Zr, R_{Zr}=0.160 nm). In compounds such as Zr_2Ni (Al_2Cu-type), substitution of Be on the Ni-site in the crystal should be accompanied by substantial lattice strain. This in turn should reduce the stability of the compound as Be is substituted in the form $Zr_2(Ni_{1-x}Be_x)$. A similar argument applies to crystalline compound $Zr(Ni_{1-x}Be_x)$. Such compounds should have limited solubility for Be. In the absence of a new ternary Zr-rich Zr-Ni-Be phase, one would expect to find a ternary eutectic feature in the Zr-rich portion of the ternary diagram. This eutectic should lie at Be concentrations exceeding the solubility limits of the above crystalline phases. Further, this ternary eutectic should lie lower at temperature than the binary eutectic of Fig. 2 or the corresponding liquidus features in the binary Zr-Cu and Zr-Ni systems. Such a deep eutectic region in fact exists in the ternary and pseudo-ternary phase diagrams. The liquidus/solidus temperatures for numerous alloys lying in the shaded region of Fig. 3 have been determined by DTA studies and found to vary between about 910 K and about 1115 K. For example, DTA scans of the melting curves of several ternary Zr-Cu-Be alloys are shown in Fig. 4. The alloys have compositions varying over a substantial region of the Zr-rich part of the ternary diagram. All four alloys have a solidus temperature of about $T_s \approx 1070$ K while the liquidus curves vary from 1080 to 1125 K. These temperatures lie well below the corresponding binary eutectic temperatures in the Zr-Be Diagram (eutectic at 35 at. % Be with T_e= 1238), or the Zr-Cu system (eutectics at 28 and 46 at. % Cu with T_e= 1273 K, and 1201 K respectively). In higher order alloys, this broad region of low lying solidus and liquidus features becomes even more pronounced. Fig. 5 shows the DTA scan of the melting transition for the pentiary alloy $Zr_{41.2}Ti_{13.8}Ni_{10}Cu_{12.5}Be_{22.5}$. Here, T_s=937 K while $T_l \approx 985$ K. This pseudo-ternary alloy exhibits an exceedingly low melting region. In this region, one finds $t_g \approx 0.67$. This will be discussed in the next section in more detail. A list of some typical ternary and pseudo-ternary alloys studied and some of their characteristic properties is given in Table I.

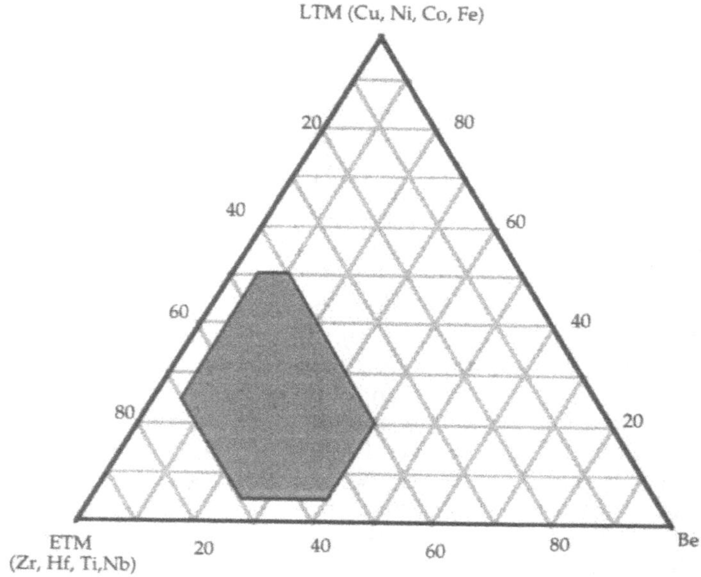

Fig. 3. Schematic ternary phase diagram showing the region in which deep lying liquidus features are expected in ETM-LTM-Be alloys.

The table lists the glass transition temperature of the amorphous alloys as measured by DSC at scanning rates of 20 K/s along with the crystallization temperature, T_x, measured under the same experimental conditions. It is important to point out that the crystallization temperature depends strongly not only on heating rate, but also on the experimental conditions (e.g. environment etc.). This is discussed in more detail in the next section.Also listed in the table are a few examples of liquidus/solidus temperatures for several selected alloys. For these pseudo-terary alloys, T_g varies from 580 K to 680 K. The variation of T_g can be most easily visualized using a contour plot. Fig. 6 shows such a contour plot for the pseudo-ternary system $Zr_{1-x-y}(NiCu)_xBe_y$. We can group these alloys on one ternary plot since the T_g is relatively insensitive to the Ni/Cu ratio. The plot is obtained by fitting data for Tg from 30 alloy compositions. Among other features, one sees that the T_g is predominantly determined by the overall Zr content of the alloy with T_g gradually increasing with decreasing Zr concentration. The reduced glass

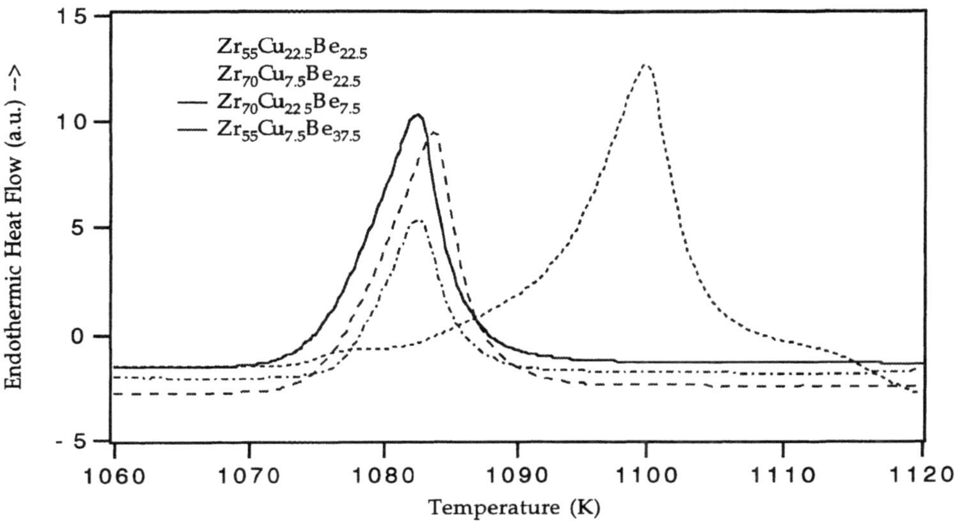

Fig. 4 DTA scans (Setarim DSC-2000) of the melting endotherms for a series of ternary Zr-Cu-Be alloys.

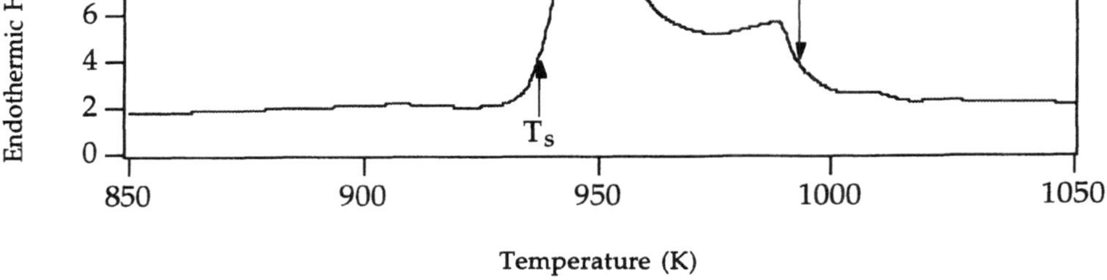

Fig. 5. DTA scan (Setarim DSC-2000) of the melting transition of a pentiary Zr-Ti-Ni-Cu-Be alloy. See text for details.

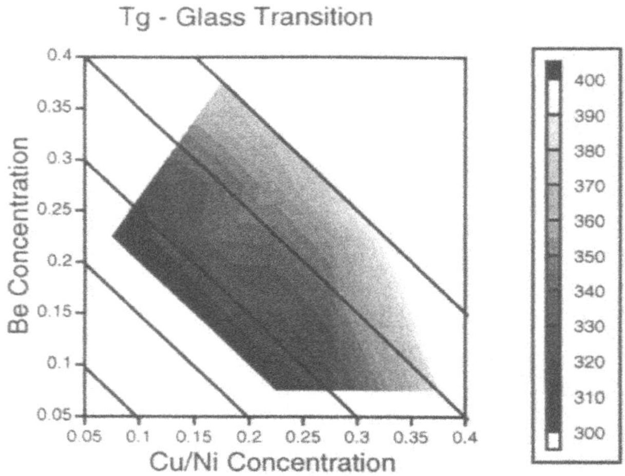

Fig. 6. Contour Plot showing the dependence of T_g on composition in the pseudo-ternary system $Zr_{1-x-y}(NiCu)_xBe_y$. The T_g is relatively independent of the Ni/Cu ratio. Data for about 30 compositions were used to construct the plot.

transition temperature of these alloys is typically 0.60-0.68. Further, such values of t_g are obtained over a rather broad range of pseudo-ternary compositions.

Based on Turnbull's criterion, one expects these alloys to be excellent glass formers. According to Turnbull, the rate of homogeneous nucleation of crystalline phases in such alloys should become very small relative to laboratory time scales [3]. This has been experimentally found to be the case as discussed in the next section.

3. Bulk Glass Formation, Undercooling, and Critical Cooling Rates

In the present section, we will focus attention on a subset of the pseudo-ternary alloy systems described above, those having the greatest t_g values. In particular, we have extensively studied alloys of the type $(Zr\text{-}Ti)_{1-x-y}(Ni\text{-}Cu)_xBe_y$. We have found that such alloys have t_g values as high as 0.68 and form glasses when cooled from the melt at cooling rates of 1 K/s or less. A typical glass forming composition is $Zr_{.412}Ti_{.138}Cu_{.125}Ni_{0.1}Be_{0.225}$. In terms of the pseudo-ternary formula, this corresponds to x=.225, y=.225, with a Zr/Ti ratio of 3:1 and a Cu/Ni ratio of 5:4. The glassy alloy has T_g= 625 K (as measured at a heating rate of 20 K/m), a solidus temperature of T_s= 937 K, and a liquidus temperature of T_l= 980 K as shown in the above figure. The

Table I. Properties of some typical pseudo-ternary alloys used in this study.

Alloy	Tg (K)	T_x (K)	T_s/T_l (K)
$Zr_{70}Ni_{7.5}Be_{22.5}$	578	606	1100/1130
$Zr_{55}Cu_{22.5}Be_{22.5}$			1075/1120
$Zr_{70}Cu_{7.5}Be_{22.5}$			1075/1095
$Zr_{70}Cu_{22.5}Be_{7.5}$			1075/1090
$Zr_{50}Cu_{17.5}Ni_{10}Be_{22.5}$	634	726	
$Zr_{70}Cu_{12.5}Ni_{10}Be_{7.5}$	584	654	
$Zr_{45}Cu_{12.5}Ni_{10}Be_{32.5}$	648	733	
$Zr_{41.2}Ti_{13.8}Cu_{12.5}Ni_{10}Be_{22.5}$	625	700	937/985
$Zr_{34.5}Nb_9Ti_{11.5}Cu_{12.5}Ni_{10}Be_{22.5}$	650	705	

alloy lies very close to a eutectic composition with eutectic temperature of $T_e = T_s = 937$ K.

Samples of this material have been prepared in a variety of ways. Ingots ranging in size from 1 cc to 4 cc (6-25 grams of material) have been made by successively induction melting the constituents together on a water cooled silver (or copper) boat under a Ti-gettered Ar atmosphere. The liquid sample is levitated during melting. Upon obtaining a homogeneous liquid alloy, it is observed that the sample freezes to a homogeneous glassy ingot when the rf power is shut off. During cooling, the sample is in contact with the silver or copper boat. The estimated cooling rate is 10-30 K/s. Fig.7 shows a 25 g (4cc) ingot of metallic glass obtained in this manner. The glassy structure of the ingot was verified by x-ray diffraction scans on cross sectioned surfaces of the ingot. No evidence of any crystallinity was detected. Schwarz, at Los Alamos Laboratory, has used plasma melting on a water cooled copper hearth to form fully glassy ingots up to 0.2 Kg (33 cc). The estimated cooling rate is less than 10 K/s. To determine the minimum cooling rate necessary to form the glass, Kim et. al. [21], have been performed experiments using an electrostatic levitation device where a 3mm diameter sphere of the material can be levitated in high vacuum, heated and melted with a focussed quartz lamp, then undercooled in a controlled manner. These experiments suggest that cooling rates of 0.1 K/s are sufficient to supress crystallization. This suggests that ingots having dimensions of 10 cm or more can be cast to the glassy state.

Other methods for preparing glassy samples of various sizes and shapes have been used. The alloy can be sealed under inert gas in a silica tube, induction melted, then water quenched (in the silica tube). A 1/2" rod of metallic glass prepared by this method is shown in Fig. 7. An x-ray diffration pattern of the cross sectional surface of the rod in shown below in Fig.8. No evidence of any crystallinity is detected anywhere in the rod. Such rods have

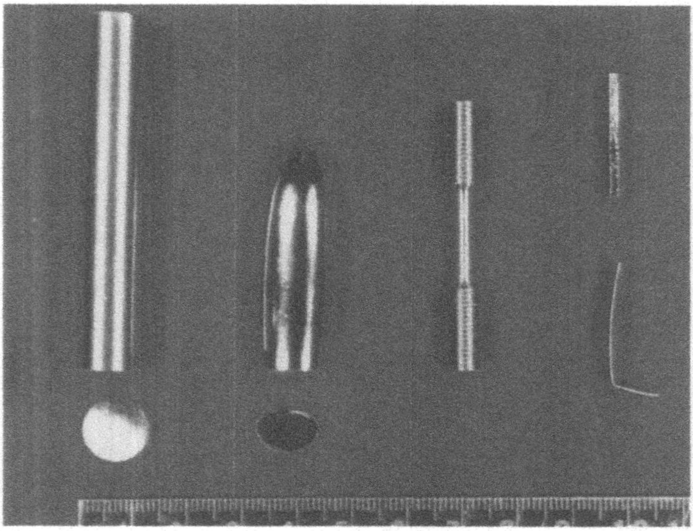

Fig. 7. Glassy samples of the Zr-Ti-Ni-Cu-Be alloy referred to in the text. From left to right: a) a 1/2" diameter rod made by melting the alloy in a silica tube and water quenching, (b) a 25 g ingot made by levitation melting on a copper boat, (c) a standard tensile bar machined from a rod, (d) a 1.5 mm thick, 8mm wide strip (side view) which was subsequently cold rolled.

been used for machining samples of the metallic glass for mechanical testing. Fig. 7 shows a standard ASTM E8-68 tensile bar which was machined from a 7mm rod. Also shown is a cross section of a 1.5 mm thick strip of the the metallic glass which was prepared by injecting the melt into a copper strip mould. This stip was used to test the ductility of the metallic glassy under cold rolling conditions. The strip was cold rolled down to an ultimate thickness of about 10 μm by multiple passes in a rolling mill at ambient temperature. A section of the rolled strip is shown in the figure just below the original cast strip.

Apparently, the homogeneous nucleation of crystalline phases in these glass forming alloys is exceedingly sluggish, if it occurs at all. In fact, we have determined that when crystallization occurs, it usually initiated at the surface of the material and is apparently brought about when the melt comes into contact with other crystalline material or is contaminated by interaction with the container or atmosphere. Oxidation of the sample surface, or contact with a crystalline container wall have been found to be mechanisms for initiating crystallization. It is natural to as whether foreign crystalline particles (or even

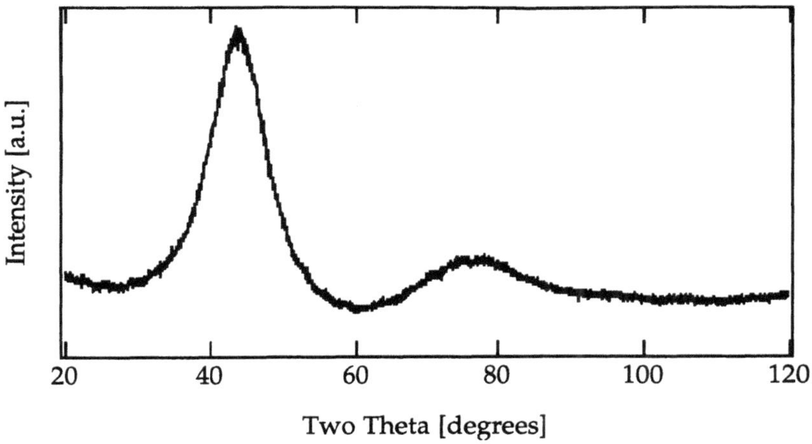

Fig. 8. X-ray diffraction scan of the cross sectioned surface of a 1/2" rod of Zr-Ti-Ni-Cu-Be glass prepared by water quenching in a silica tube. The scan was taken using Co-KÅ radiation.

crystallographic clusters) also exist within the interior of the melt and whether such particles can act as nucleation sites for crystalline phases. In the next section, we discuss experimental results which help to clarify the origins of crystal phase nucleation.

4. Metastability of the Undercooled Melt and Heterogeneous Nucleation of Crystals.

The electrostatic levitation experiments metioned above have demonstrated that when the melt is undercooled containerlessly in a high vacuum (~10^{-7} torr) under "clean" conditions, a critical cooling rate of less than 10^{-1} K/s is sufficient to bypass crystallization. Kim et. al. [21], have determined the time required initiation of crystallization at a given temperature by undercooling the molten sphere and measuring the elapsed time before recalescence occurs. This time exceeds 10 min. at all undercooling temperatures. It places an upper limit on the nucleation rate of crystals (at any degree of undercooling) of

$$dN/dt \leq 10^{-1} \text{ (crystals/cm}^3\text{-s).}$$

This is consistent with the predictions of classical theory of homogeneous crystal nucleation. For example, the rate of homogeneous nucleation in one model [3] is shown as a function of reduced undercooling $t_r = (T_l-T)/T_l$ in

Fig.9 for various values of t_g. For $t_g \geq 2/3$, the homogeneous nucleation rate should become to small to measure in the laboratory. Other models such as the Uhlmann/Davies analysis [22] vary in detail but are consistent with the same conclusion.

We have studied the crystallization of the Zr-Ti-Ni-Cu-Be glass using DSC at various heating rates. Fig.10 shows the crystallization (two crystallization events) at heating rates of 0.33 K/s and 3.33 K/s. One observes the onset of initial crystallization at about T_x = 710 K and 760 K in the two curves. The glass transition can be identified from the characteristic jump in heat capacity. This begins at T_g = 625 K. For these experiments, the sample volume is of the order of 10^{-3} cc. From the above bound on the homogeneous nucleation rate, it is apparent that the DSC samples crystallize prematurely. Apparently, heterogeneous nucleation must occur in the DSC measurements. Possible causes include reaction with the flowing gas atmosphere of the DSC or contact with the sample pan (in the case an aluminum pan). In an effort to reduce these problems, Peker [23] has sealed a DSC sample within a small thin walled sphere of pyrex glass under an inert gas atmosphere. When this is done and the DSC scan repeated, the crystallization temperature is displaced to higher temperature (at the same scanning rate) by a shift of 50 K. Encapsulating the sample in a pyrex ball reduces oxidation and eliminates contact with the aluminum DSC pan. Apparently, one of these two effects is responsible for the heterogeneous nucleation of crystals. Schneider [24], at this meeting, reports results of a study of the influence of surface oxidation on the nucleation of crystals in a similar glass forming alloy Zr-Ni-Al. She directly demonstrates how the formation of a surface oxide leads to surface initiated crystallization. Prelimary results suggest a similar mechanism in the present Zr-Ti-Ni-Cu-Be alloys.

The role of surface oxidation in triggering heterogeneous nucleation leads naturally to the question of whether oxide contamination in the melt interior also causes heterogeneous nucleation. Peker has shown that this is apparently not the case [23]. By heating a glassy ingot of the Zr-Ti-Ni-Cu-Be alloy to temperatures near the melting point (T_l) and above in an oxygen atmosphere for various times, he produced severe oxidation of the sample surface. The amount of reacted oxygen was measured by weight analysis. The surface of the sample exhibits a thick oxide layer. Following this procedure, he remelted the oxidized ingot in the induction melting system on the water cooled copper boat. Oxide debris were observed floating on the surface of the melt with oxide particles entrained in the interior by the rf stirring. In this condition, the ingot could not be significantly undercooled without rapid crystallization. Next the ingot was held above the melting point (~850 K) for an extended time period (10 m - 1 hr). The crystalline oxide debris gradually dissolved. Following this dissolution of the oxide in the melt, the ingot could

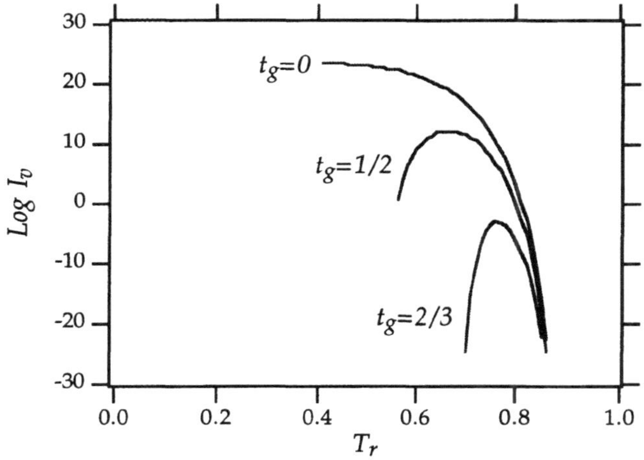

'ig. 9. Homogeneous rate of crystal nucleation as predicted by Turnbull.

ıgain readily be undercooled to the glass state. Weight analysis showed that ıll of the oxygen remained in the sample, apparently in the form of solution n the glass. Peker found that he could dissolve up to about 1.5 at. % oxygen in he liquid. Apparently, the oxides of Zr, Ti, Be, etc. which form naturally luring oxidation of the material, are soluble in the liquid alloy. The ›aturation concentration of oxygen in the liquid phase is rather high. The ›xygen is retained on undercooling and does not precipitate as oxide. This self fluxing" property of the liquid alloy with respect to oxide, must be 'esponsible in part for the apparent ease with which large glassy samples can ›e prepared in the laboratory. Other studies of bulk glass forming alloys such ıs Pd-Ni-P [4] have required surface fluxing to supress the influence of surface ›xides. A B_2O_3 flux was required in the Pd-Ni-P case. This problem is ıpparently resolved in the present case as a result of intrinsic oxide phase :quilibria considerations. We believe that this may be a general property of Zr-base and Ti-base alloys.

›. Some Properties of Bulk Zr-Ti-Ni-Cu-Be Glasses

The ability to fabricate large three dimensional samples of this family ›f pseudo-ternary metallic glasses has made possible various types of property neasurements which were previously difficult to carry out. For example, uch large specimens with well defined geometry allow one to study the nechanical behavior of the metallic glass in a variety of complex states of tress. In this section, we present some selected results.

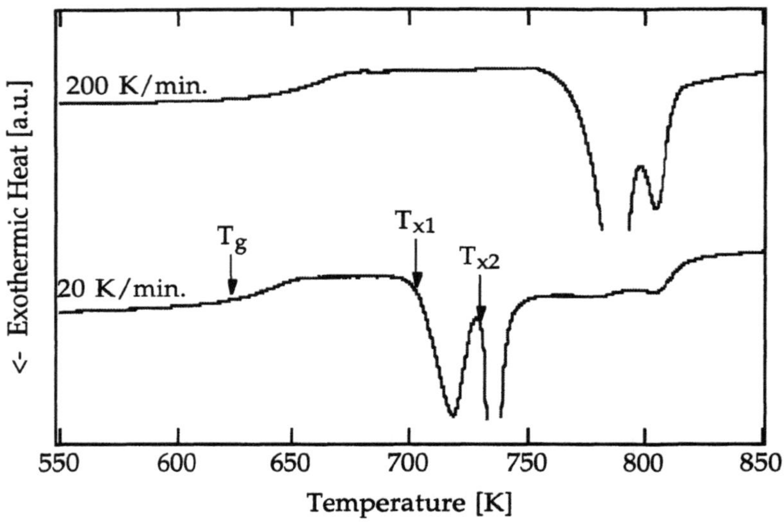

Fig. 10. DSC scans (Perkin-Elmer DSC-4) at two different heating rates showing the glass transition, and crystallization events for a Zr-Ti-Ni-Cu-Be glass. See text for details of composition etc. (Taken from ref. 11).

It has been previously demonstrated that the elastic modulii, hardness, and yield strength of metallic glasses tend to scale with one another. Fig.11, shows the composition dependence of the Vicker's hardness of a series of quaternary Zr-Ni-Cu-Be alloys. The contour plot was determined using hardness data from about ten quaternary alloy compositions. Variation of the Cu/Ni ratio as found to have little effect on hardness. As such, the data are plotted as a function of overall Cu/Ni content (x-axis) and Be content (y-axis). It can be observed that the microhardness depends primarily on the overall Zr concentration, increasing as the Zr-content of the alloy decreases. It is also apparent that the addition of Be is somewhat more effective in increasing the hardness than is the addition of either Ni or Cu. When Ti is added to the alloy, the tendency of Be addition to increase hardness becomes more pronounced. Fig. 12 shows Vickers hardness for a pentiary alloy series in which the Zr/Ti ratio is fixed at 3:1 while the Cu<Ni and Be concentrations are changed. It is apparent that Be increases hardness far more effectively in this alloy series.

Bruck et. al. [25] have carried out quasistatic mechanical testing to determine the yield behavior of the glassy alloy $Zr_{41.2}Ti_{13.8}Ni_{10}Cu_{12.5}Be_{22.5}$ in a variety of stress states (tension, compression, torsion, etc.). They used rod

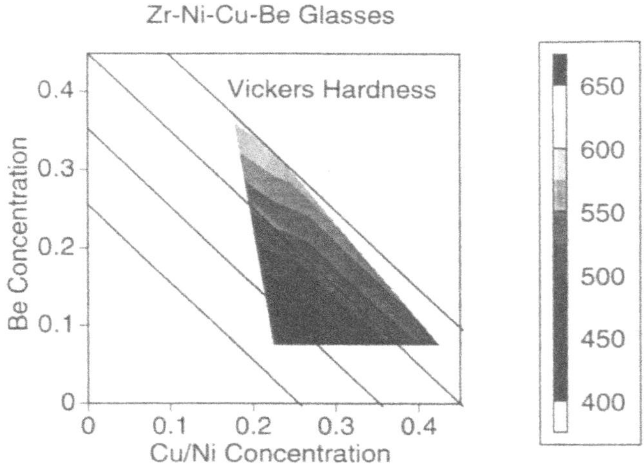

Fig. 11. Vicker's hardness of Zr-Ni-Cu-Be alloys.

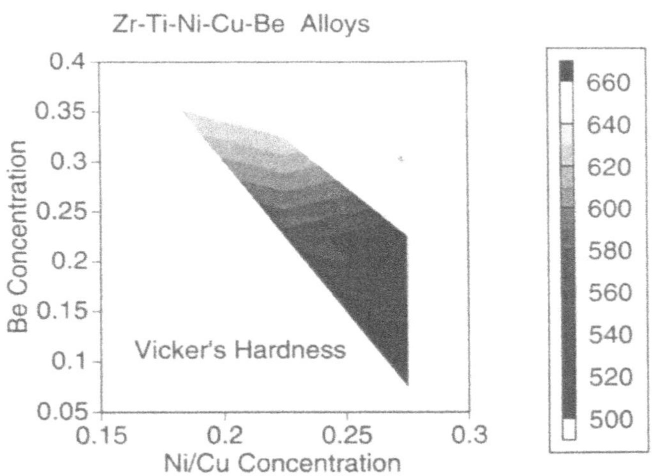

Fig. 12. Vicker's harness of ZrTi-Ni-Cu-Be alloys with Zr/Ti=3:1.

specimens for compression tests, and standard ASTM tensile bars (see Fig. 7) for tension tests. They show that this metallic glass accurately obeys the Von Mises yield criteria. The alloy shows a yield strength of $\acute{I}_y$ = 1.91 +/- 0.03 GPa in compression (3 and 7 mm rods having a height/diameter ratio of 3:2), and σ_y= 1.89 +/- 0.03 GPa in tension. The ratio of the Vicker's Hardness = 575 Kg/mm^2 to the yield strength in tension (194 Kg/mm^2) is 2.96, much the same as for other rapidly quenched metallic glasses [26]. The failure stress in pure shear (torsion) was found to be 1.10 +/- 0.03 GPa. According to the Von Mises criterion, this ratio of the tensile to the torsional yield strength should be 1.73. in excellent agreement with the experimental value of 1.74. Ultrasonic measuements on large rod samples are relatively easy. Measurements of the shear and longitudinal sound velocity using the pulse echo method in this material gave 2337 m/s and 4913 m/s repectively. The density of the glassy alloy is 6.11 g/cc. This gives Young's modulus to be 93 GPa and a Poisson ratio of 0.387. The elastic strain limit of the material is this 2.1 % in tension and compression.

These results illustrate how the ability to form large three dimensional samples of metallic glass aids in carrying out fundamental properties studies. The thermal stability of these metallic glasses in the undercooled liquid region above T_g offer addtional new opportunities for experimental investigation. For instance, diffusion and viscosity measurements become possible throughout the undercooling region. The thermodynamic properties of the undercooled melt , such as specific heat, can also be studied throughout the undercooled region enabling one to determine the entropy and free energy functions of the melt as T_g is approached from above. Such measurements have been previously difficult or impossible to carry out due to the lack of thermal stability of glass forming alloys with respect to crystallization. As such, these new metallic glasses offer the promise of significant advances in our understanding of undercooled liquid metals.

References

1. W. Klement, R. Willens, and P. Duwez, Nature, **187,** 869 (1960)
2. D. Turnbull and M.H. Cohen, Nature, **189**, 131 (1961)
3. D. Turnbull, Contemp. Phys., **10**, 473 (1969)
4. A.J. Drehman and A.L. Greer, Acta Metall., **32**, 323 (1984): also H.W. Kui, A.L. Greer, and D. Turnbull, Appl. Phys. Lett., **45**, 615 (1984)
5. M.C. Lee, J.M. Kendall, and W.L. Johnson, Appl. Phys. Lett., **40,** 382 (1982)
6. A. Inoue, T. Zhang, and T. Masumoto, Mater. Trans. JIM, **31**, 425 (1990)

7. A. Inoue, T. Nakamura, N. Nishiyama, and T. Masumoto, Mater. Trans. JIM, **33**, 937 (1992)
8. A. Inoue, T. Zhang, and T. Masumoto, Mater. Trans. JIM, **31**, 177 (1990)
9. T. Zhang, A. Inoue, and T. Masumoto, Mater. Trans. JIM, **32**, 1005 (1991)
10. A. Inoue, T. Zhang, N. Nishiyama, K. Ohba, and T. Masumoto, Mater. Trans. JIM, **34**, 1234 (1993)
11. A. Peker and W.L. Johnson, Appl. Phys. Lett., **63**, 2342 (1993): also in Proc. 8th Int. Conf. on Rapidly Quenched Metals, to appear in Mat. Sci. & Eng., (1994)
12. R.B. Schwarz, Los Alamos National Laboratory, private communication,
13. T.B. Massalski, Rapidly Quenched Metals VI, ed. by S. Steeb and H. Warlimont, (Elsevier, Amsterdam, 1985), p.171
14. L. Battezzati and A.L. Greer, Z. Phys. Chem., **156**, 103 (1987)
15. Z. Altounian and J.O. Strom Olsen, Phys. Rev. B, **27**, 4149 (1983)
16. C.N.J. Wagner, D. Lee, L. Keller, and L.E. Tanner, Rapidly Quenched Metals IV, ed. by T. Masumoto and K. Suzuki, (Japan Inst. of Metals, Sendai, 1981), Vol. **1**, 331 (1981)
17. F. Gillessen and D.M. Herlach, Mat. Sci. & Eng., **97**, 147 (1988)
18. T.B. Massalski, Binary Alloy Phase Diagrams, (ASM International, Metals Park, Ohio) p.717
19. L. Tanner and R. Ray, Acta Metall., **27**, 1727 (1979)
20. A. Peker and W.L. Johnson, "Be Bearing Amorphous Metallic Alloys Formed at Low Cooling Rates", U.S. Patent No. 5,288,344 (1993)
21. Y.J. Kim, W.Q. Rhim, and W.L. Johnson, unpublished results
22. H. Davies, Phys. Chem. of Glasses, **17**, 159 (1976)
23. A. Peker, Ph.D. Thesis, California Institute of Technology, (1994)
24. S. Schneider and W.L. Johnson, this proceedings
25. H. Bruck, T. Christman, A.J. Rosakis, and W.L. Johnson, Scripta Met., **30**, 429 (1994)
26. L. Davis, in Metallic Glasses, (American Society for Metals, Metals Park, Ohio, 1978), Chapter 9.

SYNTHESIS AND PROPERTIES OF BULK AMORPHOUS AND NANO-CRYSTALLINE ALLOYS

S.J. POON,[1] Y. HE,[1] G.J. SHIFLET,[2] AND G.M. DOUGHERTY[2]
[1]Department of Physics, [2]Department of Materials Science and Engineering, University of Virginia, Charlottesville, VA 22903 U.S.A.

ABSTRACT. High-strength Aℓ-based amorphous alloys have been produced in sub-millimeter thick ribbons (sheets) by single-roller melt-spinning. The excellent glass formability of Aℓ-alloys has been studied by thermal and structural experiments. Amorphous Aℓ-alloy powders can be consolidated to form bulk amorphous or nanocrystalline Aℓ-alloys as high-strength, low-density engineering materials. Submillimeter Mg-(Cu,Ni)-Sn amorphous ribbons with good mechanical properties can also be produced. Recently, refractory metal based rod-shaped samples containing mixtures of amorphous and nanocrystalline phases have been produced by mold casting, and their properties have been studied. Based on these results as well as those from the variety of bulk-glass forming systems reported recently, several key factors for easy vitrification are suggested.

1. Introduction

It is well-known that metallic glasses exhibit high mechanical strengths and corrosion resistance relative to crystalline alloys at comparable compositions [1]. Unlike the brittle oxide glasses used in composite materials, metallic glasses have not only high strengths but also good ductility, resulting from the nature of their metallic bonds. For metallic glasses to be employed as structural and other strategic materials they must be fabricated in bulk form. Until recently, with the exception of a few alloys based on noble metals [2], metallic glasses could only be produced in the form of ribbons or sheets with thickness of ~20-50 μm, which hinders their applications. In the past three years, a number of new alloy systems, not based on noble metals, have been found to be favorable for the formation of bulk metallic glasses via casting, they include Aℓ-Ni-Co-Y [3], Aℓ-Ni-Fe-Gd[4], La-Aℓ-Ni[5], Mg-Cu-Y[6], Zr-Aℓ-Ni-Cu [7], and Zr-Ti-Cu-Ni-Be [8]. For the Zr-Ti-Cu-Ni-Be alloys, amorphous rods with diameters up to 14 mm using a water-quenching method at a critical cooling rate less than 10 K/s were reported. This preliminary success in vitrifying a variety of alloy melts by moderate to slow cooling reminisces the early days of metallic glasses when

M. A. Otooni (ed.), Science and Technology of Rapid Solidification and Processing, 43–52.

an increasing number of glass-forming systems gradually emerged. Thus, bulk-glass formation in many more alloy systems can be expected. These new alloys and the study of them, in particular their structural and electronic properties, will provide important clues to the formation of bulk metallic glasses for scientific research and application.

In this paper, we first summarize and discuss results on the unique metallic glass formability and ultra-high tensile strength in submillimeter thick Aℓ-Ni-Fe-Gd ribbons. We will also discuss bulk-glass formation in Mg-based systems. The mechanical properties of consolidated amorphous Aℓ-alloy powders will be discussed. Finally, we report on the formation of bulk nanocrystalline and glassy rods in Ti- and Ni-based alloys. Based on our investigation of unconventional glass formation in Aℓ-alloys, as well as, recent bulk-glass formation in a number of alloys that are in all actuality extensions of previous binary glass-forming systems, several favorable factors for easy vitrification of metallic melts are pointed out.

2. Experimental

The making of alloy ingots, melt spinning by use of a copper single-roller, characterization of structure, study of thermal properties, and evaluation of mechanical properties follow those procedures described in our previous papers [4,9]. Mold-casting experiments were performed by induction-melting the ingot in a partial helium atmosphere, then injecting the melt by an over-pressure of ~0.3 MPa into a water-cooled copper mold with inner diameter varying from 3 to 6 mm. To prepare TEM specimens for microscopy study, the rods were sliced into small disks about 500 μm thick, followed by mechanical polishing to ~80 μm. The disks were then electropolished in 3:1 methanol to nitric acid solution at 243K, followed by ion milling for several hours, also at 243K. Consolidation of amorphous Aℓ-alloy powders was carried out by first cold-pressing in a die at pressures of 250 psi (1725 MPa), followed by treating the pressed buttons in a hot isostatic press (HIP) at 20 ksi (138 MPa) hydrostatic pressure and 300 C for 30 hours to test for further consolidation [10]. Because of the high strength of the amorphous powder it proved impossible to consolidate to 100% density using these methods, so microhardness testing was done by using a high-precision microhardness tester with applied Vickers loads of 200 to 500 grams.

3. Results and Discussion

3.1 THICK Aℓ-GLASS RIBBONS

Aℓ-Ni-Fe-Gd alloy melts that contain 85-87 at.% Aℓ and 5-8 at.% Gd can be vitrified to form ribbons with thickness up to ~0.25 mm, obtained at a tangential speed of the wheel of about 3 m/s by the single-roller melt-spinning technique [4,11]. Table 1 summarizes the critical (largest) thickness of ribbons for different Aℓ content. Tensile strengths as high as 1280 MPa and Young's modulus of up to 75 GPa for these ribbons have been measured. It is evident that these tensile strengths are at least two times higher than those found in commercial Aℓ-alloys, while the densities are only 20-30 % higher than a commercial 2219 alloy. The specific strength (tensile strength/density) of Aℓ-Ni-Fe-Gd glasses ~40 is approaching that of some of the strongest metallic glasses (~50).

Table 1. Compositional dependence of the lowest wheel speed required to make amorphous alloys. x_c is the critical thickness.

Composition	Lowest wheel speed (rpm)	x_c (μm)
$A\ell_{85}Ni_5Fe_2Gd_8$	400	200
$A\ell_{85}Ni_6Fe_3Gd_6$	300	250
$A\ell_{85}Ni_7Fe_2Gd_6$	600	150-175
$A\ell_{87}Ni_6FeGd_6$	400	250
$A\ell_{83}Ni_6Fe_3Gd_8$	600[a]	150-175
$A\ell_{83}Ni_8Fe_2Gd_7$	600[a]	150-175
$A\ell_{83}Ni_{10}Fe_2Gd_5$	600[a]	150-175

[a]Mainly amorphous with some trace of crystalline phases when quenched at this speed

Since the initial discovery of Aℓ-glasses in Aℓ-TM-RE (TM=transition metal, RE=rare earth) ternary alloys [9,12], it was immediately pointed out by the present authors that glass formability in these alloys was unique. This is because for conventional amorphous alloys, the glass formation range often coincides with a deep eutectic region where the liquid structure is stable to a significantly lower temperature than its components [13]. On the contrary, the liquidus temperature of Aℓ melt increases upon alloying with TM or RE. Figure 1 is a DTA trace for $A\ell_{85}Ni_6Fe_3Gd_6$ alloy. It shows that this alloy begins to melt at T_e=630C (eut-ectic melting), and its liquidus temperature is T_ℓ=900C (1173K). DSC measurement, on the other hand, showed that the glass temperatures for these quaternary Aℓ-glasses are around 500-600K, as illustrated in figure 2. From these results, the reduced glass temperature $T_{rg}=T_g/T_\ell$ is estimated to be ~0.49, which is much lower than that of conventional

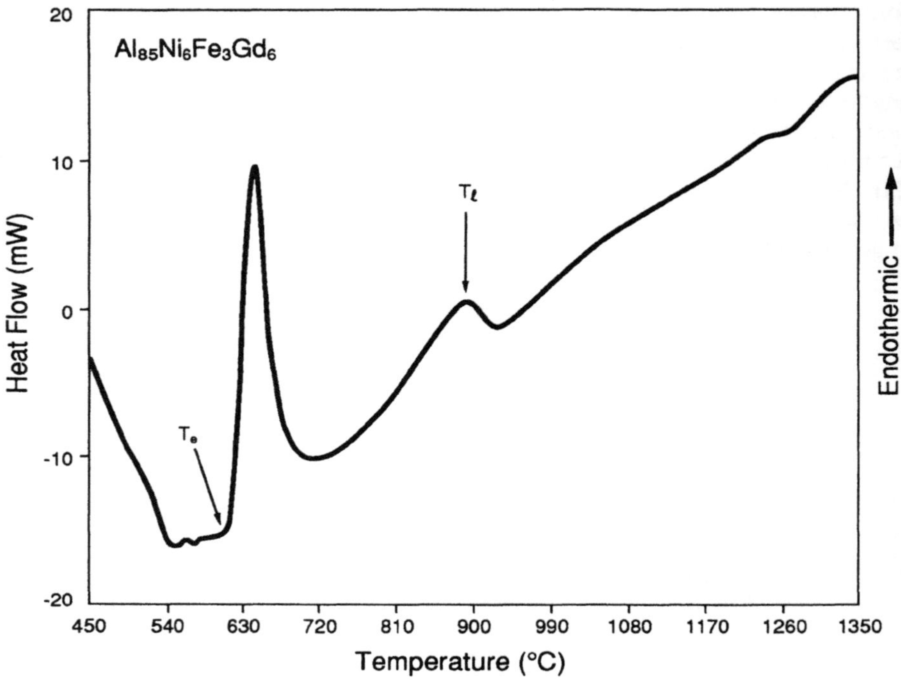

Fig. 1. DTA trace for $A\ell_{85}Ni_6Fe_3Gd_6$ metallic glass. T_e = 900°C is the liquidus temperature. T_e is the eutectic melting temperature.

good glass formers whose T_{rg} ~0.6-0.7 [13]. In fact, T_{rg} for Aℓ-glasses are close to that of the lowest limit in glass forming systems previously observed. It is interesting to note that in chalcogenides alloys such as Ge_xSe_{1-x}, easy glass formation (via air or water quenching) has been observed in the region where the system exhibits only low T_{rg} (0.4-0.5). However, high cooling rates are required to melt-quench metallic alloys with such low T_{rg}'s into amorphous states. Unexpectedly despite the low T_{rg} for Aℓ-Ni-Fe-Gd alloys, only low cooling rates are necessary for the glass formation. This suggests that, to some degree, the glass forming behavior in Aℓ-alloys is similar to that in chalcogenide systems. It is this unique glass formability which makes it possible to produce bulk amorphous Aℓ-alloys.

3.2 CONSOLIDATION OF AMORPHOUS ALUMINUM-ALLOY POWDERS

Die-pressing of amorphous $A\ell_{80}Ni_8Fe_4Gd_8$ powders produced by ball-milling yields

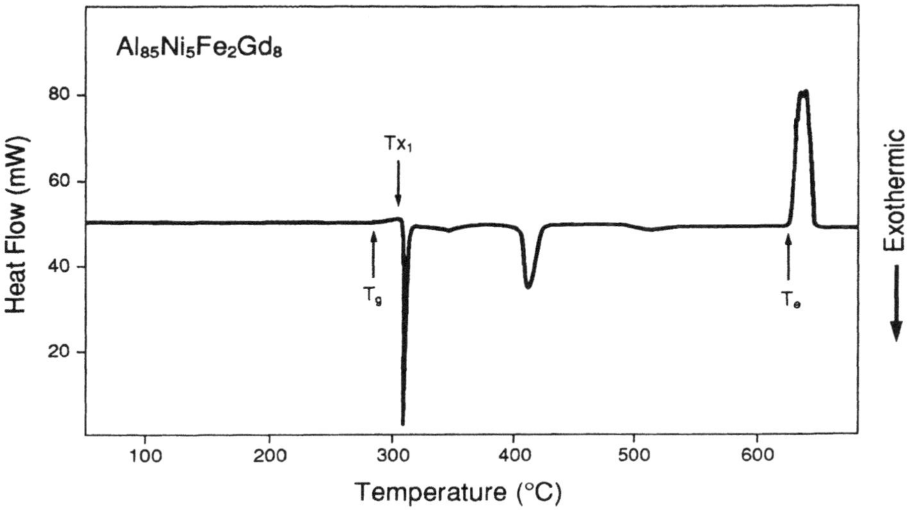

Fig. 2. DSC scan for $A\ell_{85}Ni_5Fe_2Gd_8$ metallic glass. The glass transition temperature is T_g = 570 K. T_{x1} is the first crystallization temperature.

buttons containing as little as 7% volume fraction of voids, with no strain-induced crystallization detected. The thermal stability of the product is characterized by a crystallization temperature of 360C. Following HIP treatment at ~300C, no crystallization was observed after the maximum treatment time of 30 hours. There was no further consolidation of the buttons after high-temperature HIP treatment. Clearly, higher pressures are needed for full consolidation, such as those produced during warm extrusion [14]. The Vicker's hardness is found to be ~625-650 DPN, as compared with 100-190 for structural Aℓ-alloys. From measurement of cracks produced during the indentation process, the fracture toughness can, with confidence, be said to lie between 1 and 5 MPa/m , which is clearly in the range of engineering ceramics. Thus, amorphous Aℓ-alloy powders produced by ball-milling belong to the brittle kind. However, in general, fracture toughness can be enhanced through ductile phase toughening. Buttons of powders that contain a mixture of amorphous and nanocrystalline phases produced by ball-milling for a shorter time were employed to investigate whether the presence of softer nanocrystals in the amorphous phase would increase the fracture toughness. It is found that the presence of softer phases significantly reduces the Vickers hardness to approximately 390, while the fracture toughness appears to increase.

3.3 BULK GLASS FORMATION IN MG-BASED ALLOYS

The choice of alloy systems for experiment in sections 3.3 and 3.4 is mainly motivated by the idea of bulk-glass formation to be discussed in section 4. In view of the prominent presence of deep eutectics in glass-forming binary systems, easy glass formation in Mg-Cu-Y [6] can be thought of as conventional. That is to say by adding a third element to a binary alloy to further reduce its liquidus temperature, glass formability should be enhanced. It is found that Mg-(Cu or Ni)-Sn can readily form thick amorphous ribbons of ~0.2 mm [15]. Fracture tensile strength as high as 900 MPa is recorded for $Mg_{80}Ni_{15}Sn_5$ metallic glass, which is much higher than that of conventional Mg-based crystalline alloys. T_{rg} of ~0.52-0.59 measured for Mg-glasses, even though higher than those of Aℓ-glasses, are considered to be low for easy glass forming systems. However, it is plausible that the low liquidus temperature of Mg-alloys has rendered it technically easier to solidify the melt without crystallization.

3.4 BULK METASTABLE REFRACTORY-ALLOY RODS

In view of favorable glass formability in Ti-Cu and Ni-Nb alloys, we have begun to investigate formation of bulk metallic glasses and nanocrystalline materials in higher order alloys based on these binary systems. Preliminary results are illustrated here. Structure of mold-cast Cu-Ti-Zr-TM and Ni-Nb-TM (TM=Co,Ni, or Fe) was studied by powder x-ray diffraction (figure 3). The x-ray patterns show a broad amorphous band and some broad crystalline peaks indicating the fine grain size of the samples. TEM examination revealed the coexistence of amorphous and crystalline regions in the mold-cast $Cu_{50}Ti_{20}Zr_{15}Co_5Ni_{10}$ rod sample. The crystal grain size is ~200 nm, similar to that of the amorphous region. Figure 4(a) is a selected area electron diffraction pattern taken from the amorphous region. It clearly shows the diffuse diffraction ring pattern characteristic of amorphous phase. The diffraction spots in this micrograph is probably due to overlapped crystalline region. Figure 4(b) is the bright field image of the area corresponding to figure 4(a).

The microhardness of several Cu-Ti-Zr-Co-Ni rods was measured with a load of 1 Kg. The values for $Cu_{50}Ti_{20}Zr_{15}$ Co_5Ni_{12} , $Cu_{50}Ti_{20}Zr_{13}Co_5Ni_{10}$, and $Cu_{50}Ti_{20}Zr_{15}Ni_{15}$ are H =585, 556, and 533 DPN, respectively. Compare to other metallic glasses, these values are very close to that of $Ti_{50}Cu_{50}$ (610), and they are significantly higher than Aℓ-glasses of the ductile kind (250-350), but half that of $Fe_{80}B_{20}$ (1080).

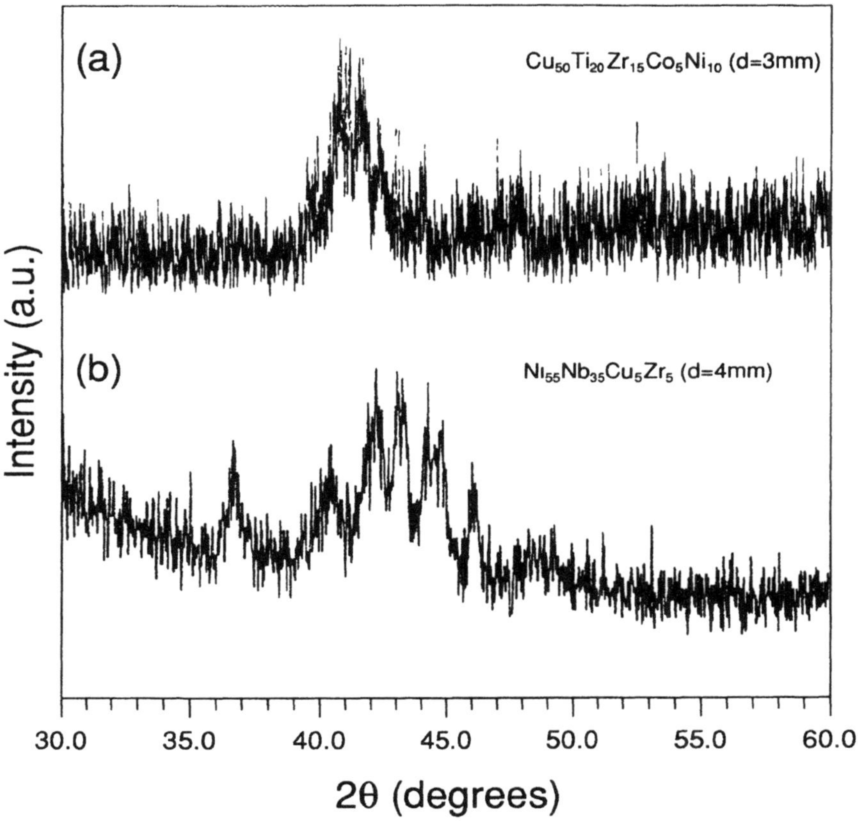

Fig. 3. Powder x-ray diffraction patterns of (a) $Cu_{50}Ti_{20}Zr_{15}Co_5Ni_{10}$ and (6) $Ni_{55}Nb_{35}Cu_5Zr_5$ rods produced by mold-casting.

4. The Origin of Easy Glass Formation

The unusual glass formability of Aℓ-alloys and also to some extent that of Mg-alloys may be understood in terms of strong interatomic interaction between the components. Structural studies by x-ray and neutron scattering measurements have shown ~7.5% shortening in the Aℓ-TM nearest-neighbor distance compared with (Aℓ+TM) atomic radii [16], which indicates strong Aℓ-TM interaction. Ab-initio investigations of electronic factors on phase stability of Aℓ-TM intermetallics [17] and crystal-analogs of quasicrystals known as approximants which have very complex structures [18] have demonstrated the importance of strong interatomic potential on phase stability. These

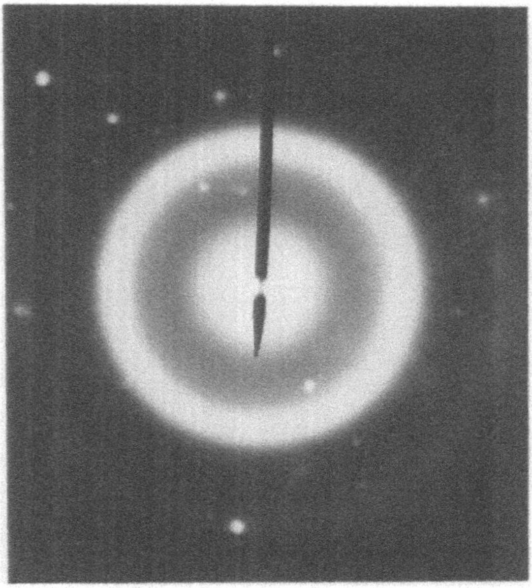

Fig. 4 (a) SAD pattern taken from thinked disk of $Cu_{50}Ti_{20}Zr_{15}Co_5Ni_{10}$ showing diffuse rings characteristic of an amorphous pattern.

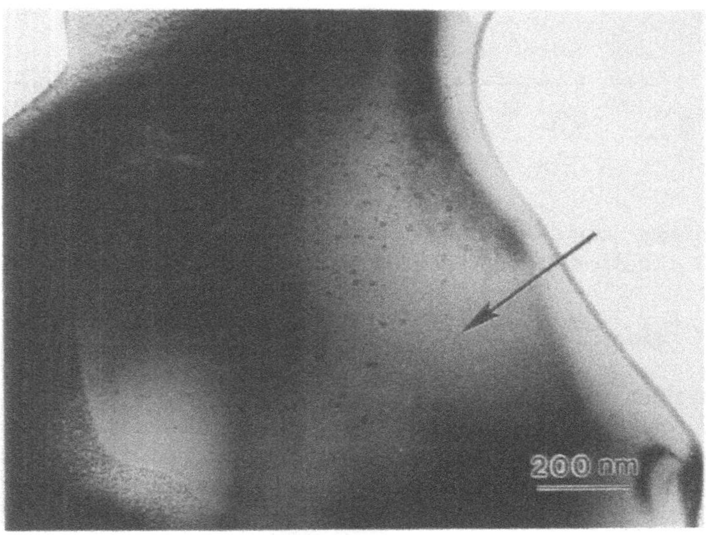

Fig. 4 (b) Bright field image of the area corresponding to (a).

results can also be viewed as realization of the Hume-Rothery rule in real space. The enhanced vitrifiability witnessed upon adding Sn to glass-forming binary alloys of Mg-Cu and Mg-Ni may also be due to the strong associaton of Mg-Sn pairs in Mg-Sn compounds [19], resulting in a 9% shortening in Mg-Sn distance.

Of course, just having strong atomic-pair interaction may be a favorable but not sufficient condition for glass formation, since strong interaction only implies compound formation but does not specify the structure. Nevertheless, in high order alloys (those that contain at least three components) derived from binary alloys with strong pair associations and also forming vastly different crystal structures, complex structures are favored upon solidification in order to accomodate the different types of pair associations as phase separation is suppressed by the strong atomic-pair interactions that are present. From the dense-packing viewpoint, diversity in crystal structure may be favored when there is a significant difference in the atomic size. Based on this discussion, it seems clear that high order alloys of elements that range in atomic size and are strongly interacting are better glass forming systems than their binary counter-parts. It should be emphasized that although the picture of glass formation depicted here does not involve the idea of deep eutectic, the existence of the latter does imply strong interaction and thus it can serve as a first-order guide to bulk-glass formation. Examples are seen in the Mg-, Zr-, and La-based high order alloys [5-8,15] as well as the more current Cu- and Ni-based alloys that form bulk glasses. In the near future, more high order alloys derived from previously studied glass forming systems can be expected to form bulk glasses. On the other hand, strong interaction does not necessarily imply the existence of a deep eutectic, as exemplified by the $A\ell$-TM systems that also tend to form bulk glasses (section 3.1). This distinction aside, the role of atomic-pair interaction and number of components as well as atomic size difference on bulk vitrification in the two classes of high order alloys is equally evident. We close by pointing out that one should extend the search for easy glass formation to alloy systems without a deep eutectic.

Acknowledgement - This research is sponsored by U.S. Army Research Office under Contract No. DAAL 03-91-G-0009.

References

1. For reviews on metallic glasses, see articles in "Amorphous Metallic Alloys", ed. F.E. Luborsky, Butterworths, Boston, 1983.
2. A.J. Drehman, A.L. Greer, and D. Turnbull, Appl. Phys. Lett. 41, 716 (1982) and references sited therein.
3. A. Inoue, N. Masumoto, and T. Masumoto, Mater. Trans. JIM 31, 493 (1990).
4. Y. He, G.M. Dougherty, G.J. Shiflet, and S.J. Poon, Acta Metall. Mater. 41, 337 (1993).

5. A. Inoue, T. Zhang, and T. Masumoto, Mater. Trans. JIM 31, 425 (1990).
6. A. Inoue, A. Kato, A. Zhang, S.G. Kim, and T. Masumoto, Mater. Trans. JIM 32, 609 (1991).
7. T. Zhang, A. Inoue, and T. Masumoto, Mater. Trans. JIM 32, 1005 (1991).
8. A. Peker and W.L. Johnson, Appl. Phys. Lett. 63, 2342 (1993).
9. Y. He, S.J. Poon, and G.J. Shiflet, Science 241, 1640 (1988); G.J. Shiflet, Y. He, and S.J. Poon, Scripta Metall. 22, 1661 (1988); Y. He, S.J. Poon, and G.J. Shiflet, Scripta Metall. 22, 1813 (1988).
10. G.M. Dougherty, G.J. Shiflet, and S.J. Poon, to be published in Acta Metall. Mater.
11. G.M. Dougherty, Y. He, G.J. Shiflet, and S.J. Poon, Scripta Metall. 30, 101 (1994).
12. A. Inoue, K. Ohtera, A.P. Tsai, and T. Masumoto, Jpn. J. Appl. Phys. 27, L479 (1988).
13. H.A. Davies, Chapter 2 in ref. 1, p. 8; F. Spaepen and D. Turnbull, Ann. Rev. Phys. Chem. 35, 241 (1984).
14. K. Ohtera et al., Mater. Trans. JIM 32, 331 (1991).
15. S.Y. Su, Y. He, G.J. Shiflet, and S.J. Poon, to be published in Mater. Sci. Engng.
16. H.Y. Hsieh, B.H. Toby, T. Egami, Y. He, S.J. Poon, and G.J. Shiflet, J. Mater. Res. 5, 2807 (1990); H.Y. Hsieh et al., J. Non-Cryst. Solids 135, 248 (1991).
17. J. Zou, Ph.D. Dissertation, Dept. of Physics, Washington University, (1993); J. Zou, R. Phillips, and A. Carlsson, to be published in Phys. Rev. B.
18. J. Hafner and M. Krajci, Europhys. Lett. 17, 145 (1992) and references cited therein.
19. In Pearson's Handbook of Crystallographic Data for Intermetallic Phases, ed. by P. Villars and L.D. Calvert, American Society for Metals, 1985, New York.

ALUMINUM-RICH METALLIC GLASSES

G.J. SHIFLET, Y. HE[+], S.J. POON[+], G.M. DOUGHERTY and H. CHEN*
Department of Materials Science and Engineering, [+]Department of Physics
University of Virginia, Charlottesville, VA 22903
**Center for Advanced Materials, Department of Chemistry*
University of Massachusetts, Lowell, Massachusettes 01854

ABSTRACT. This paper will review work done in our laboratory on the formation, thermodynamics and crystallization of aluminum based metallic glasses. The results demonstrate that the metallic glasses are truly amorphous and not microcrystalline and that they are unusual in their formation because the alloys contain up to 90 at.% aluminum, there is no deep eutectic present in the phase diagram and the reduced glass temperatures are low (0.44) relative to the ease of glass formability in this system. Tensile strengths that exceed 1200 MPa are obtained for several of the compositions and reveal increased strength values after partial crystallization. Various methods to form nanocrystals of aluminum in an amorphous matrix are discussed and include mechanical and thermal processing.

1. INTRODUCTION

Although metallic glasses have been studied for more than two decades very few were based on aluminum (1). Aluminum-based glasses were actively sought both for their scientific interest but also because of their possible technological importance. Only small fractions of glassy phase have been observed in various Al-binary alloys produced by rapid solidification, most notably Al-X (X=Si, Ge, Cu, Ni, Cr, and Pd) alloys. The first successful production of a homogeneous Al-based amorphous phase was achieved in Al-(Fe,Co)-B (2) and Al-Fe-(Si,Ge) ternary alloys (3). However, these amorphous alloys are extremely brittle, and their tensile strength is as low as about 50-150 MPa, therefore, they have no practical applications. Then in 1988 a new family of metallic glasses containing up to 90 at.% Al was discovered (4,5). In our group, these Al-rich metallic glasses were discovered during the search for aluminum-rich quasicrystals in the $Al_{20}Cr_2Ce$-type alloys ($Al_{18}Cr_2Mg_3$ structure, 184 atoms per unit cell) (6). In this process, an Al-rich metallic glass of composition $Al_{20}Fe_2Ce$ was obtained, which was produced in continuous ribbons by single-roller melt spinning. This has led to the synthesis of many metallic glasses containing up to more than 90 at.% aluminum. Many of these metallic glasses were found to be flexible with good local ductility, and the amorphous ribbons can be bent 180 degrees without fracture. It was also found that the formation range in this new family of metallic glasses can be as wide as 20 at.% or more. They exhibit unique mechanical and corrosion-resistant behavior resulting from their amorphous structure (4,7). The glasses were found to have excellent strength per weight, or specific strength (8-10). Typical ex-

M. A. Otooni (ed.), Science and Technology of Rapid Solidification and Processing, 53–71.

amples of these glasses are Al-TM-RE (TM=Fe, Co, Ni, etc.; RE=Ce, Y, Gd, etc.). Quaternary alloys include Al-TM1-TM2-RE. The tensile strength of $Al_{90}Fe_5Ce_5$ reaches as high as 940 MPa, while its physical density is only 3.21 g/cm^3. These strengths greatly exceed those of conventional aluminum alloys.

The formation of aluminum rich metallic glasses is anomalous in that the glasses are formed in the absence of a eutectic in the equilibrium phase diagram (11). Also, it is unusual that metallic glasses have such a high concentration of the primary element, in this case Al. Thus, from both scientific and technological points of view, it is important to characterize the atomic scale structures, thermodynamics and properties of the aluminum based metallic glasses and to understand their unusual formability. This paper will review some of the results found in our laboratory over the past few years.

2. EXPERIMENTAL

The alloys were prepared by arc melting nominal amounts of elements in an argon atmosphere. Melt spinning was carried out in a partial helium atmosphere using a 0.20 m diameter copper wheel with variable circumferential velocity. Samples are formed with dimensions of 15-250 μm thick, 1-2 mm wide, and up to several meters long. Partially transformed materials were secured by first encapsulating in an argon atmosphere and then isothermally heat treated. Ball milling was performed with a SPEX-8000 laboratory ball mill. Further details can be found in our original papers.

3. RESULTS AND DISCUSSION

3.1. FORMATION AND STABILITY

3.1.1 *Wholly Glassy Material.* We have followed two routes to produce aluminum glass, *viz.*, melt-spinning and mechanical alloying. With the formation of Al-Fe-Ce glassy alloys, by melt-spinning, as the basis for obtaining other aluminum glasses, many other compositions were studied by replacing Fe with Co, Ni and Rh; and Ce with other lanthanides, Y, and Hf (Table 1). Within an individual system, i.e., Al-Fe-Gd, the effect of alloy composition of the transition and rare earth elements on crystallization temperature, onset of melting and enthalpy is given in Table 2. All of the ternary systems could be produced in glassy form only if the ribbon thickness was maintained less than about 20 μm, that is, at high spinning rates. The effect of adding a second TM was examined in the Al-Fe-Gd system by adding Ni (12). The system was then optimized (Fig. 1) to produce higher strengths (discussed in Section 3.2) and ribbon thicknesses approaching 0.25 mm. Further details are given in the paper by Poon *et al.* in this volume (13).

One of the most important parameters influencing the glass formation ability of the alloy system is the reduced glass temperature T_{rg} (14-16), which is the ratio of the glass transition temperature T_g to the liquidus temperature T_l. The glass transition temperatures and liquidus temperatures of Al-Ni-Fe-Gd alloys were measured by DSC and DTA respectively.

TABLE 1. Mechanical Properties of Amorphous and Partially Crystallized (PC) Metallic Glasses (from refs. 8,9,10)

Alloys	Structure	Heat Treatment	Tensile Strength	Elastic Modulus
$Al_{90}Fe_5Gd_5$	PC	445 K, 20 min.	1010 MPa	49 GPa
$Al_{90.5}Fe_{4.75}Gd_{4.75}$	PC	as-spun	938	45
$Al_{90}Fe_5Gd_5$	amorphous	as-spun	735	51
$Al_{87}Fe_{8.7}Gd_{4.3}$	amorphous	as-spun	835	53
$Al_{87}Fe_{6.6}Gd_{6.3}$	amorphous	as-spun	810	58
$Al_{87}Fe_{7.7}Gd_{5.3}$	amorphous	as-spun	850	51
$Al_{90}Fe_5Ce_5$	amorphous	as-spun	940	66
$Al_{87}Ni_{8.7}Ce_{4.3}$	amorphous	as-spun	500	46
$Al_{87}Ni_{8.7}Y_{4.3}$	amorphous	as-spun	880	50
$Al_{87}Co_{8.7}Ce_{4.3}$	amorphous	as-spun	790	63
$Al_{87}Rh_{8.7}Ce_{4.3}$	amorphous	as-spun	730	63

A differential scanning calorimetry trace for amorphous $Al_{25}Fe_2Gd$ is shown in Fig. 2. Four exothermic peaks can be seen. The first exothermic peak, at 245 °C, is associated with the primary crystallization of aluminum. When as-quenched ribbons were also annealed for five minutes at the temperature indicated by point A, the X-ray diffraction pattern showed only aluminum peaks and an amorphous background. The volume fraction of aluminum is estimated to be about 30%.

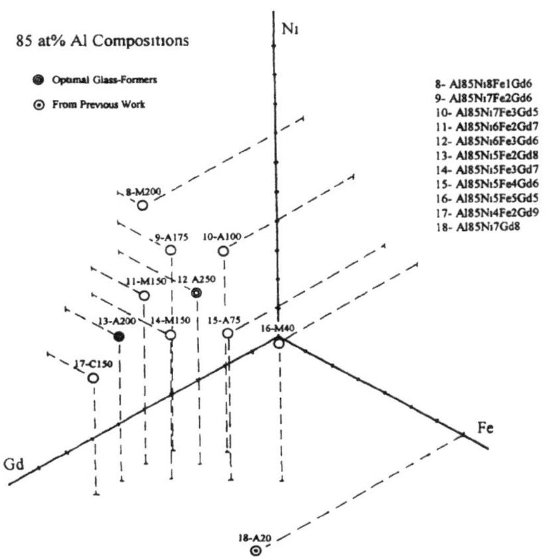

FIGURE 1. Optimization map for producing thick metallic glass ribbons (12).

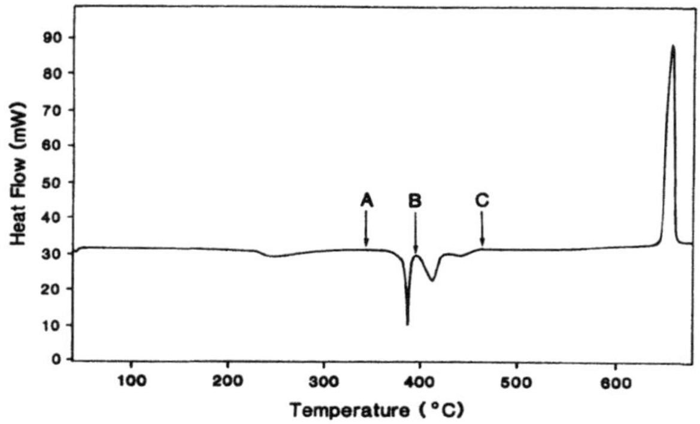

FIGURE 2. The DSC trace for $Al_{25}Fe_2Gd$ metallic glass (11).

The second exothermic peak at 387 °C, which is much sharper, is associated with the crystallization of the remaining glassy matrix. The amorphous sample annealed at point B was found to contain mainly f.c.c. Al and $Al_{10}Fe_2Gd$ phase, as well as a trace amount of the $Al_2(Fe,Gd)$ phase (of the Cu_2Mg structure in this case). The majority phases were also formed upon crystallizing the other Al_xFe_2Gd glasses. Samples annealed at point C contained the same phases as those at point B. Therefore, the third exothermic peak may be related to some yet unknown changes in the microstructure. Finally, the endothermic peak at 658 °C represents the beginning of the solid-to-liquid transformation. The latter is also confirmed by direct observation of the sample melting while encapsulated in a pyrex tube. In Table 2 the onset of melting is denoted as T_e and represents the beginning of the endothermic curve. Upon complete crystallization, aluminum and the $Al_2(Fe,Gd)$ phase are formed in the Al_xFeGd alloys. For $Al_{40}Fe_3Gd$, $Al_{30}Fe_2Gd$ and $Al_{18}FeGd$ where aluminum concentration exceeds 90 at.%, the primary crystallization of Al occurs at a lower temperature and extends over a wider temperature range. However, reducing Al concentration to 85 at.% and below, primary crystallization of Al is no longer observed. Other thermodynamic data obtained from DSC scans listed in Table 2 include T_x , the temperature where the sample completely transforms to crystalline phase, corresponding to the on-

TABLE 2. Thermodynamic Data of Al-Fe-Gd Metallic Glasses Obtained From DSC (ref. 11)

Alloy	T_x(°C)	T_e(°C)	ΔH_x(kJ/mol)
$Al_{40}Fe_2Gd$	366	649	-3.25
$Al_{30}Fe_2Gd$	374	647	-4.15
$Al_{25}Fe_2Gd$	383	648	-4.75
$Al_{20}Fe_2Gd$	373	638	-5.87
$Al_{17}Fe2Gd$	331	636	-5.48
$Al_{15}Fe_2Gd$	346	648	-4.54
$Al_{18}FeGd$	371	638	-4.25
$Al_{15}FeGd$	365	638	-4.62
$Al_{13.3}FeGd$	369	637	-4.80
$Al_{11.3}FeGd$	340	637	-4.97

set of the second exothermic peak in Fig. 2. ΔH_x is the change of enthalpy between the initial amorphous state and the final crystallized state prior to melting, which is the integral of all exothermic peaks.

It is well known that to obtain an amorphous alloy, the metal melt must be cooled fast enough from its liquidus temperature T_l, to below the glass transition temperature T_g where the supercooled liquid solidifies. Thus, a reduced glass temperature can be defined as $T_{rg}=T_g/T_l$ which represents a measure of glass forming ability. The basis for this is that the higher the T_g and the lower the T_l, the easier to supercool the metal melt to a glassy state. Conventional theory predicts T_{rg}=0.65 ~0.7 for good glass formers (e.g. T_{rg} ~ 0.63 for $Pd_{77}Cu_6Si_{17}$) where a quenching rate as low as 100 ~ 1000 K/s is sufficient to produce the glassy state (14-16). However, substantially smaller T_{rg} values were observed for Al-Ni-Fe-Gd metallic glasses, in fact, we have T_{rg}~0.44 for $Al_{85}Ni_5Fe_2Gd_8$ and T_{rg}~0.49 for $Al_{85}Ni_6Fe_3Gd_6$, and these values are close to the lowest limit of T_{rg} in glass forming metallic systems previously observed. High cooling rates (10^9-10^{12} K/s), however, are normally required to melt-quench metallic alloys with such low T_{rg}'s into amorphous states. Unexpectedly despite the low T_{rg} for Al-Ni-Fe-Gd alloys, only low cooling rates are necessary for the glass formation. The observed good glass formability combined with a low reduced glass transition temperature once again strongly suggest that the formation of metallic glass in Al-Ni-Fe-Gd alloys is unique, and to some degree, the glass forming behavior in Al-Ni-Fe-Gd alloys is similar to that in chalcogenide systems. It is this unique glass formability which makes it possible to produce bulk amorphous alloys based on Al.

Information that can help elucidate the unique glass formability in Al-Ni-Fe-Gd alloys must ultimately be related to the atomic interactions and the structure of the amorphous state. The atomic structure of amorphous $Al_{90}Fe_xCe_{10-x}$ has recently been studied using neutron and X-ray scattering (17,18). It was found that in these amorphous alloys, the distance of an Fe-Al pair is 0.02 nm shorter than the sum of the atomic radii of Al and Fe, suggesting strong interaction between Al and Fe atoms. Further structural analysis of these data reveals that Fe atoms are surrounded by approximately ten Al atoms within 0.3 nm, and of these six are in close contact with Fe. It is also shown that the rare earth atoms in Al-Fe-Ce metallic glasses form a dilute dense random packing substructure and they repel each other, while the substructure of Fe atoms and the surrounding Al atoms are substantially different from the random packing. The strong Al-Fe interactions may be explained from an electronic structure point of view. The *d*-orbitals of Fe and the *s-p* orbitals of Al should hybridize strongly to form a bonding *d*-band, which will be filled by the transfer of electrons from *s-p* orbitals of Al as a result of high Fermi level of Al relative to that of Fe. This results in the increased covalency in the bonding. The strong Al-Fe interaction suggested from structural studies is consistent with the large negative volume additivity of the Al-Fe alloys in the liquid or solid state (19). Gebhardt and co-workers reported that an Al melt actually contracts with small addition of Fe, and more remarkably, the shear viscosity of molten Al was sharply increased by dissolution of small amount of Fe (20). These results can be attributed to the strong atomic interaction between Al and Fe in the melt. Based on these analyses, the unusual glass formability in Al-TM-RE may be explained: the shear viscosity of the molten alloy is increased through Al-TM interactions, resulting in a high resistance to nucleation and crystallization, while the amorphous arrangement of Al-TM clusters is further stabilized by the randomly distributed rare-earth atoms. This behavior occurs through the whole supercooling process, to the temperatures lower than glass temperature T_g, assuring the glass formation.

Given the unusual formation behavior of rapidly solidified Al-based material, an area of some concern, then, is whether metallic glasses are indeed amorphous (21). Amorphous materials are generally characterized by a broad diffuse X-ray or electron diffraction pattern. However, it has been a subject of intensive debate whether a material exhibiting a broad diffuse X-ray diffraction pattern has a truly amorphous structure or whether it consists of randomly oriented microcrystallites or microquasicrystallites. Thus, techniques other than diffraction experiments are necessary to distinguish between the two. High-resolution electron microscopy (HREM) and differential scanning calorimetry can provide direct and indirect observations of the structural nature of metallic glasses. Modern medium-voltage high-resolution electron microscopy with a resolution better than 0.2 nm enables identification of microcrystallinity in a sample. Using calorimetric techniques to study the isothermal crystallization kinetics allows a differentiation to be made between amorphous and microcrystalline structures (22). Figs. 3-5 show amorphous aluminum glass and microcrystalline glass (23). The $Al_{90}Fe_5Gd_5$ samples observed in HREM do not reveal any crystalline structure in the sample (fig. 3). The Fourier transform of the HREM image shows structureless features. This provides direct evidence that the glass is truly amorphous. Annealed samples illustrate crystallization in the glass (fig. 4). A HREM image of one of the crystallites (fig. 5) reveals the periodic structure as an aluminum crystal embedded in a glassy matrix. X-ray's would not differentiate crystallites this small.

FIGURE 3. Underfocused HREM image of the rapidly solidified $Al_{90}Fe_5Gd_5$ alloy. No crystalline phases can be found at the nanometer scale (23).

To confirm that the matrix is amorphous following solidification, isothermal DSC experiments can be performed. The character of the isothermal transformation kinetics makes it possible to distinguish a microcrystalline structure from an amorphous structure assuming that the rate of heat released *dH/dt* in an exothermic transformation is proportional to the transformation rate *dx/dt*, where *H* is the transformation enthalpy and *x(t)* is the trans-

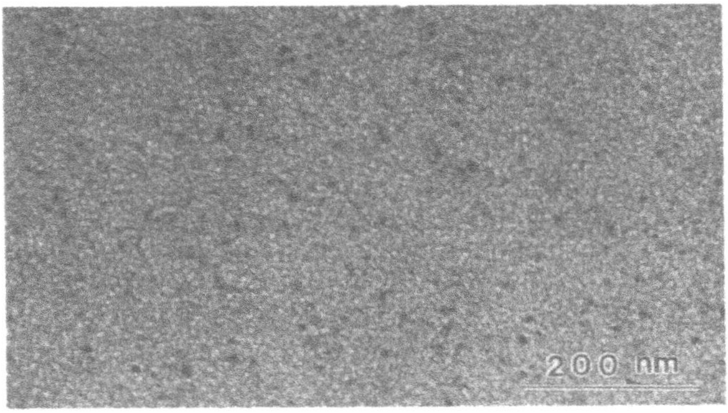

FIGURE 4. TEM image of a partially crystallized $Al_{90}Fe_5Gd_5$ metallic glass which has been annealed at 445K for 20 minutes. The various shades of contrast of the crystallites are due to random orientations of the aluminum crystallites (10).

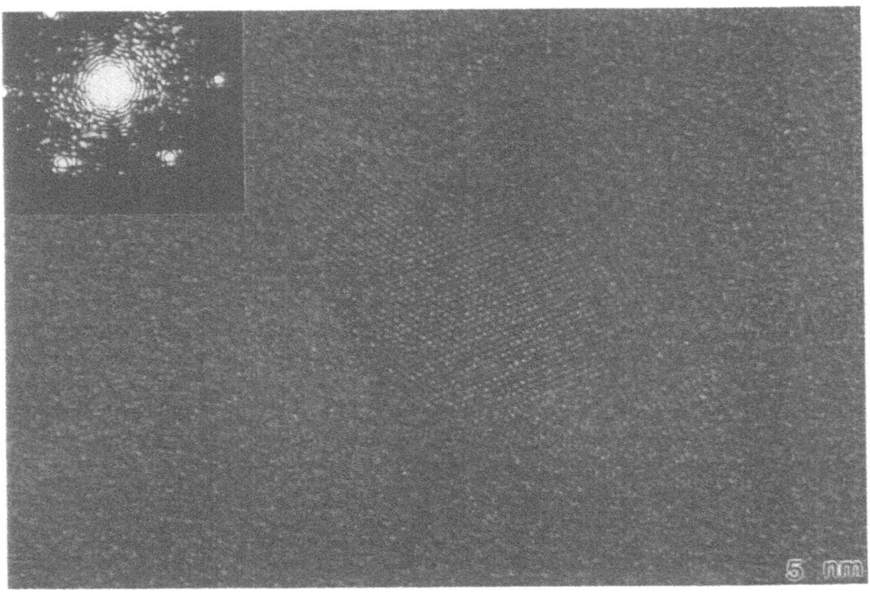

FIGURE 5. HREM image of the rapidly solidified and annealed $Al_{90}Fe_5Gd_5$ alloy. The inset represents the optical diffractogram, which is equivalent to a Fourier transformation and shows that the crystallite in the center of the micrograph is oriented along the 110 direction (23).

formed volume fraction at time t. If microcrystals do exist in a grain growth process, the isothermal calorimetric signal *dH/dt* is proportional to $1/r^{m+2}$, where r is the average grain size and *m* is a positive exponent (22). The heat release rate decreases monotonically with time, which is the case in a sputtered Al-Mn film studied by Chen and Spaepen (22).
In an amorphous material, the alloy, when heated to a constant, isothermal, temperature and maintained, will show a DSC trace as in Fig. 6 (23). This figure is not a characteristic of microcrystalline growth, but rather, can be well described by an isothermal nucleation-and-growth process based on the Johnson-Mehl-Avrami (JMA) transformation theory (24). The transformed volume fraction at time t can be written as

$$x = 1 - \exp\{-[K(t-\tau)]^n\},\ t > \tau,$$

where $K = K_0\exp(-E_x/RT)$ is the reaction rate constant, E_x is the activation energy in the crystallization process, which incorporates the activation energy for both nucleation and growth of the new phases. E_x generally has a value between 40 and 400 kJ/mol. T is the absolute temperature and R is the gas constant, with n>0 the Avrami exponent which varies between 1.5 to 4.0 (24).

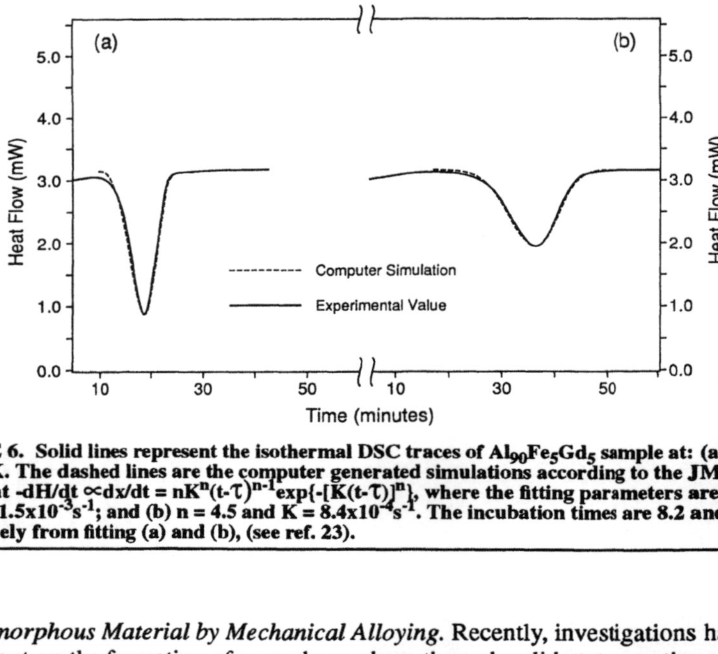

FIGURE 6. Solid lines represent the isothermal DSC traces of $Al_{90}Fe_5Gd_5$ sample at: (a) 605 K; and (b) 598 K. The dashed lines are the computer generated simulations according to the JMA model, using that $-dH/dt \propto dx/dt = nK^n(t-\tau)^{n-1}\exp\{-[K(t-\tau)]^n\}$, where the fitting parameters are: (a) n=4.5 and $K = 1.5\times10^{-3}s^{-1}$; and (b) n = 4.5 and $K = 8.4\times10^{-4}s^{-1}$. The incubation times are 8.2 and 17.9 min. respectively from fitting (a) and (b), (see ref. 23).

3.1.2 *Amorphous Material by Mechanical Alloying.* Recently, investigations have been carried out on the formation of amorphous phase through solid-state reactions, typically by the use of mechanical alloying and interdiffusion between thin films. Because these solid-state transformations do not rely on high-heat transfer rates, this suggests the possibility of forming bulk amorphous material through powder metallurgy with much larger dimensions than by the rapid-solidification techniques. Mechanical alloying can be performed by any of several methods, but the most common method is ball milling. Using this method, alloys are formed by placing the constituent elemental powders into a hardened metal jar with a number of hard metal balls, then sealing the jar (usually under some inert atmosphere to prevent powder oxidation) and placing the jar into a machine that shakes it for a long period of time, until the elemental powders are so intimately cold-welded and interdeformed that they transform into powdered alloy. In the case of easily amorphizable alloy systems, the powder product is often seen to be the amorphous phase. Never during the process does any material melt, therefore eliminating all consideration of rapid-solidification phenomena. At short milling times, when the elemental powders have not yet interdeformed to the extent that they have lost their individual identities, a fine structure of highly deformed cold-welded elemental layers is observed. Schultz was able to show that amorphous powders could be formed simply by mechanically milling (and thereby mechanically alloying) the elemental powders of an appropriate binary metallic system such as Ni-Zr (25). Particularly useful is the fact that the milling process itself produces enough localized thermal energy to drive the interdiffusion and lead to amorphization. The structure of the amorphous phase formed in this way is essentially the same as that produced through rapid-quenching (25). Application of this technique to $Al_{80}Ni_8Fe_4Gd_8$ (26) where the elements are mixed and ball-milled up to 80 hours is shown in Fig. 7 where X-ray diffraction traces, as a function of milling time, illustrates the amorphizing of this alloy. At shorter times, some amorphous phase is evident in conjunction with the crystalline peaks. Observations at various milling times reveal that after 1 hour of ball milling, the elemental powders have deformed to produce a well-ordered layered structure (Fig. 8a). The dark layers are pure Al and the bright layers are pure Gd. The harder Ni and Fe powder particles show much less plastic deformation, persisting as fairly distinct particles within the layered structure. The layers are of varying thickness, with an average value of a few microns. After two hours of milling, the layers have been folded upon one another many times and in the process have thinned considerably (Fig. 8b).

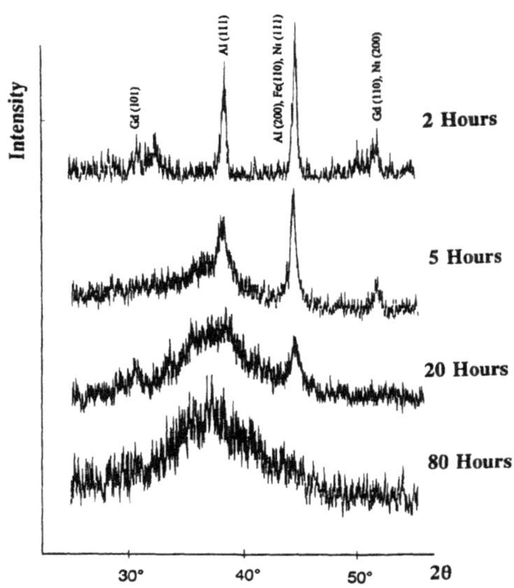

FIGURE 7. $Al_{80}Ni_8Fe_4Gd_8$ X-ray diffraction traces vs. milling time. The locations of the principal elemental crystalline peaks for the system are indicated.

Some of the bright areas are Ni and Fe particles that have now become more distorted with the increased milling time. TEM (Fig. 9) show that the 2 hour structure contains the initial formation of the amorphous phase. The structure consists almost entirely of layers of Al and Gd of varying thicknesses, with some extremely fine (<20 nm) while others are comparatively quite thick (>100 nm). Highly deformed regions containing the lower-concentration elements Fe and Ni can be found, lightly dispersed throughout the structure. EDS analysis shows that while the Al layers can be elementally pure to the resolution of the EDS system, the Gd layers invariably show the presence of some Al. This observation is central to the model of solid-state amorphization that these studies suggest. In addition, a number of layers are neither as dark as the Gd layers nor as light as the Al layers, show no extinction contours, and have a diffuse gray appearance in the images. These are layers of Al-Gd amorphous phase (indicated by arrows).

A more highly magnified TEM image of a well ordered region where the amorphization mechanism is apparent is shown in Fig. 10. The figure shows several layers of f.c.c.-Al, as confirmed by narrow beam CBED and EDS analysis, and two layers of Al-Gd amorphous phase. CBED patterns for an Al-Gd amorphous layer and for an Al crystalline layer are given in Fig. 10b and c, and the EDS trace for one amorphous layer is given in Fig. 10d. There are no crystalline Gd layers, suggesting that the amorphous layers have formed in place of existing Gd layers in the same locations.

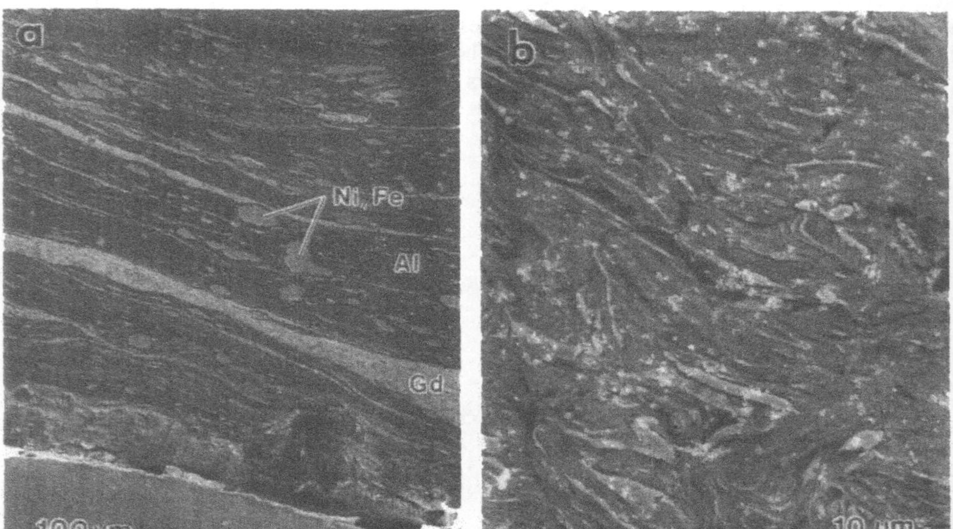

FIGURE 8. (a) SEM image of 1-hour and (b) 2-hour microstructures (26). $Al_{80}Ni_8Fe_4Gd_8$

At 80 hours (Fig. 11) the microstructure is approaching its maximum refinement, with only a few very small nanocrystals remaining (generally less than 10 nm in size). The diffraction pattern is a characteristic amorphous pattern. Dougherty *et al.* (26) described the microstructural evolution of the amorphous phase to indicate that the initial amorphization reaction takes place between critically thin layers of aluminum and gadolinium after roughly 2 hours of milling. Iron and nickel do not participate in the initial formation of amorphous phase. Aluminum diffuses into neighboring layers of crystalline gadolinium, and when the aluminum concentration becomes a significant fraction of the total composition layer, an amorphous character begins to be seen. The degree of amorphous character appears to be proportional to the aluminum concentration in the gadolinium parent layers.

In the case of thick gadolinium layers, the aluminum concentration does not easily reach a level high enough for appreciable amorphous character to appear, while in layers

FIGURE 9. TEM of 2-hour microstructure with arrows indicating amorphous regions, $Al_{80}Ni_8Fe_4Gd_8$ (26).

below the critical thickness interdiffusion is very thorough and the entire resultant gadolinium-aluminum layer transforms into amorphous Al-Gd phase. The critical thickness is on the order of 50 nm for the studied system.

3.1.3 Nanocrystals in an Amorphous Matrix. Crystallization need not always be deleterious to properties such as strength. In Al-based glasses, partially crystallized material actually increases the fracture strength by ~30%. A wholly amorphous $Al_{90}Fe_5Gd_5$ material obtained by melt-spinning and then annealed at 445 K for 20 min. resulted in the nucleation and growth of nanocrystals (figs. 4,5)(10). Because of the very high density (10^{23} per cubic meter) and uniform distribution, the efficacy of the nanocrystalline-amorphous matrix material in preventing the onset of fracture is high in this system.

We have found four different methods of producing nanocrystals. They are: 1) melt-spinning into wholly amorphous material followed by a thermal anneal (10); 2) melt-spinning with an alloy content rich in aluminum (10); 3) mechanical alloying (27); and 4) deformation induced crystallization (28).

The first two methods produce a microstructure similar to Fig. 4. Here the partially crystallized metallic glasses (whether as-spun $Al_{90}Fe_5Gd_5$, annealed at 445 K for 20 minutes or as-quenched $Al_{90.5}Fe_{4.75}Gd_{4.75}$) have the crystalline particles homogeneously embedded in the amorphous matrix. The particle density was estimated from both TEM and X-ray experiments to be about $4x10^{23}$ per cubic meter, which is typical for homogeneous nucleation. Fig. 5 is a HREM image of one of the crystallites. Tilting reveals the crystal structure to be f.c.c. with a lattice parameter of 0.4 nm, suggesting that the crystallites are aluminum. The size of the crystalline particles is about 8 nm and their morphology is roughly spherical.The volume fraction of crystallites for both the annealed alloy and the as-quenched, partially crystallized metallic glass is about 28%.

The final microstructure of an $Al_{90.5}Fe_{4.75}Gd_{4.75}$ alloy, when produce by melt-spinning only, i.e., no subsequent thermal treatment, is identical to Fig. 4. Indeed, the liquid has sufficient aluminum content that precludes a wholly amorphous material. This results in aluminum precipitates forming uniformly upon quenching.

The formation of localized shear bands and the deformation mechanism of amorphous alloys have been the subject of much theoretical discussion (29-31) and computer simulations (32,33). Localization of plastic flow requires that the material within the shear bands undergoes some structural changes, causing a local softening of the material, so the material inside the shear bands deforms more easily than the rest of the sample. In our experiments we used four different aluminum-based glasses, namely $Al_{90}Fe_5Ce_5$, $Al_{90}Fe_5Gd_5$, $Al_{87}Ni_{8.7}Y_{4.3}$ and $Al_{85}Ni_{10}Ce_5$. The ribbons were bent through 180 degrees and the area of the shear band examined with conventional and high resolution TEM. The seemingly "straight" shear bands are shown with SEM in Fig. 12. TEM shows the shear bands not to be straight (as expected) and to contain small precipitates Fig. 13. The apparent width of the shear band is due to tilting the sample in the microscope; tilting for minimum thickness reveals the band to be less than 10 nm. A high resolution image of the shear band precipitates shows they are identical to Fig. 5.

The structural changes in Al-based glasses caused by ball milling were investigated to further understand the deformation-induced crystallization observed in some aluminum-based glasses (27). In an amorphous $Al_{90}Fe_5Gd_5$ alloy, ball milling for 2 min. induced the formation of nanocrystalline Al particles (Fig. 14). These nanocrystals are roughly spherical, with a diameter of 10-20 nm, and the crystallographic changes with milling time can be followed employing X-ray analysis (Fig. 15). The results are consistent with the observation that nanocrystal formation can be induced within the shear bands of some Al-based glasses after bending, as discussed above. SEM micrographs of ball-milled $Al_{90}Fe_5Gd_5$ amorphous ribbons revealed that the resulting shear bands are similar to those produced by bending. After ball milling for a very short time (30 sec to 2 min), severe shear deformation is readily observed on the ribbon surfaces. The ribbons break into shorter pieces, and cracks are formed. Generally, three types of surface morphology of the deformed ribbons are observed. In some areas narrow shear bands are found which confine the plastic deformation. These shear bands are quite straight and similar to those produced by bending. Different areas reveal shear steps that can be rather wide and more irregular. Another different area where the plastic deformation is still more irregular and severe is also observed. This latter area has features similar to the vein or river pattern commonly observed

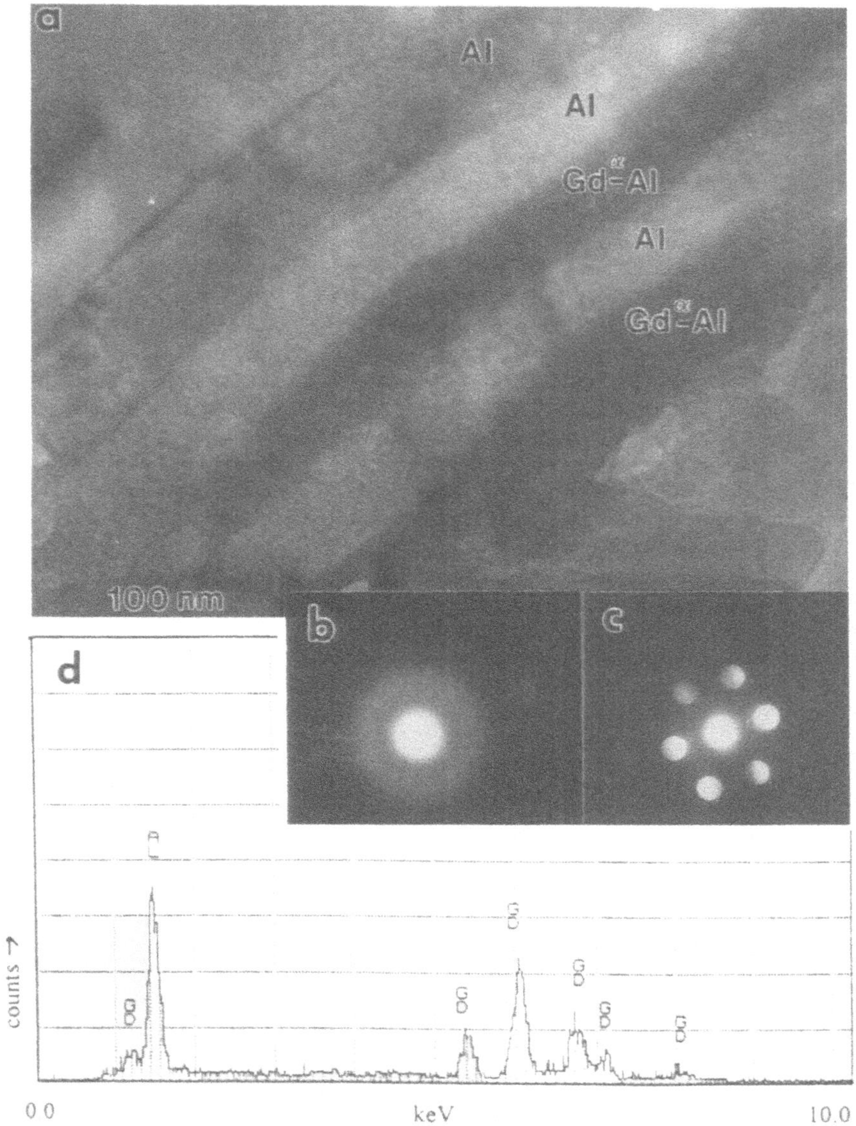

FIGURE 10. Formation of the amorphous phase from gadolinium thin layers. The TEM image in (a) shows three aluminum crystalline layers of Al-Gd amorphous phase. The symbol α denotes amorphous phase. CBED patterns for Al-Gd amorphous and Al crystalline layers respectively are given in (b) and (c). The EDS trace of an amorphous layer is given in (d), $Al_{80}Ni_8Fe_4Gd_8$.

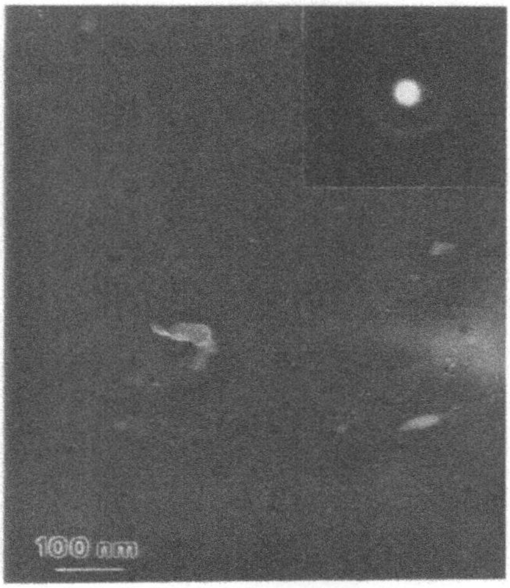

FIGURE 11. TEM image and selected area diffraction (SAD) pattern for $Al_{80}Ni_8Fe_4Gd_8$ sample ball milled for 80 hours (26).

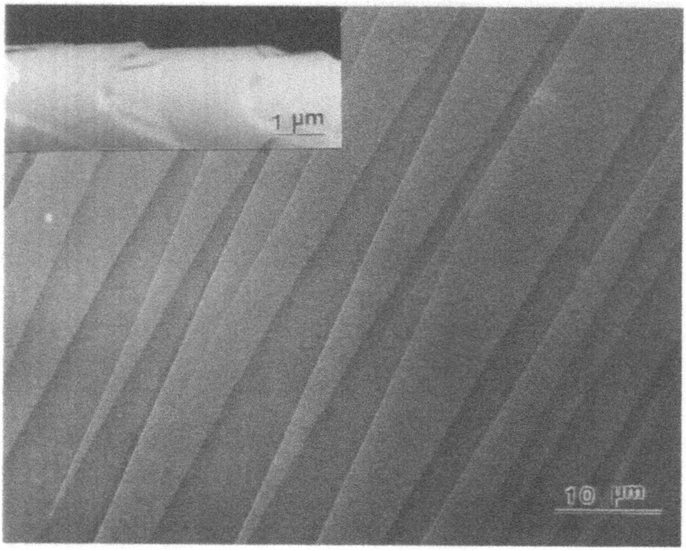

FIGURE 12. SEM picture of the surface of $Al_{90}Fe_5Gd_5$ metallic glass ribbon that has been bent through 180 degrees at room temperature. The straight lines are slip steps which are terminations of the shear bands on the surface. The inset picture is a profile of the surface steps (28).

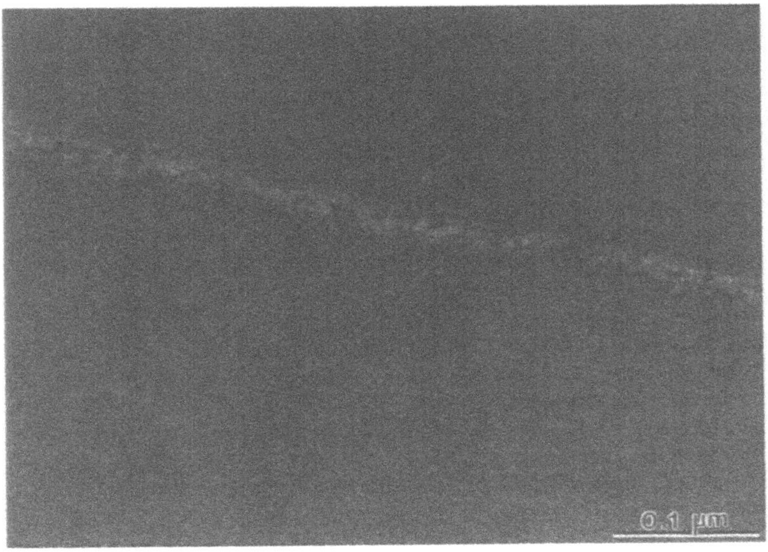

FIGURE 13. TEM micrograph of a slip band shown in Fig. 12. The white and black circles inside the shear band are the crystalline precipitates. The surrounding area is amorphous matrix. HREM images look identical to Fig. 5 (28).

on the tensile fracture surface of an amorphous alloy. The SEM patterns reveal that under ball milling the Al-based metallic glasses can have large plastic strain and the ribbons do not fracture in a brittle fashion. When subjected to prolonged milling (greater than 10 min.) the ribbons become flakes and eventually powders, and the typical particle size after milling for several hours in 10-20 microns. In other samples, with different chemistries, no such ball milling induced crystallization was observed, indicating that the mechanical deformation-induced structural changes are sensitive to the chemical composition and thus the atomic structure and the bonding nature of the metallic glasses. DSC measurements showed that certain structural change occurs in these amorphous alloys after milling even though no crystallization was observed (27). The formation of nanocrystals in the ball-milled amorphous alloys in attributed to the atomic displacements under high stresses during the deformation process. Further structural studies using X-ray and neutron scattering may help to elucidate the mechanism of such structural changes caused by mechanical deformation. Our results also confirm that the crystallization induced by mechanical deformation is different from that due to a thermal process. Thus, the mechanical process provides a new method of preparing amorphous powders embedded with a nanocrystalline phase. High-strength composite materials containing nanocrystals and amorphous matrix can therefore be synthesized by subsequent consolidation of these powders, using warm pressing, extrusion or explosive compaction processes.

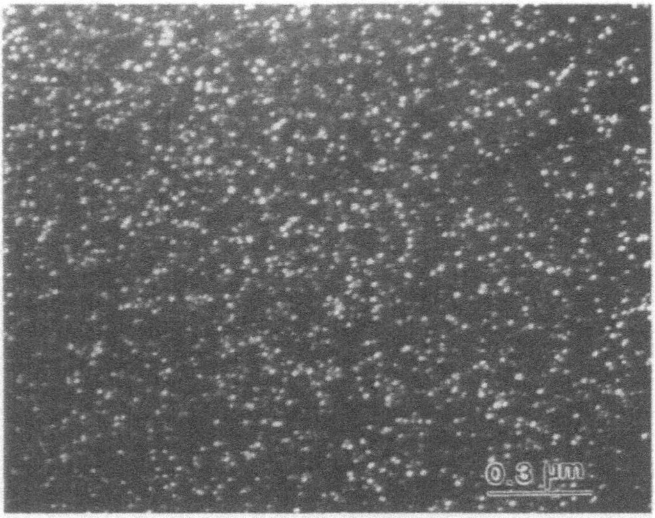

FIGURE 14. TEM dark-field image of an $Al_{90}Fe_5Gd_5$ ribbon after ball milling for 2 min (27).

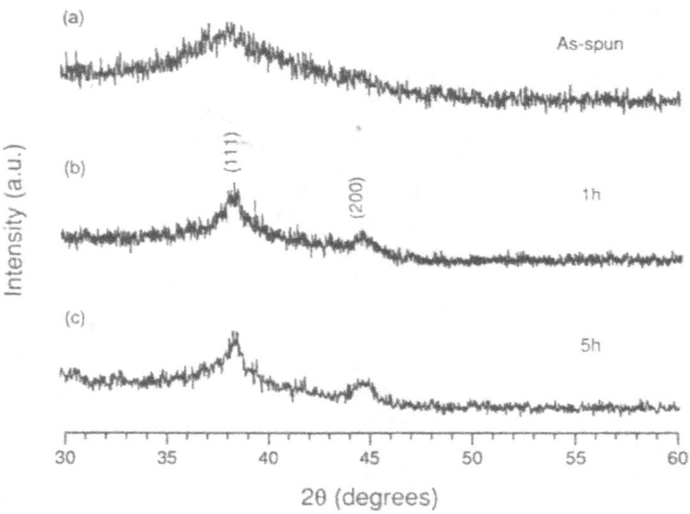

FIGURE 15. X-ray diffraction patterns of (a) as-spun $Al_{90}Fe_5Gd_5$, (b) as-spun ribbon followed by ball milling for 1 hour, (c) after 5 hours ball milling (27).

3.2 MECHANICAL PROPERTIES

It is well documented that metallic glasses possess high strength with good bending ductility. The fracture tensile strength of a metallic glass is usually much higher than that of its crystalline state with the same chemical composition. Most metallic glasses lose their high fracture tensile strength and ductility upon crystallization, although there are reports that some metallic glasses still exhibit high fracture strengths and adequate ductility after partial crystallization by quenching or isothermal annealing near their glass transition temperature. Such a sudden loss of ductility and strength is a severe disadvantage of metallic glasses as engineering materials; the retention of these properties after partial crystallization would obviously be very attractive. Aluminum glasses have been examined in their "pure" state as well as partially crystallized microstructure (2-5).

When examined with SEM, the fracture surfaces are fairly typical of metallic glass material. Fig. 16 is an image of a fracture surface. The typical vein pattern of metallic glasses is clear seen. The fracture surface has two distinct features: a smooth area is due to slip along the "plane" of maximum resolved shear stress, while the vein pattern is evidence of plastic deformation in the metallic glass, and this fracture morphology is the result of an instability that occurs during the propagation of the fracturing crack front [21].

TABLE 3. Mechanical Properties of Al-Ni-Fe-Gd Metallic Glasses (from ref. 34)

Alloys	Tensile Strength	Density	Specific Strength	Elastic Modulus
$Al_{85}Ni_6Fe_3Gd_6$	1280 MPa	3.51 Mg/m^3	37.2	72.7 GPa
$Al_{85}Ni_5Fe_2Gd_8$	1210	3.71	33.3	75.3
$Al_{87}Ni_6FeGd_6$	1140	3.47	33.5	67.1
Al 2219 alloy	475	2.84	17.1	

Systematic invesitgations of adding a second TM to the basic ternary Al-TM-RE composition resulted in both a dramatic increase in ribbon thickness (13,34) that remained completely amorphous and in ribbon fracture tensile strength (Table 3).

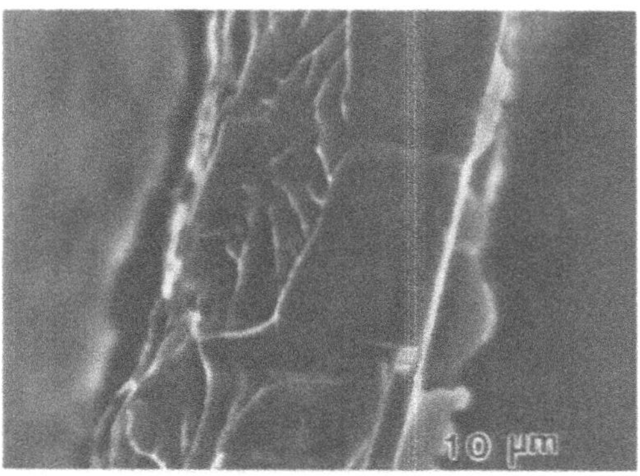

FIGURE 16. Fracture surface of the $Al_{90}Fe_5Ce_5$ metallic glass

3.2.1 Partially Crystallized Glass. Partially crystallized material was prepared two different ways. In the first a slightly aluminum enriched alloy was prepared ($Al_{90}Fe_{4.75}Gd_{4.75}$). This results in a partially crystallized alloy directly from melt-spinning without any further heat treatment. The second approach taken was to first obtain wholly amorphous $Al_{90}Fe_5Gd_5$ material through melt spinning, followed by isothermal annealing at 445K, which is just below the crystallization temperature determined by DSC. The mechanical properties of the two alloys are included in Table 1. The strength of the isothermally heat treated and partially crystallized $Al_{90}Fe_5Cd_5$ alloy has a fracture tensile strength of 1010 MPA which is much higher than its pure metallic glass form.The fracture strength of the quenched, partially crystallized $Al_{90.5}Fe_{4.75}Gd_{4.75}$ metallic glass also exceeds that of the as-spun $Al_{90}Fe_5Gd_5$ metallic glass.

An $Al_{90}Fe_5Gd_5$ metallic glass was annealed at 170 °C for increasing times to examine the uniaxial tensile strength as a function of crystallization (fig. 17, see also fig. 2 and Table 1). The fracture tensile strength increases from the as-spun value of 720 MPa to 920 MPa during the first 20 min. of annealing. This increase in strength corresponds with nanocrystalline precipitation. As annealing time increases from 20 min to 90 min., the strength decreases dramatically to 330 MPa. A reduced rate of decreasing strength occurs from 90 min. to 350 min. As annealing time exceeds 400 min. the fracture tensile strength remains at about 200 MPa.

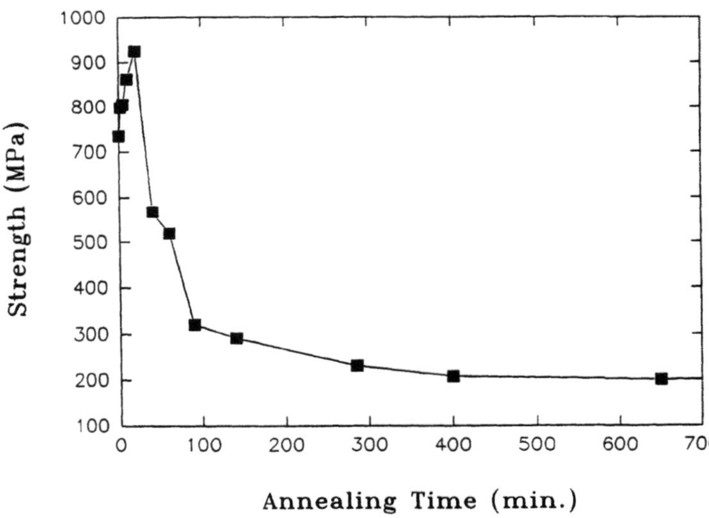

FIGURE 17. Fracture tensile strength of a partially crystallized $Al_{90}Fe_5Gd_5$ metallic glass as a function of annealing time. The annealing temperature is 170 °C.

Acknowledgment

The authors acknowledge support from the Army Research Office under contract No. DAAL03-91G-0009.

References

1. R.W. Cahn, *Nature* **341**, 183 (1989).
2. A. Inoue, A. Kitamura and T. Masumoto, *J. Mater. Sci.*, **16**, 1895 (1981).
3. R.O.Suzuki, Y. Komatsu, K.E. Kobayashi and P.H. Shingu, *J .Mater. Sci.*, **18**, 1195 (1983).
4. Y. He, S.J. Poon, and G.J. Shiflet, *Science*, **241**, 1640 (1988).
5. A. Inoue, K. Ohtera, A.P. Tsai and T. Masumoto, *Jpn. J. Appl. Phys.*, **27**, L479 (1988).
6. S. Samson, *Acta Crystall.*, **11**, 851 (1958).
7. F.E. Luborsky, editor, *Amorphous Metallic Alloys*, Boston, Butterworths (1983).
8. G.J. Shiflet, Y. He, and S.J. Poon, *Scripta Met.*, **22**, 1661 (1988).
9. G.J. Shiflet, Y. He, and S.J. Poon, *J. Appl. Phys.*, **64**, 6863 (1988).
10. H. Chen, Y. He, G.J. Shiflet and S.J. Poon, *Scripta Met.* **25**, 1421 (1991).
11. Y. He, S.J. Poon, and G.J. Shiflet, *Scripta Met.*, **22**, 1813 (1988).
12. G.M. Dougherty, Y. He, G.J. Shiflet, and S.J. Poon, Scripta Metall. et Materl., 30, 101 (1994).
13. S.J. Poon, Y. He, G.J. Shiflet and G.M. Dougherty, this volume.
14. H. A. Davies, *Phys. Chem. Glasses* **17**, 159 (1976).
15. H. A. Davies, in *Amorphous Metallic Alloys* (edited by F.E. Luborsky), p. 8. Butterworths, London (1983).
16. F. Spaepen and D. Turnbull, *Ann. Rev. Phys. Chem.* **35**, 241 (1984).
17. H.Y. Hsieh, B.H. Toby, T. Egami, Y. He, S.J. Poon, and G.J. Shiflet, *J. Mater. Res.*, **5** (1990) 2807.
18. H.Y. Hsieh, T. Egami, Y. He, S.J. Poon, and G.J. Shiflet, *J. Non-Cryst. Solids*, **135**, 248 (1991).
19. D. Turnbull, *Acta Metall. et Mater.* **38**, 243 (1990).
20. E. Gebhardt, M. Becker and S. Dorner, *Z. Metallk.* **44**, 510 (1953), ibid. 44, 573 (1953).
21. G.S. Cargill III, *Solid St. Phys.* **30**, 227 (1975).
22. L.C. Chen and F. Spaepen, *Nature*, **336**, 366 (1988).
23. Y. He, H. Chen, G.J. Shiflet, S.J. Poon, *Phil. Mag. Lett.*, **61**, 297 (1990).
24. J. W. Christian, *The Theory of Transformations in Metals and Alloys* (New York, Pergamon, 1975).
25. L. Shultz, *Phil. Mag. B*, **61**, 453 (1990).
26. G. M. Dougherty, G.J. Shiflet and S.J. Poon, *Acta Metall. Mater.* (1994) in press.
27. Y. He, G.J. Shiflet and S.J. Poon, *Acta Metall. Mater.*, (1994) in press.
28. H. Chen, Y. He, G.J. Shiflet and S.J. Poon, *Nature*, **367**, 541 (1994).
29. F. Spaepen, *Acta Metall.*, **25**, 407 (1977).
30. A.S. Argon, *Acta Metall.* **27**, 47 (1979).
31. P. S. Steif, F. Spaepen and J. W. Hutchinson, *Acta Metall.* **30**, 447 (1982).
32. D. Srolovitz, V. Vitek and T. Egami, *Acta Metall.*, **31**, 335 (1983).
33. S. Kobayashi, K. Maeda and S. Takeuchi, *Acta Metall.*, **28**, 1641 (1980).
34. Y. He, G.M. Dougherty, G.J. Shiflet and S.J. Poon, *Acta Metall. Mater.*, **41**, 337 (1994).

NUMERICAL AND EXPERIMENTAL STUDY OF POROSITY EVOLUTION DURING PLASMA SPRAY DEPOSITION OF W

HUIMIN LIU
WEIDONG CAI
ROGER H. RANGEL
ENRIQUE J. LAVERNIA
Department of Mechanical and Aerospace Engineering
University of California, Irvine
Irvine, CA 92717, U.S.A.

ABSTRACT. The porosity that is commonly associated with discrete droplet processes, such as plasma spraying and spray deposition, effectively degrades the quality of the sprayed material. In the present paper, numerical and experimental studies on porosity evolution in plasma spray deposition of W are undertaken to provide insight into the formation and evolution of porosity. In the numerical study, deformation, interaction and solidification of molten droplets impinging onto a flat and non-flat substrate during plasma spraying are investigated. The full Navier-Stokes equations coupled with the Volume Of Fluid (VOF) function are solved to determine the exact movement and interaction of droplets. A 2-domain method is employed for the treatment of the thermal field and solidification problem within the flattening droplet to track the moving solid/liquid interface. A two-phase flow continuum model is employed for the simulation of the flow problem with a growing solid layer during droplet impingement. On the basis of the VOF function and the two-phase flow continuum model the micro-porosity is quantitatively calculated. In the experimental study, a W deposit of 2 - 3 mm in thickness is prepared using low pressure plasma spraying (LPPS). The microstructure of the deposit is characterized in detail, paying particular attention to the presence of porosity. The mechanisms that govern the formation of porosity during LPPS are proposed in light of numerical and experimental results. On the basis of the mechanisms, some fundamental trends and effects of important processing parameters on micro-porosity may be reasonably explained and optimal processing conditions for reducing micro-porosity may be determined.

1. Introduction

Plasma spraying has been widely used to process thin protective coatings on preshaped parts [1-2]. Recently, new applications, such as self-standing structures, have received more attention [3]. Studies focusing on fundamental phenomena during the stage of melting and traveling of particles in plasma spray processes have been conducted by a number of investigators [4-10]. In contrast, studies related to the heat and fluid flow phenomena during the impact and solidification of droplets are indeed scarce [11-13]. While these phenomena are attracting an increasing attention in numerical [14-16] and experimental [17-18] studies, reports on mechanisms governing the formation of micro- or macro-pores in plasma spray processed coatings or deposits are conspicuously absent

M. A. Otooni (ed.), Science and Technology of Rapid Solidification and Processing, 73–107.

in the relevant literature. In manufacturing processes involving the consolidation of numerous solidifying droplets, micro- and macro-porosity is an inherent problem. Porosity typically reduces useful properties such as bond strength, hardness and corrosion resistance, degrading the deposited coatings or bulk materials. Therefore, it is of critical importance to investigate the transient deformation and solidification behavior of multiple droplets impinging onto a target substrate. Accordingly, the primary objective of the present paper is to elucidate the mechanisms governing the formation of micro- and macro-pores during plasma spraying of W due to its important application in plasma spray processes as a protective coating. The fundamental problem simulation described here is also applicable to other materials and discrete droplet processes, such as thermal spraying and spray deposition. Ultimately, the numerical and experimental results will be used to determine the processing conditions that are required in order to minimize the porosity in plasma spray processed coatings or bulk materials.

2. Numerical Model And Computational Method

Two challenging problems existing in the numerical study are: a) the treatment of complex interfacial phenomena induced by the solidification and the formation of free surfaces with arbitrary geometrical shape, and b) the simulation of three phases, i.e., liquid, solid and gas (void) phases present and/or generated during the deformation and solidification of droplets. The transient nature of the moving solid/liquid interface is one of the most difficult factors inherent in the modeling. Since the details of the model formulation and computational methodology have been given elsewhere [15], only a brief description will be presented here.

The governing equations describing the motion and interaction during droplet flattening include the incompressibility condition, the full Navier-Stokes equations for transient, axisymmetric, viscous, incompressible fluid flow, and the transport equation for VOF function. Considering solidification during the flattening, the governing equations are modified on the basis of a two-phase flow continuum model [15] as

$$\nabla \cdot (\Theta \vec{v}) = 0 \tag{1}$$

for continuity,

$$\Theta \frac{\partial \vec{v}}{\partial t} + \nabla \cdot (\Theta \vec{v}\vec{v}) = -\frac{\Theta}{\rho}\nabla p + \frac{\Theta}{\rho}\nabla \cdot \tau + \Theta \vec{g} + \Theta \vec{F}_b \tag{2}$$

for momentum transport, and

$$\frac{\partial}{\partial t}(\Theta F) + \nabla \cdot (\Theta F \vec{v}) = 0 \tag{3}$$

for VOF transport. In this continuum model, the flow is modeled as a special case of two-phase flow by introducing a volume fraction Θ, which is a continuous average quantity defined to be unity in any cell fully occupied by fluid, zero in any cell fully occupied by solid, and between zero and unity in any cell containing the solid/liquid interface

$$\Theta = \begin{cases} 1, & \text{fluid cell} \\ >0, <1, & \text{interface} - \text{containing cell} \\ 0, & \text{solid cell} \end{cases} \tag{4}$$

In order to calculate Θ, the solid/liquid interface position at any time is determined by employing the 2-domain method [19]. The thickness of the solidified layer in a splat is estimated here using the 1-D solution of the Stefan problem of solidification, given by [12, 13, 20]

$$Y = U\sqrt{a_s(t - t_i)} \tag{5}$$

with the freezing constant U [12, 13, 20]

$$U = \frac{2}{\sqrt{\pi}}\left\{\frac{T_s^*}{\left[\operatorname{erf}\frac{U}{2}\right]\exp\left(\frac{U^4}{4}\right)} - \frac{T_l^*\sqrt{\frac{(kc\rho)_l}{(kc\rho)_s}}}{\operatorname{erfc}\left[\frac{U}{2}\sqrt{\frac{a_s}{a_l}}\right]\exp\left(-\frac{U^2 a_s}{4a_l}\right)}\right\} \tag{6}$$

where T_s^* and T_l^* are defined as [12, 13, 20]

$$T_s^* = \frac{k_s\left(T_m - T_s\right)}{a_s\rho_s\Delta h}, \qquad T_l^* = \frac{k_s(T_l - T_m)}{a_s\rho_s\Delta h} \tag{7}$$

The governing equations, Eqs. (1) - (3), in primitive variables are solved on an Eulerian, rectangular, staggered mesh in cylindrical geometry with the extended SOLA-VOF method [21]. At the bottom boundary and the solid/liquid interface, a no-slip boundary condition is imposed, whereas at other boundaries and at the symmetry axis, a free-slip boundary condition is employed. The splat temperatures at the substrate surface, at the solid/liquid interface and in the liquid phase are assumed to equal the substrate temperature, the melting temperature of droplet materials, and the initial droplet temperature, respectively. Free surfaces are determined with the data of the VOF function on the mesh and surface tension at free surfaces is simulated using a continuum surface force model [21]. The solid/liquid interface is tracked using Eqs. (4) - (7). The numerical computations are performed by modifying a program called RIPPLE [21] on a DEC 3000/400 AXP workstation. The details of numerical schemes and solution techniques have been completely documented in [21] and will not be repeated herein.

In order to quantitatively investigate the effects of processing parameters on the formation of micro-pores, micro-porosity is calculated using the data of the VOF function. The micro-porosity is defined as the volume fraction of the micro-voids which are entrapped into the splat during deformation and eventually contained within the solidified layer due to rapid solidification. The micro-porosity is calculated using the following equation (Figure 1)

$$\text{Porosity} = \frac{\sum_{i=1,N} \sum_{t=t_1,t_s} [z_s(i,t+dt) - z_s(i,t)]\pi\left[r(i+1)^2 - r(i)^2\right](1 - F(i,t))}{\sum_{i=1,N} z_s(i,t_s)\pi\left[r(i+1)^2 - r(i)^2\right]} \quad (8)$$

The calculation is conducted at each time step for the annular volume i in the cell where the solidified layer thickness has increased from $z_s(i, t)$ to $z_s(i, t+dt)$ within the time step dt. This calculation is complete when the solidification front reaches the top of volume i. The corresponding solidification time and solidified layer thickness in volume i is t_s and $z_s(i, t_s)$. Such calculation is conducted for each annular volume (i = 1, 2, 3, ... N) once solidification starts in that volume. When the entire splat is solidified, the calculated volume of voids in each annular volume is added together (i = 1, N) to get the total volume of voids in the entire splat. This total volume of voids is then divided by the total volume of the splat. The result is the micro-porosity of the splat.

3. Experimental Procedures

The plasma spraying of W was conducted at Electro-Plasma Inc. (Irvine, CA) using an EPI-LPPS™ system and EPI-03CP plasma gun. The experimental procedure is described as follows. First, a thin Mo coating of about 0.2 mm in thickness was plasma-sprayed on a cylindrical graphite substrate. The objective of this step was to form a diffusion barrier between W and graphite, and thus to prevent the formation of tungsten carbide in the deposit. Second, pure W powders (purity ~99.9% and size -45+15 µm) were then plasma sprayed and deposited on the Mo coated graphite substrate, using the primary processing parameters listed in Table 1. The substrate was heated to a temperature of 1500°C before and following plasma spraying.

Table 1. LPPS Processing Parameters

Power (kW)	135
Plasma Gas Ratio: Ar/H_2	10 : 1
Powder feed rate (g/min)	75
Spray distance (mm)	250
Atmosphere (torr)	80

To determine the density of the deposited materials, the Mo layer and the interaction zone (approximately 125 µm in thickness) that formed between Mo and W were dissolved using a solution of 2% hydrogen peroxide. The density of the deposited W was then determined on the basis of Archimedes' principle. Scanning electron microscopy (SEM) was used to document the morphology and location of pores in the deposited W with a Hitachi-S-500 instrument. Furthermore, to provide insight into the mechanisms that govern the formation of porosity during LPPS, the as-received W powders were examined using SEM.

4. Results And Discussion

4.1. NUMERICAL RESULTS AND DISCUSSION

The processing parameters are chosen to cover a large range of operation conditions in plasma spray processes. Droplets of initial diameter 30 μm with impact velocities of 100, 400 and 900 m/s, respectively, are taken into consideration. Two droplet temperatures, i.e., 3650 and 3750 K, and different substrate temperatures, ranging from 1000, 1500, 2000, 2900 to 3400 K, are employed since in typical LPPS procedures the plasma torch is used to clean the substrate prior to deposition, thus significantly heating the substrate. The physical properties of liquid and solid W, as well as the temperature-dependence of the freezing constant U are given in [15]. In the following, the numerical results will be presented and discussed in detail.

4.1.1. *Formation Of Micro-Pores During Deformation Of A Single Droplet On Flat Surface.* The calculated deformation and solidification sequence of a single W droplet impinging onto a flat substrate is shown in Figure 2. The solid/liquid interface is indicated by the dark line and the splat contour is represented by the thin line. As observed in Figure 2, the droplet does not splash upwards during impact, but instead spreads uniformly in the radial direction. With increasing time, the splat height decreases and its diameter increases. Finally, it solidifies into a very thin splat at about 0.21 μs after impact. At $t = 0.045$ μs, the edge of the splat separates from the solid/liquid interface. In the present study, we refer to this phenomenon as separation phenomenon. This phenomenon, as schematically depicted in Figure 3, will lead to the formation of voids near the splat edge which may develop into micro-pores in as-sprayed materials. Hence, in the present paper, the voids which are entrapped into the splat during separation are considered as a type of micro-porosity and quantitatively calculated using Eq. (8).

4.1.2. *Effects of Droplet and Substrate Temperatures on Formation of Micro-Pores.* The calculated micro-porosity within a solidified splat vs. the substrate temperature is summarized in Figure 4. Also shown are the effects of the droplet temperature. The initial droplet diameter and velocity are maintained constant. The results shown in Figure 4 reveal that for a constant droplet temperature the micro-porosity decreases with increasing substrate temperature; and for a constant substrate temperature the micro-porosity decreases with increasing droplet temperature. However, there exists a clear transition in the variations of the micro-porosity with temperature. For example, at a droplet temperature of 3650 K and a substrate temperature lower than 2000 K (or at a droplet temperature of 3750 K and a substrate temperature lower than 1500 K), the micro-porosity decreases with decreasing substrate temperature. This observation suggests that there may be two opposite factors influencing the formation of micro-pores. In order to determine the nature of these factors, the initial rising velocity of the solid/liquid interface and the time corresponding to the separation, are plotted vs. the substrate temperature in Figures 5 and 6, respectively.

The rising velocity of the solid/liquid interface is defined as the derivative of the thickness of the solidified layer with time and is calculated using Eq. (5). The results in Figure 5 indicate that the initial rising velocity of the solid/liquid interface at 0.015 μs after impact is indeed very high, and it increases rapidly with decreasing substrate temperature, especially for the lower droplet temperature (comparing the solid line to the dash line in Figure 5). The results in Figure 6 show that for a constant droplet temperature, the time, at which the separation begins, decreases with decreasing substrate temperature; and for a constant substrate temperature, the separation time decreases with decreasing droplet temperature. This trend is a direct result of the increase in the solid/liquid interface rising velocity with decreasing substrate temperature and/or decreasing droplet temperature. Accordingly, the faster the rising velocity of the solid/liquid interface, the earlier the separation occurs. On the basis of these results, the influence of the droplet and substrate temperatures on the formation of micro-pores may be then rationalized.

For higher substrate temperatures, the rising velocity of the interface is still not very high and the separation occurs at a later time. This implies that fewer voids are formed as a result of later separation, and the droplet experiences extensive deformation before significant solidification occurs. Hence, the voids formed during deformation are filled in by the molten metal before they are fixed by solidification. In contrast, for lower substrate temperatures, two mechanisms may act. First, solidification is more rapid than deformation, so that the molten metal is solidified before voids are generated. Alternatively, fewer voids are contained in the solidified layer as a result of the decreased total solidification time with decreasing substrate temperature [15]. This is the reason why the micro-porosity turns to decrease when the substrate temperature decreases from 2000 to 1500 K for T_l=3650 K and from 1500 to 1000 K for T_l=3750 K.

At the transition temperature (T_s=2000 K for T_l=3650 K and T_s=1500 K for T_l=3750 K), the solid/liquid interface grows rapidly upwards, and simultaneously, deformation also occurs rapidly. Accordingly, the large radial velocities of the remaining liquid above the interface lead to a rapid spreading of the fringe of the splat, while the underlying solidified portion of the splat is fixed on the substrate. As a result of the separation of the liquid from the interface, voids are formed and fixed within the solidified layer due to its rapid growth. This suggestion is supported by the result that more pores are found in the fringe region of the splat. Additional support to this suggestion is the nearly identical values of the micro-porosity for the droplet temperatures of 3650 K and 3750 K at a substrate temperature of 3400 K. This interesting phenomenon indicates that droplet deformation is indeed completed before significant solidification occurs due to the relatively low rising velocity of the interface present at this high substrate temperature, regardless of the droplet temperatures.

The relatively low micro-porosity corresponding to a high viscosity value (the square symbol in Figure 4) will be discussed below.

4.1.3. *Effect of Droplet Viscosity on Formation of Micro-Pores.* In order to examine the effect of viscosity, a larger value of viscosity is assumed, the magnitude of which is two orders higher than that in the liquid state. The micro-porosity calculated under this mushy droplet condition is plotted in Figure 4 (the square symbol). Clearly, the micro-porosity in this case is lower than that at the same droplet temperature (3750 K) and the same substrate temperature (1500 K). This is not unexpected, since the rising velocity of the solid/liquid interface is the same in the two cases (due to the same thermal condition), but the deformation under the mushy droplet condition is slower (due to higher viscous consumption). Moreover, the flattening splat exhibits a negligible edge separation only at 0.045 μs after impact, and no violent separation is detected at subsequent time. Nevertheless, under the mushy droplet condition, a high initial impact velocity is absolutely necessary to achieve a large flattening extent.

4.1.4. *Effect of Initial Impact Velocity on Formation of Micro-Pores*. The calculated micro-porosity within a solidified splat vs. the initial impact velocity is summarized in Figure 7. The initial droplet diameter, droplet temperature and substrate temperature are maintained constant.

The calculated results reveal that the violent extent of the separation and hence the formation of micro-pores is critically influenced by the initial impact velocity, as evident from the results shown in Figure 7. The trend in this figure again suggests that the mechanism governing the formation of micro-pores is intimately coupled to the relative magnitudes of the solidification and deformation velocities. Accordingly, at a high initial impact velocity (e.g. 900 m/s), significant flattening takes place before significant solidification has occurred. Hence the micro-porosity is low (see the point on the right hand side in Figure 7). In contrast, at a low initial impact velocity (e.g. 100 m/s), the flattening is slow. The solid/liquid interface moves into the droplet liquid phase which does not contain many voids due to the slow deformation. Consequently, the micro-porosity is also low (see the point on the left hand side in Figure 7). However, it should be noted, that in spite of the low micro-porosity corresponding to the low initial impact velocity, the diameter of the final splat formed at such low initial impact velocity will also be small. A small splat diameter, on the one hand, reduces the contact area between the splat and the substrate, and on the other hand, may lead to the formation of macro-pores between adjacent droplets. Therefore, a high initial impact velocity should be preferred, when a small porosity is desired. However, an initial impact velocity of 400 m/s leads to the highest micro-porosity among the velocities considered under the thermal conditions of T_l=3750 K and T_s=1500 K (see the point in the middle of Figure 7). In this case, the deformation and solidification velocities are comparable. As a result, a large portion of the voids generated during droplet flattening are contained within the solidified layer, leading to the high micro-porosity.

4.1.5. *Formation of Micro-Pores during Interaction of multiple Droplets on Flat Surface.* The calculated sequence of collision, interaction and solidification of two droplets in tandem is shown in Figure 8. As illustrated in Figure 8, when the second droplet impinges the flattening splat beneath it (t = 0.120 μs after impact), its interaction with the radial sheet jet of the first droplet produces upward and outward ejection. With increasing time, e.g., from t = 0.120 to 0.255 μs after impact, the outward ejection enlarges itself gradually. As a result of solidification, the ejection is largely depressed relative to that in a fully liquid case. Therefore, a mushy surface condition should be preferred in order to reduce ejection and improve adhesion between droplets and/or splats. Moreover, it is also clear in the figure, the interaction between the two droplets intensifies the separation, leading to an increase in the micro-porosity.

The calculated sequence of collision, interaction and solidification of an axisymmetric toroidal ring impinging onto the splat of an earlier droplet is plotted in Figure 9. As shown in Figure 9, when the ring impinges the flattening splat below it, the outside fringe of the ring spreads outwards, while the interaction between the inside fringe of the ring and the radial sheet jet of the earlier droplet causes inward and upward bounce of the liquid in the interaction region. This phenomenon results from the opposite flow directions of the splat liquid (outwards) and the ring liquid (inwards). This interaction leads to the formation of voids at the locations of vortices within the liquid (t = 0.165 to 0.255 μs after impact). Under the present computation conditions, however, some of such voids are filled in by the remaining liquid during the subsequent deformation, since the rising velocity of the solid/liquid interface reduces with increasing time, and hence, is not as rapid so to freeze these voids within the solidified layer. At the same time, some new voids form during the subsequent vortex movement. In addition, the inwards flow of the ring liquid contributes to fill in the voids in the fringe region of the earlier splat, leading to a decrease in the micro-porosity.

4.1.6. *Formation of Macro- and Micro-Pores during Deformation of Droplets on Non-Flat Surface.* Figure 10 shows the deformation sequence of a single droplet impinging onto a solid particle on a substrate. As observed from Figure 10, the droplet liquid flows in good contact with the upper hemisphere of the particle upon impact, and separates from the lower hemisphere of the particle when $t > 0.060$ μs. Beyond this time point, break-up of the liquid occurs. When the liquid impacts onto the substrate surface (for example at $t > 0.120$ μs), it forms two lateral sheet jets, one rushes outwards and the other inwards toward the symmetry axis ($t = 0.210$ μs). The inwards flowing liquid consumes most of its kinetic energy due to the viscous shear stress and wall adhesion forces when it reaches the corner between the particle and the substrate ($t \geq 0.21$ μs). The flow then changes its direction due to stagnation at the corner. However, it is evident that this corner will never be completely filled by the liquid as a result of the reduced kinetic energy and the further break-up of the liquid (compare $t = 0.210$ μs to $t = 0.345$ μs). In addition, the numerical simulation of the heat transfer and solidification phenomena [15] revealed that solidification of a liquid splat is completed within a time of 0.23 μs at an initial impact velocity of 400 m/s. Therefore, solidification will further prevent the voids in the corner from being filled by the liquid. The voids eventually develop into large pores, which reduce density and deteriorate bonding among as-sprayed deposits or coatings. In view of the size scale of these pores with respect to the droplet size, we refer to them hereafter as macro-pores, in order to distinguish these from the micro-pores which develop as a result of the separation phenomenon and the interactions of vortices described above.

Figure 11 shows the deformation sequence of an axisymmetric toroidal liquid ring impinging onto a substrate with a solid particle. The ring liquid spreads uniformly in the radial direction upon impact onto the substrate ($t > 0.060$ to $t = 0.120$ μs). When the inwards flowing liquid reaches the particle surface, a portion of liquid rebounds upwards and breaks up subsequently, while the remaining liquid fills in the corner between the particle and the substrate during the spreading of the ring splat ($t = 0.120$ to $t = 0.210$ μs). At this impact velocity, no macro-voids will form in the corner between the particle and the substrate. A comparison of this result to that in Figure 10 suggests that increasing the fraction of liquid droplets in the spray is of critical importance for reducing the macro-porosity.

Figure 12 shows the deformation sequence of an axisymmetric toroidal liquid ring, co-existing in the same space level with a droplet at the symmetry axis, and impinging onto a solid particle on a substrate. At $t < 0.060$ μs, the spreading behavior is essentially same as those illustrated in Figures 10 and 11. From $t \geq 0.120$ μs, the interactions between the downwards flowing droplet liquid and the inwards flowing ring liquid, as well as the upwards rebounding ring liquid, cause the formation of ring-like vortices. The rebounding liquid breaks up subsequently ($t = 0.210$ μs). If solidification occurs at $t \geq 0.210$ μs, some micro-pores may form as a result of this type of interactions among multiple droplets.

Figure 13 shows the deformation sequence of an axisymmetric toroidal liquid ring, co-existing in the same space level with a droplet at the symmetry axis, and impinging onto a solid particle ring on a substrate. During impingement, the break-up of the liquid occurs at an earlier time ($t = 0.060$ μs). The subsequent interactions between the droplet and the ring liquid lead to the formation of inner micro-pores when $t \geq 0.120$ μs. Under this surface condition, it is also difficult to avoid the formation of the macro-pores as a result of the increased fraction of solid particles in the spray (more solid particles arrive on the substrate surface). These results reveal that the fraction of unmelted particles in plasma spraying or solidified particles in spray deposition processing is a significant factor influencing the spreading behavior of droplets and the formation of micro- and macro-pores among deposits. Since these solid particles essentially do not deform or flatten only to a small extent during impact, they will effectively increase porosity and reduce bonding among deposits or between coatings and substrate. These results are consistent with related experiments. Jackson, et al. [22], for example, note that incomplete melting of coarse W powders leads to excessive porosity in plasma sprayed W. It should be indicated that the flow patterns shown at the top in this figure result from the limit of the calculation domain, where the liquid flow reaches the top boundary and hence changes its direction from vertical upwards to lateral outwards due to the free-slip condition. Therefore, the flow patterns in this region are not representative of any physical processes during thermal spraying.

The surface conditions used in the above-described calculations represent some extreme cases where solid particles do not deform at all upon impact. Microstructural studies on sprayed deposits (for example in [23, 24]) demonstrate that in spray deposition processing, particularly in plasma spraying, the deposition surface generally exhibits a non-flat, rough morphology in a microscopic scale, typified of a wave-like geometry, which may be characterized with a sinusoidal form surface pattern. In the following, we present and discuss the calculated results for this type of surface geometry condition.

Figure 14 shows the deformation sequence of a single droplet impinging onto a solid, waved surface layer on a substrate. The spreading behavior is distinctly different from that on a solid flat substrate or on a solid particle discussed above. The droplet does not splash upwards at impact, but instead spreads in the radial direction and in a relatively good contact with the waved surface. With increasing time, the splat height decreases and its diameter increases. Eventually, it becomes a very thin, non-flat splat. The break-up of the liquid begins at $t = 0.060$ μs, and becomes significant only at $t = 0.345$ μs. At this time, solidification is expected to be completed according to the numerical studies on the heat transfer and solidification in [15]. Therefore, no macro-pores form during the spreading of a droplet on this type of waved surface. This deformation behavior suggests that a better adhesion between the splat and the surface may be obtained, in the sense of the even distribution of the liquid sheet (no macro-pores) and the good contact, if the roughness of the solid surface at impact is not as large as in Figure 10, and significant solidification begins after complete deformation. From this viewpoint, the waved surface condition is favorable relative to those involving some undeformed solid particles. These results further demonstrate the importance of the extent of droplet flattening to the ultimate microstructure and quality of spray processed materials.

Figure 15 shows the deformation sequence of an axisymmetric toroidal liquid ring, co-existing in the same space level with a droplet at the symmetry axis, and impinging onto a solid, waved surface layer on a substrate. Significant break-up of the liquid occurs at $t \geq 0.120$ μs. Since the voids between the surface and the ejected liquid (in the fringe region, $t = 0.120$ μs) are still open to the flow of the liquid (not closed by solid or solidified materials), no macro-pores will be generated in this case. A comparison of the results under this surface condition to those in Figure 13 reveals that an improvement of the surface condition is helpful for reducing the porosity.

Figure 16 shows the deformation sequence of a single droplet impinging onto a solid, waved surface layer on a substrate at an impact velocity lower than that in Figure 14. Comparing Figure 16 to Figure 14 reveals that the deformation behavior at the two impact velocities is essentially similar. As expected, the spreading of the droplet is slower at the lower impact velocity and the break-up of the liquid occurs at a later time.

Figures 17 and 18 show the deformation sequence of a single droplet impinging onto a solid, waved surface layer on a substrate with a dimensionless roughness height of 0.66 and 1, respectively. Other calculation conditions are the same as those used for the results shown in Figure 16. A comparison of the results shown in Figures 16, 17 and 18 indicates that the dimensionless roughness height critically influences the deformation behavior of droplets. An increase in the dimensionless roughness height leads to a decrease in the lateral projected diameter of the splat. In Figure 16, no voids exist in the vicinity of the surface; while in Figures 17 and 18, voids in the liquid above the surface are evident ($t \geq 0.210$ μs). These voids result from the break-up of the liquid and the concomitant entrapment of gas. The voids are surrounded by the liquid and will be closed to the flow when solidification occurs. Therefore, they may eventually develop into micro-pores. These results demonstrate the tendency that the extent of void formation increases with increasing dimensionless roughness height.

4.2. Experimental Results And Discussion

The measured density of the deposited W is 17.507 g/cm^3. When compared to the theoretical density, 19.250 g/cm^3 [25], the measured density suggests a porosity of 9.1%. This result is consistent with SEM studies of the fracture surface which revealed that an elevated concentration of micrometer-sized pores is present in plasma sprayed W (Figure 19). Figure 19a shows a fracture surface of the plasma-deposited W, clearly revealing the presence of pores both intragranularly and transgranularly. Small intragranular pores with an equivalent diameter that is generally smaller than 2 μm (0.1 to 2 μm) are observed on cleavage surfaces (Figure 19b). The relatively large pores (in the range of 1 to 5 μm) are generally present on the grain boundaries (Figure 19c).

Some investigators have studied the effects of processing parameters on the density of the deposited materials [26-29]. For example, Smith and Mutasim [28] have studied the concomitant influences of gas flow rate, nozzle type, spray distance and other variables on the density of the deposit. It is evident that the formation of the pores during plasma spraying and deposition is influenced by a large number of factors. However, from the results presented above we can propose some possible mechanisms of pore formation. Accordingly, we classify pores into five major types:

(i) macro-pores, whose size scale is comparable to droplet size and which are formed primarily at locations around solidified particles in spray deposition processing or unmelted or partially melted particles in plasma spraying. These solid particles do not flatten or flatten only to a small extent. The presence of such particles hinders the spreading and flow of subsequent droplets. Therefore, some of the voids in the vicinity of solid particles cannot be completely filled by liquid due to flow stagnation, leading to the formation of the macro-pores.

(ii) micro-pores, whose size scale is smaller relative to droplet size and resulting essentially from droplet/droplet and droplet/liquid-surface-layer interactions. The interactions lead to the formation of voids at the locations of vortices. The voids will eventually form pores when solidification is sufficiently fast and the subsequent solidification shrinkage occurs.

(iii) micro-pores formed as a result of the separation of liquid from the solid/liquid interface or the break-up (ejection or rebounding) of liquid. When the detached liquid subsequently falls down to the surface, some voids will be entrapped, eventually developing into pores as solidification proceeds.

(iv) micro-pores resulting from the presence of cavities which already exist in droplets before impact. Such cavities may be generated during atomization in spray deposition processes or are present in the powders which are used in plasma spraying, and do not vanish during flight in spray chamber as a result of short flight time at high velocities. These cavities may develop into the micro-pores depending on the relative magnitude of the deformation rate and the solidification rate.

(v) micro-pores resulting from extraneous inclusions and dispersoids (carbides, nitrides and oxides) that may be present in droplets. These defects pre-existing within droplets will lead to the formation of cavities or gaps during solidification.

The size of the macro-pores should be in the order of a few ten micrometers. The dimensions of the micro-pores of the types (ii), (iii), and (iv) might be several micrometers, several hundred nanometers, and a few nanometers, respectively. The ultimate geometry and size of the micro-pores of the type (v) will depend on changes in molar volume during solidification, and on differences in wetting characteristics and other physical properties.

On the basis of these mechanisms, our experimental results and some results from available microstructural studies on plasma sprayed coatings may be examined.

Some investigators [4, 8, 29, 30] have indicated that in order to achieve high densities during plasma deposition, it is necessary that a large proportion of the powders is completely molten during impingement with the deposition surface. Therefore, the presence of partially molten or unmolten powders normally results in a low deposit density. The degree of melting during plasma spraying is primarily associated with the particle-plasma interactions [4]. During plasma processing, an extremely sharp temperature gradient exists in the plasma jet. For example, the work done by McKelliget, et al. [31] indicated that the temperature varies from 1000 K to 12000K in the plasma jet for argon/hydrogen plasma. Therefore, depending on powder type, size, shape and injection velocity, various powder states are produced as the powders travel through this gradient. These should include fully molten droplets, partially molten powders and/or unmolten powders. Consequently, the spray may consist of fully molten, partially molten or unmolten powders before impinging the substrate. In the case of W, it may be anticipated that there will be a significant proportion of partially molten or unmolten powders, as a result of the high melting temperature of this material. Thus, when these partially molten or unmolten powders impinge, first on the substrate and subsequently on each other, they do not deform extensively, leading to the formation of interstitial pores (type (i)). This mechanism is thought to be responsible for the pores that are present on the grain boundaries as shown in Figure 19c.

Figure 20 shows the fracture surface of the deposited W, in which solidification shrinkage is observed and apparently is one reason of formation of a fraction of the pores in the deposit (type (ii)). Therefore, non-uniform shrinkage during solidification of both completely molten powders and partially molten powders, would certainly contribute to the porosity in the deposited material. The pores resulted from the shrinkage could be present both intragranularly or transgranularly.

The SEM studies revealed that some pores are also present in the starting powders (Figure 21). Most of these pores appeared to be located in the interior of powders (Figure 21a) and the surface seldom revealed any porosity (Figure 21b). Since some of powders are partially molten during deposition, it is plausible that the original pores that are present in the powders eventually become entranced in the deposited W. Clearly, such pores would correspond to those observed intragranularly as shown in Figure 19b and may be interpreted to be micro-pores of the type (iv).

Figure 22 shows a micrograph and a corresponding schematic diagram of the microstructure of a low-pressure plasma sprayed W deposited on graphite, obtained from [23]. The experimental conditions and processing parameters are given in [23]. As demonstrated in the schematic diagram, macro-pores of the type (i) are primarily located in the vicinity of partially melted or unmelted particles. Some micro-pores are also present, and may be interpreted to be of the types (ii) and (iii). In addition, it can be observed that most of the splats exhibit a lamellar structure, the shape and dimension of which are critically limited by the surface morphology composed of the prior splat or particle boundaries. As shown in the figure, a comparison of the splat configuration and void geometry in this schematic diagram to that calculated suggests a relatively good agreement between the calculation and the experiment.

Figure 23 illustrates the microstructure of a Ni-20wt.% Cr coating which is sprayed at a substrate temperature of 650 °C, obtained from [24]. The experimental conditions and processing parameters are described in detail elsewhere [24]. This figure demonstrates a typical non-flat, waved substrate surface and the resultant waved, lamellar structure in the coating. Pores of the types (i), (ii), and (iii) can also be observed in the figure. As illustrated in the figure, a qualitative comparison of the splat configuration and void geometry in this figure to those calculated suggests a relatively good agreement between the experiment and the numerical simulation presented in this paper.

5. Conclusions

1. During impingement on a flat substrate, a single droplet spreads uniformly in the radial direction and eventually solidifies into a thin splat. Accompanying the flattening and solidification, the splat edge may separate from the solid/liquid interface depending on processing parameters. The separation phenomenon causes the formation of micro-pores primarily in the fringe region of the splat.

2. The rising velocity of the solid/liquid interfaces in the deposit and the deformation velocity of the droplet are two critical factors that influence the formation of micro-pores in the deposit. These two velocities in turn critically depend on the droplet impact velocity, droplet temperature and substrate temperature. When the initial impact velocity or the substrate temperature is sufficiently high so that significant deformation occurs prior to significant solidification, the separation takes place to a less violent extent, or occurs at a later time. Hence the micro-porosity is low. When the initial impact velocity or the substrate temperature is sufficiently low that significant solidification is completed prior to significant deformation, the micro-porosity is also relatively low. This is attributable to the decreased amount of voids or to the decreased total solidification time, despite the fact that the separation occurs at an earlier time for a lower substrate temperature. At low impact velocities or low temperatures, the contact and eventual adhesion of the splat with the substrate may be worse as a result of the small flattening extent. When the initial impact velocity or the substrate temperature is between these two extremes, the deformation velocity and the solidification velocity become comparable and most of the voids formed due to the separation phenomenon are fixed in the solidified layer. As a result, the micro-porosity becomes large.

3. A droplet with a high viscosity impinging onto a solid, flat substrate leads to a relatively small micro-porosity, but may produce macro-pores as a result of the small flattening extent. A fully liquid droplet impinging onto a flattening, solidifying splat produces relatively large micro-porosity, while multiple fully liquid droplets colliding with a flattening, solidifying splat cause a relatively small micro-porosity and at the same time the formation of vortices. The ejection and bounce of the liquid may be largely suppressed by solidification.

4. A fully liquid droplet impinging onto a solid particle on a flat substrate leads to the formation of macro-pores between the particle boundary and the substrate; multiple fully liquid droplets striking onto a solid particle on a flat substrate substantially eliminate the macro-pores, but simultaneously produce vortices within the liquid, as well as ejection and break-up of the liquid; multiple fully liquid droplets striking onto multiple solid particles on a flat substrate, not only lead to vortices and ejection of the liquid, but also to the formation of the macro-pores. Therefore, a decrease in the proportion of solid

particles contained in the spray may reduce the macro-porosity significantly. These results are in good agreement with available metallographic observations [22-24].

5. Decreasing the roughness height of the deposition surface and increasing the roughness spacing may improve the extent of droplet flattening and reduce the break-up of the liquid and the micro-porosity.

6. The porosity in spray processed materials may be classified into five major types. Macro-pores, whose size scale is comparable to droplet size (type (i)) occur primarily at interparticle boundaries. These pores may be substantially eliminated by reducing the population of unmelted particles in plasma spraying or decreasing the fraction of completely solidified particles in spray deposition processing. Micro-pores, whose size scale is smaller relative to droplet size, result from the droplet/droplet and droplet/liquid-surface-layer interactions and subsequent solidification shrinkage (type (ii)), the separation of the liquid from the solid/liquid interface and the break-up of the liquid (type (iii)), and the cavities pre-existing in droplets (type (iv)), as well as the defects pre-existing in droplets, such as extraneous inclusions and dispersoids (type (v)), respectively.

Overall, in the practical plasma spray processes, the porosity can be controlled through a suitable combination of droplet temperature, droplet impact velocity and substrate temperature by adjusting various operating parameters.

Acknowledgments

The authors would like to thank Electro Plasma, Inc. (Irvine, CA 92714), the Army Research Office (DAALO3-92-0181) and the University of California (Universitywide Energy Research Group, UERG-17875) for financial support. This research is also supported in part by the University of California, Irvine through an allocation of computer resources.

NOMENCLATURE

a_l, a_s	thermal diffusivity of liquid and solid, respectively;
c_l, c_s	heat capacity of liquid and solid, respectively;
d_0	initial droplet diameter;
F	volume of fluid function;
$\vec{F}_b$	body force;
$\vec{g}$	gravitational acceleration vector;
Δh	enthalpy of fusion;
i	subscript denoting annular volumes;
k_l, k_s	thermal conductivity of liquid and solid, respectively;
N	total number of annular volumes within solidified splat;
p	scalar pressure;
r	radial coordinate;
Re	Reynolds number, $Re = \frac{d_0 u_0}{\upsilon}$;
t	time;
t_i	time that solidification begins at some radial position r;
t_s	time that solidification terminates at some radial position r;
T_l^*, T_s^*	dimensionless temperature of droplet and substrate, respectively;
T_l, T_m, T_s	droplet temperature, melting temperature and substrate temperature, respectively;
u_0	initial impact velocity of droplet in axial direction;
U	freezing constant;
$\vec{v}$	velocity vector;
We	Weber number, $We = \frac{d_0 u_0^2 \rho}{\sigma}$;
Y	thickness of the solidified layer at any real time t after impact;
z	axial coordinate;
$z_s(i,t)$	axial coordinate of solid-liquid interface in annular volume i at time t;
ρ, ρ_s	density of liquid and solid, respectively;
τ	viscous stress tensor;
Θ	volume fraction of liquid;
ε	amplitude of sinusoidal wave function for deposition surface shape; ε/d_0 is dimensionless roughness height of deposition surface;
λ	wavelength of sinusoidal wave function for deposition surface shape; λ/d_0 is dimensionless roughness spacing of deposition surface.

REFERENCES

1. J. Musil and J. Fiala, Plasma Spray Deposition of Graded Metal Ceramic Coatings, *Surface & Coatings Technology*, vol. 52, no. 3, pp. 211-220, 1992.
2. J. Chen, G. Craig, E. Farley and A. Sanjurjo, Titanium-based Coatings on Steel-DIP Coating and Plasma Spray, *Surface & Coatings Technology*, vol. 49, no. 1-3, pp. 116-120, 1991.
3. R. Tiwari, H. Herman, S. Sampath and B. Gudmundsson, Plasma Spray Consolidation of High Temperature Composites, *Mater. Sci. Eng. A*, vol. 144, pp. 127-131, 1991.
4. D. Apelian, M. Paliwal, R.W. Smith and W.F. Schilling, Melting and Solidification in Plasma Spray Deposition --- Phenomenological Review, *Int. Metals Review*, vol. 28, pp. 271-294, 1983.
5. D. Wei, B. Farouk and D. Apelian, Effects of Particle Loading on a Reduced Pressure Inductively Coupled Radio Frequency Plasma Torch, *Metall. Trans. B*, vol. 20, no. 6, pp. 949-958, 1989.
6. D. Wei, D. Apelian and B. Farouk, Particle Melting in High Temperature Supersonic Low Pressure Plasma Jets, *Metall. Trans. B* , vol. 20, no. 2, pp. 251-262, 1989.
7. R.C. Dykhuizen and M.F. Smith, Investigations into the Plasma Spray Process, *Surface & Coatings Technology*, vol. 37, no. 4, pp. 349-358, 1989.
8. R.W. Smith and D. Apelian, Plasma Spray Consolidation of Materials, *Pure and Applied Chemistry*, vol. 62, no. 9, pp. 1825-1832, 1990.
9. R. Knight, R.W. Smith and D. Apelian, Application of Plasma ARC Melting Technology to Processing of Reactive Metals, *Int. Mater. Reviews*, vol. 36, no. 6, pp. 221-252, 1991.
10. R. Westhoff, G. Trapaga and J. Szekely, Plasma-Particle Interactions in Plasma Spraying Systems, *Metall. Trans. B*, vol. 23, no. 6, pp. 683-693, 1992.
11. H. Jones, Cooling, Freezing and Substrate Impact of Droplets Formed by Rotary Atomization, *J. Phys. D.: Applied Physics*, vol. 4, pp. 1657-1660, 1971.
12. J. Madejski, Solidification of Droplets on a Cold Surface, *Int. J. Heat Mass Transfer*, 19, pp. 1009-1013, 1976.
13. C. San Marchi, H. Liu, A. Sickinger, E. Mühlberger, E.J. Lavernia and R.H. Rangel, Numerical Analysis of the Deformation and Solidification of a Single Droplet Impinging onto a Flat Substrate, *J. Mater. Sci.*, vol. 28, pp. 3313-3321, 1993.
14. H. Liu, E.J. Lavernia and R. H. Rangel, Numerical Simulation of Impingement of Molten Ti, Ni and W Droplets on a Flat Substrate, *J. Thermal Spray Technology*, vol. 2, no. 4, pp. 369-378, 1993.
15. H. Liu, E.J. Lavernia and R.H. Rangel, Numerical Simulation of Substrate Impact and Freezing of Droplets in Plasma Spray Processes, *J. Phys. D: Appl. Phys.*, vol. 26, pp. 1900-1908, 1993.
16. G. Trapaga and J. Szekely, Mathematical Modeling of the Isothermal Impingement of Liquid Droplets in Spraying Processes, *Metall. Trans. B*, vol. 22B, pp. 901-914, 1991.
17. G. Trapaga, E.F. Matthys, J.J. Valencia and J. Szekely, Fluid flow, Heat Transfer, and Solidification of Molten Metal Droplets Impinging on Substrates --

Comparison of Numerical and Experimental Results, *Metall. Trans. B*, vol. 23B, pp. 701-718, 1992.
18. S. Fantassi, M. Vardelle, A. Vardelle, and P. Fauchais, Influence of the Velocity of Plasma Sprayed Particles on the Splat Formation, *Proc. of the 1993 National Thermal Spray Conference,* Anaheim, CA, 7-11 June 1993, pp. 1-6.
19. M. Rappaz, Modeling of Microstructure Formation in Solidification Processes, *Int. Mater. Rev.*, vol. 34, no. 3, pp. 93-123, 1989.
20. J.M. Hill, *One-dimensional Stefan Problems: an Introduction*, Longman Scientific & Technical, John Wiley & Sons, Inc., New York, 1987.
21. D.B. Kothe, R.C. Mjolsness and M.D. Torrey, RIPPLE: A Computer Program for Incompressible Flows with Free Surfaces, *LA-12007-MS, UC-000*, 1991.
22. M. R. Jackson, P. A. Siemers, S. F. Rutkowski and G. Frind, Refractory Metal Structures Produced by Low Pressure Plasma Deposition, *Inter. J. Refrac. & Hard Metals*, vol. 8, no. 3, pp. 196-200, 1989.
23. R.A. Neiser, R.D. Watson, G.R. Smolik and K.J. Hollis, An Evaluation of Plasma Sprayed Tungsten for Fusion Reactors, in *Proc. of the 1993 National Thermal Spray Conference*, Anaheim, CA, 7-11 June 1993, pp. 303-308.
24. A. Itoh, M. Hirata and M. Ayagaki, Effects of Substrate Temperature During Spraying on the Properties of Sprayed Coatings, in *Proc. of the 1993 National Thermal Spray Conference*, Anaheim, CA, 7-11 June 1993, pp. 593-600.
25. J. Shackelford and W. Alexander, *The CRC Materials Science and Engineering Handbook*, CRC Press, 1992.
26. D. Apelian, R. W. Smith and D. Wei, Particle Melting and Droplet Consolidation During Low Pressure Plasma Deposition, *Powder Metall. Int.*, vol. 20, no. 2, pp. 7-10, 1988.
27. Z. Z. Mutasim and R. W. Smith, Low Pressure Plasma Spray Deposition of W-Ni-Fe Alloy, *Tungsten and Tungsten Alloys - Recent Advances*, ed. A. Crowson and E. S. Chen, TMS, 1991, pp. 69-73.
28. R. W. Smith and Z. Z. Mutasim, Plasma Sprayed Refractory Metal Structures and Properties, *Thermal Spray Research and Applications*, ed. T. F. Bernecki, ASM International, 1990, pp. 369-374.
29. H. Gruner, Vacuum Plasma Spray Quality Control, *Thin Solid Films*, vol. 118, pp. 409-420, 1984.
30. H. Liu, E. J. Lavernia and R. H. Rangel, Numerical Investigation of Micro-Pore Formation During Substrate Impact of Molten Droplets in Plasma Spray Processes, *Atomization and Sprays*, accepted, 1994.
31. J. McKelliget, J. Szekely, M. Vardelle and P. Fauchais, Temperature and Velocity Fields in a Gas Stream Exiting a Plasma Torch. A Mathematical Model and Its Experimental Verification, *Plasma Chem. & Plasma Proc*, vol. 2, no. 3, pp. 316-332, 1982.
32. W. Cai, H. Liu, E.J. Lavernia, A. Sickinger, E. Mühlberger and D. Bailey: Low Pressure Plasma Deposition of W, *J. Thermal Spray Technology*, in press, 1994.

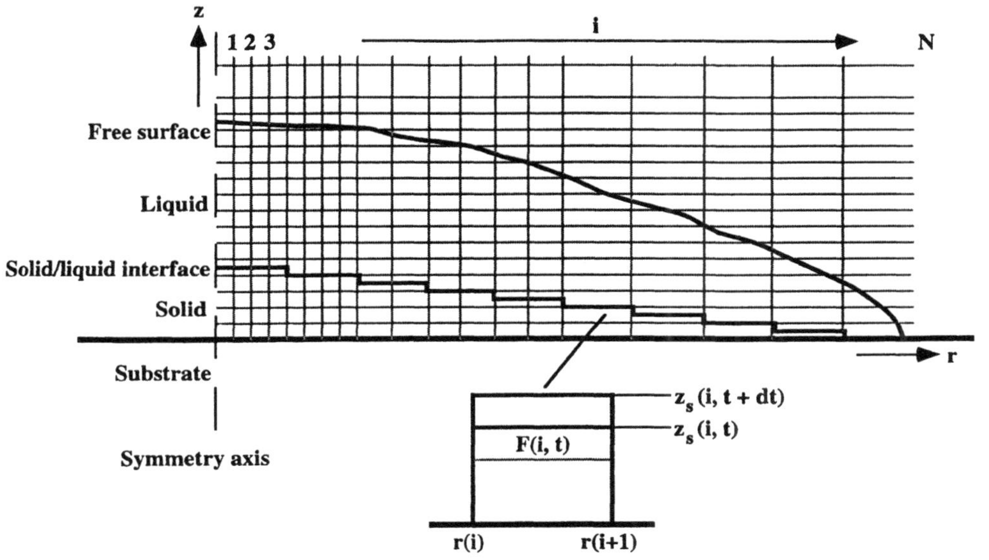

Figure 1. Schematic showing computation method of micro-porosity (from ref. [30]).

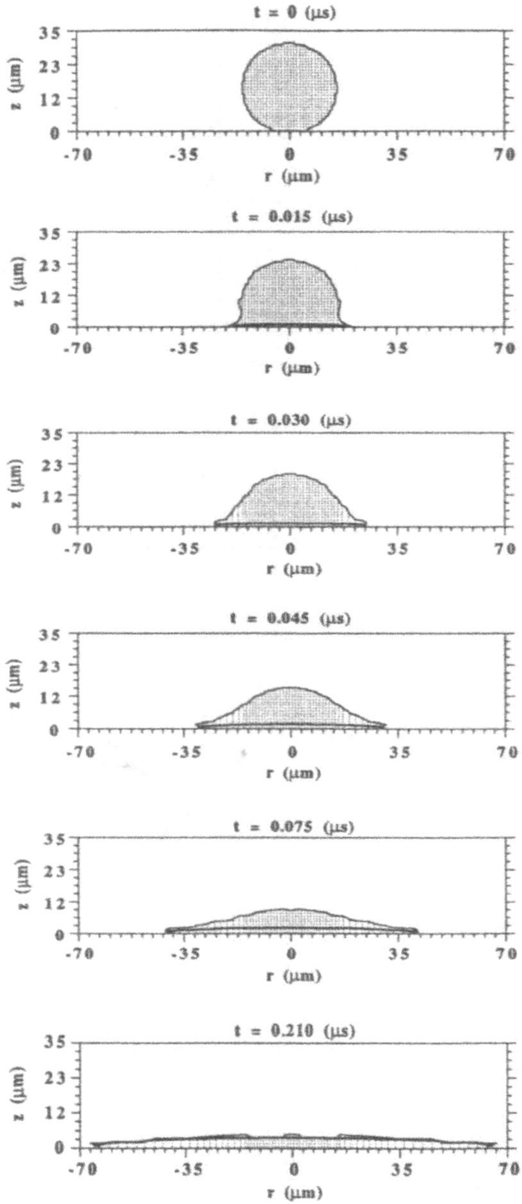

Figure 2. Deformation and solidification sequence of a single W droplet, u_0=400 m/s, T_l=3750 K, T_s=1500 K (from ref. [30]).

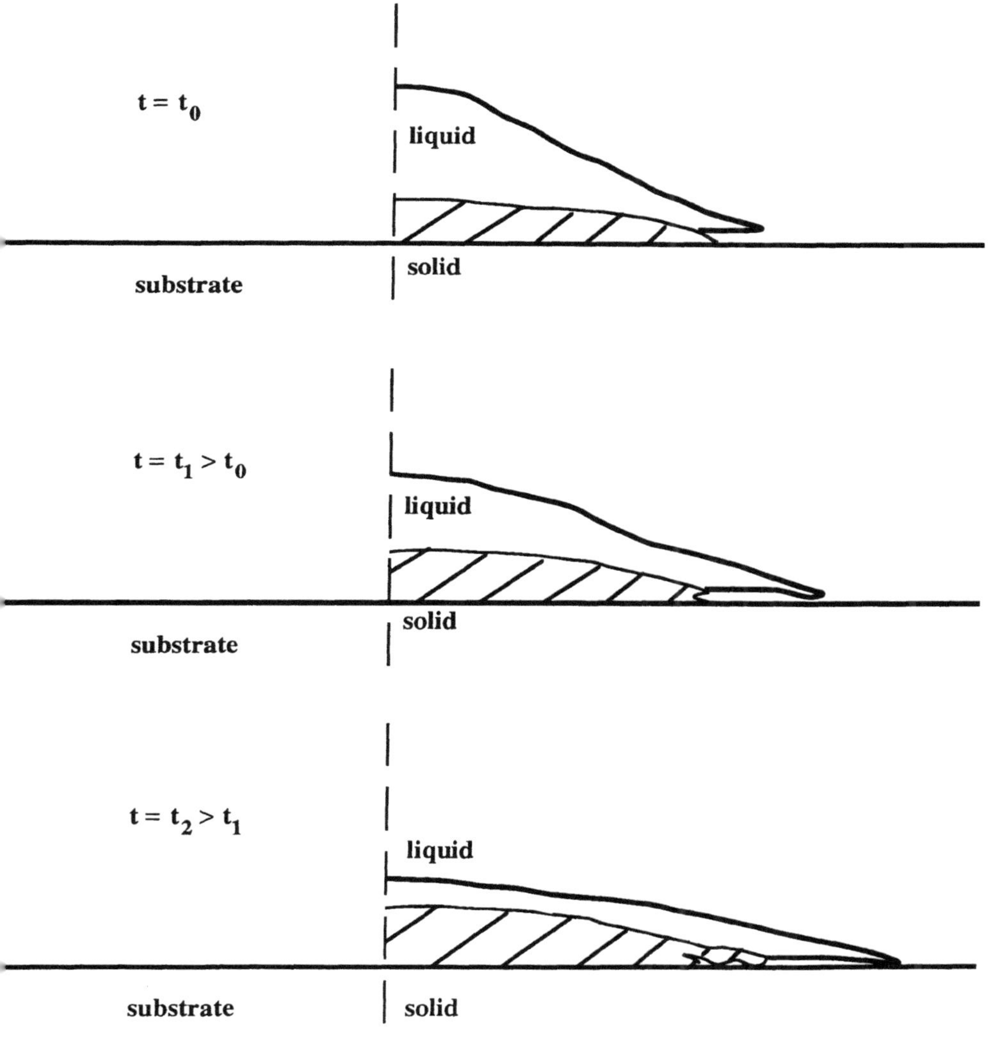

Figure 3. Schematic diagram showing formation mechanism of micro-pores (from ref. [30]).

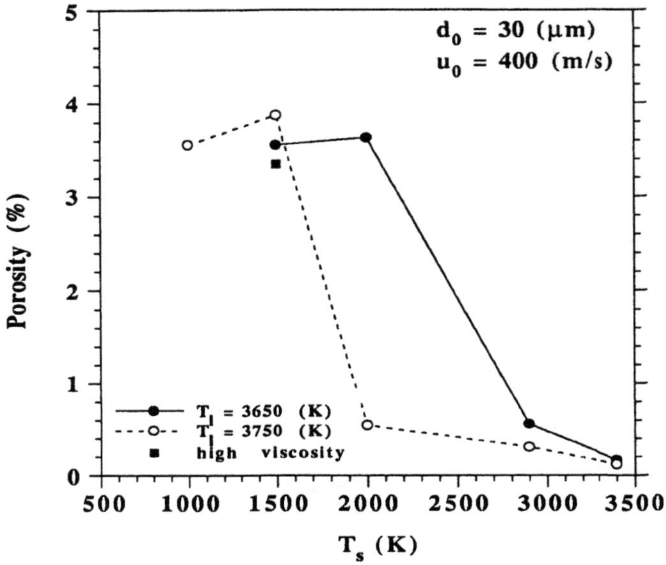

Figure 4. Micro-porosity formed during droplet deformation and solidification and its dependence on substrate temperature and droplet temperature (from ref. [30]).

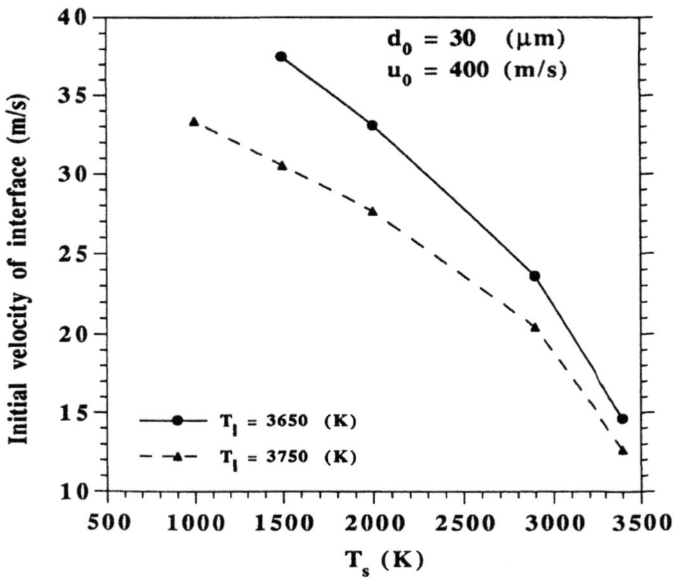

Figure 5. Initial rising velocity of solid/liquid interface as a function of substrate temperature and droplet temperature (from ref. [30]).

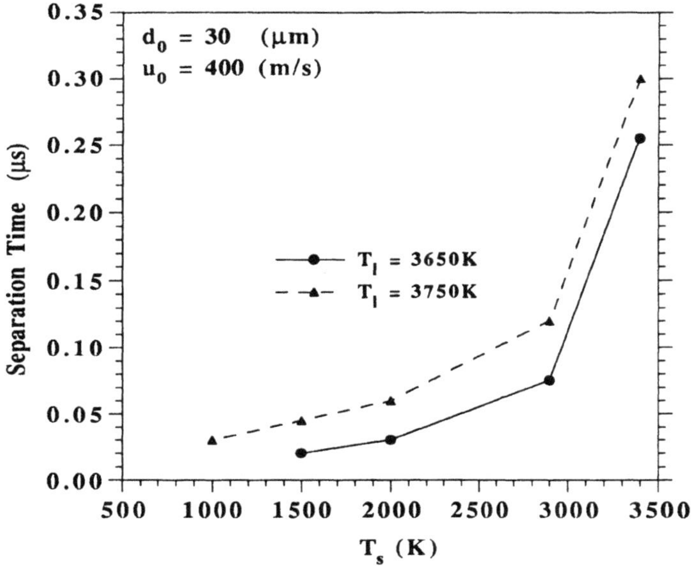

Figure 6. Separation time as a function of substrate temperature and droplet temperature (from ref. [30]).

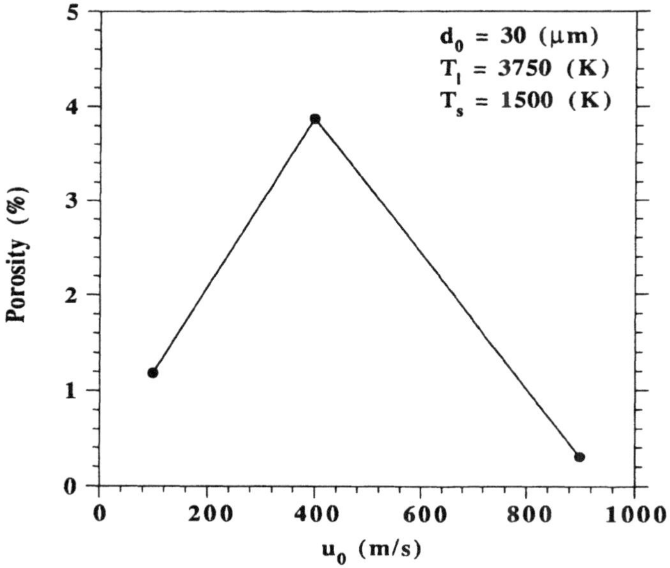

Figure 7. Micro-porosity formed during droplet deformation and solidification and its dependence on the initial impact velocity (from ref. [30]).

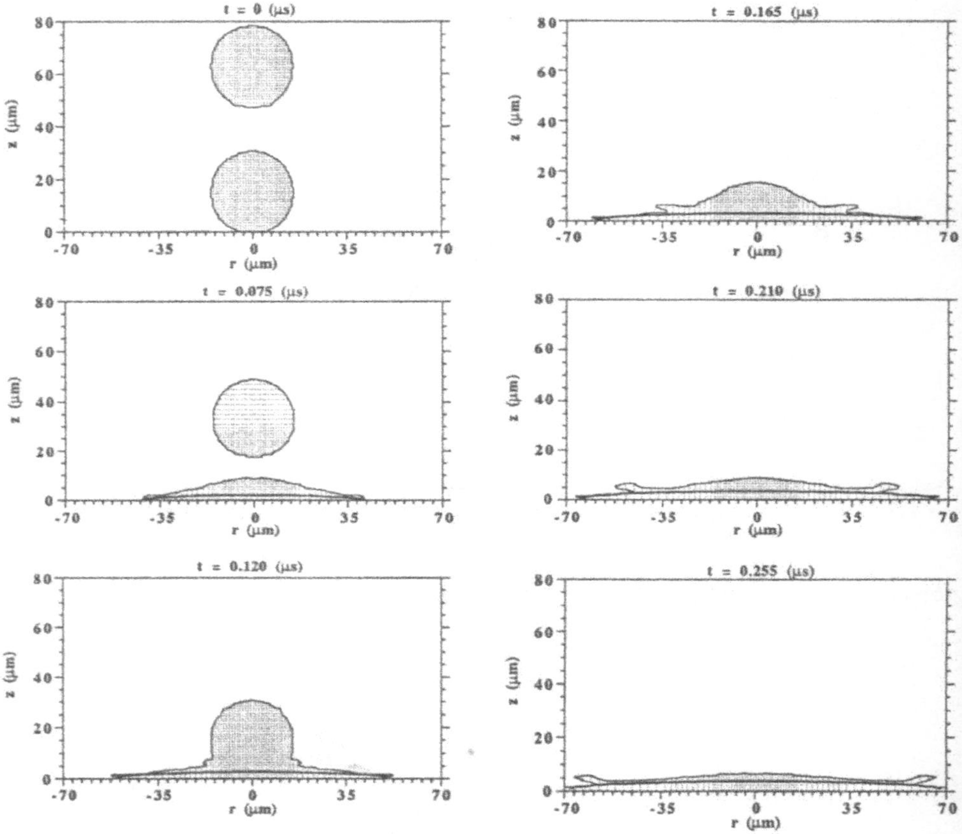

Figure 8. Deformation and solidification sequence during interaction of two droplets in tandem, u_0=400 m/s, T_l=3750 K, T_s=1500 K (from ref. [30]).

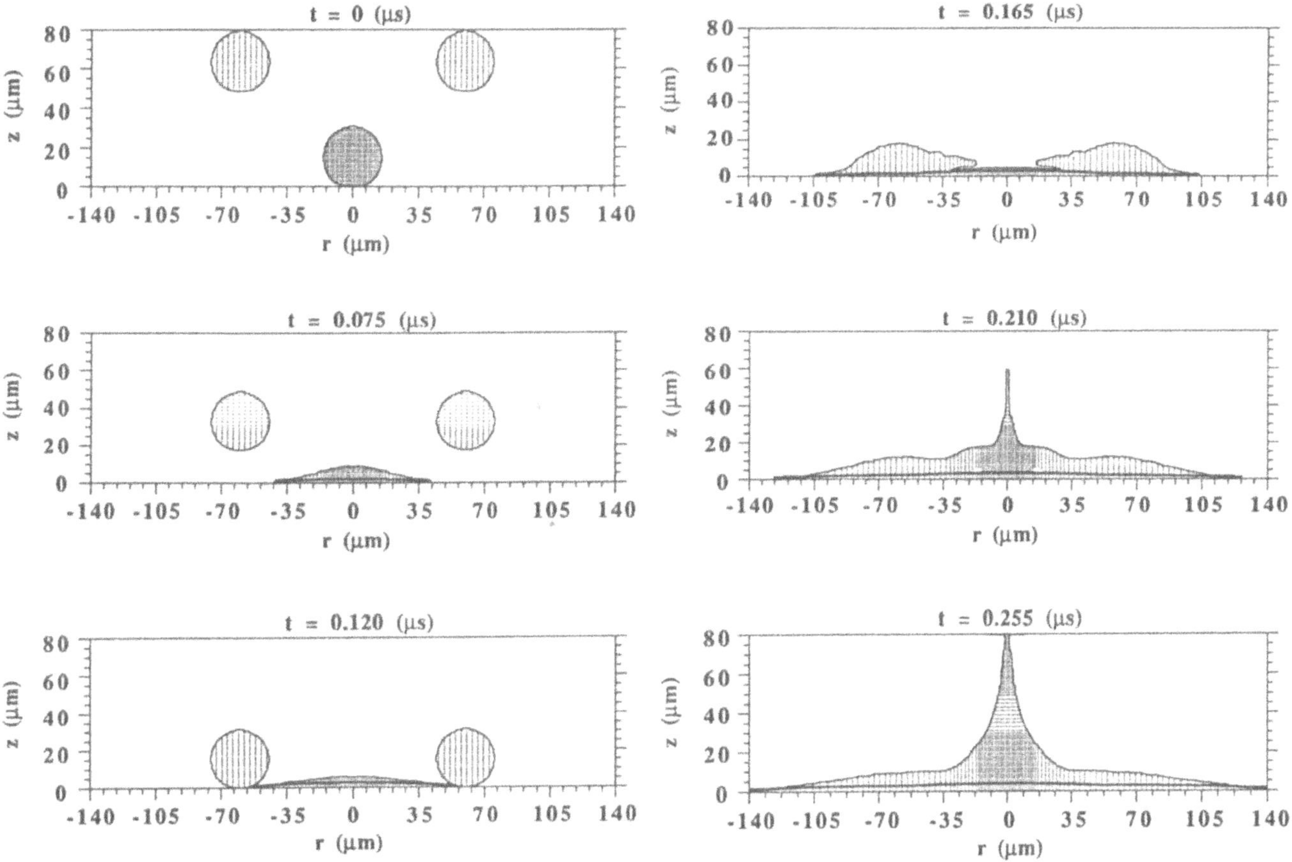

Figure 9. Deformation and solidification sequence during interaction of an axisymmetric toroidal ring and a droplet beneath it, u_0=400 m/s, T_l=3750 K, T_s=1500 K (from ref. [30]).

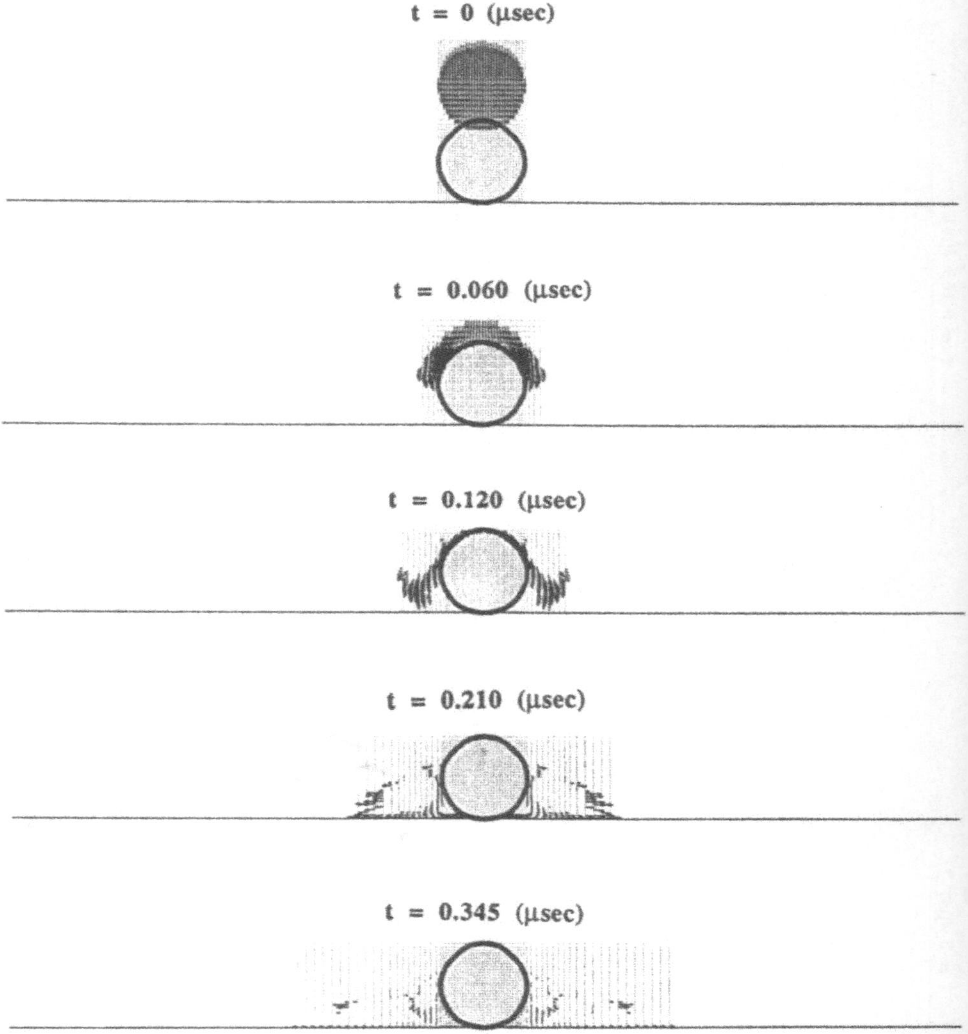

Figure 10. Deformation sequence of a single droplet impinging onto a solid particle on a substrate, u_0=400 m/s, Re=76229, We=33897.

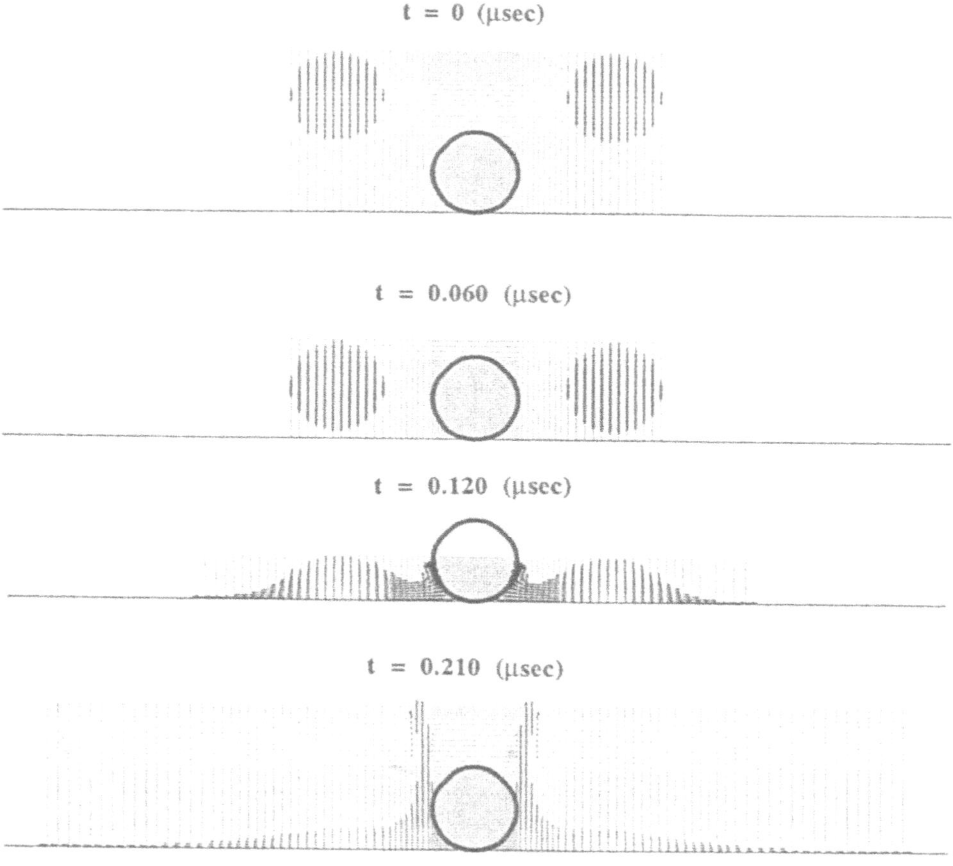

Figure 11 Deformation sequence of an axisymmetric toroidal liquid ring impinging onto a solid particle on a substrate, u_0=400 m/s, Re=76229, We=33897.

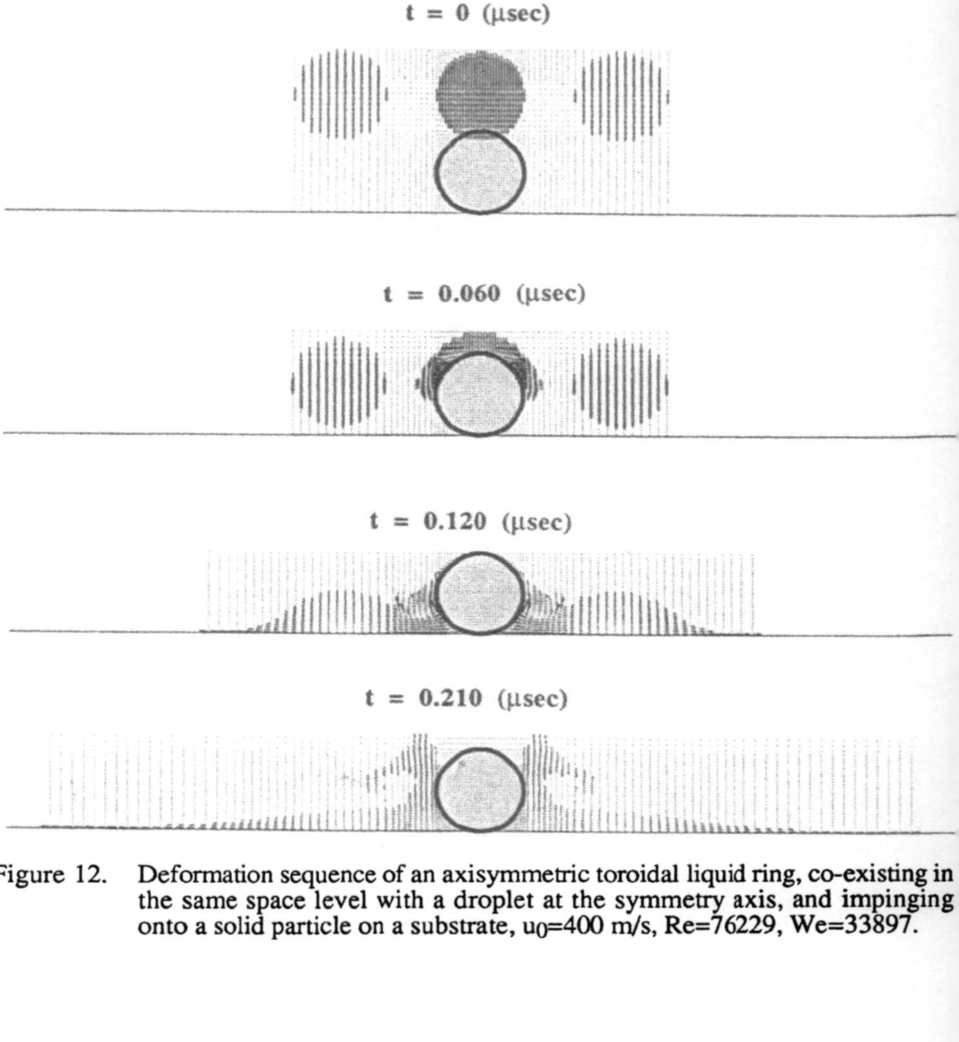

Figure 12. Deformation sequence of an axisymmetric toroidal liquid ring, co-existing in the same space level with a droplet at the symmetry axis, and impinging onto a solid particle on a substrate, u_0=400 m/s, Re=76229, We=33897.

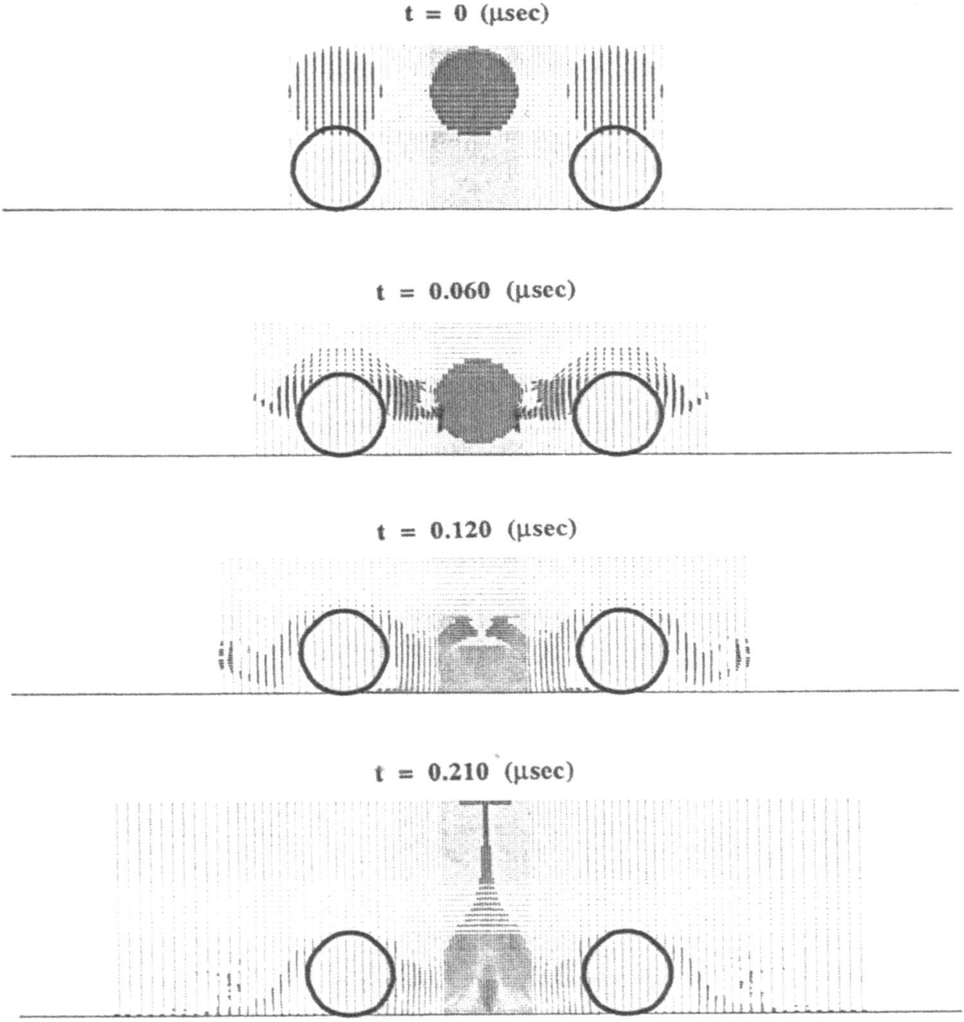

Figure 13. Deformation sequence of an axisymmetric toroidal liquid ring, co-existing in the same space level with a droplet at the symmetry axis, and impinging onto two solid particles on a substrate, u_0=400 m/s, Re=76229, We=33897.

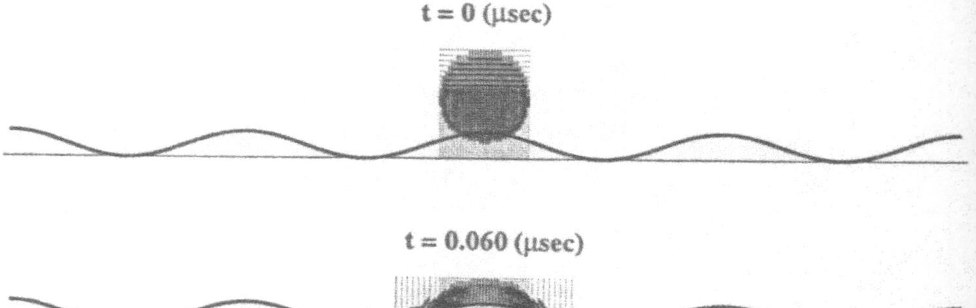

Figure 14. Deformation sequence of a single droplet impinging onto a solid, waved surface layer on a substrate, u_0=400 m/s, Re=76229, We=33897, ε/d_0=0.33, λ/d_0=2.8.

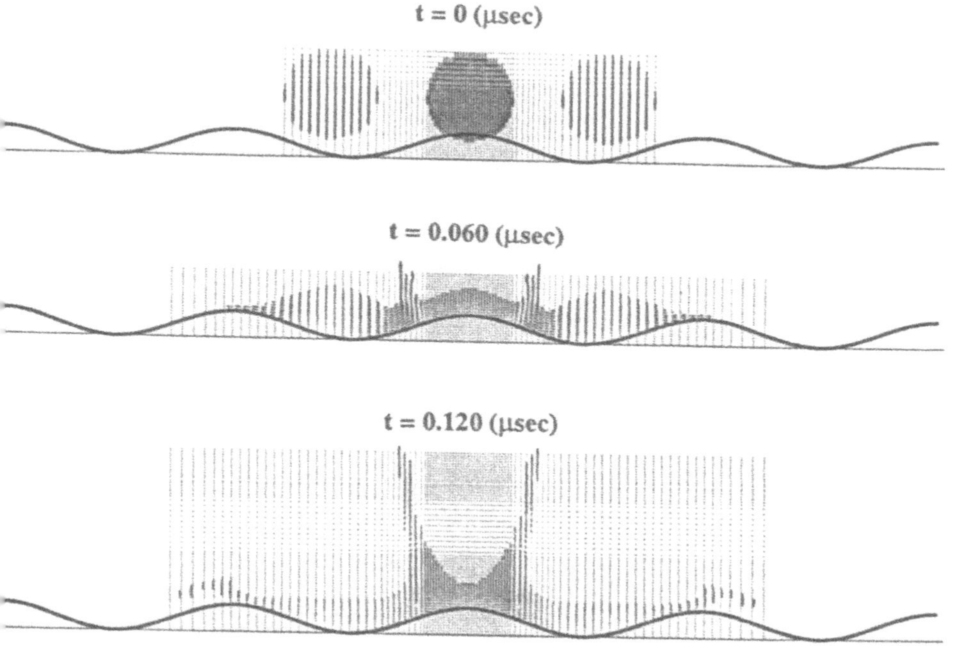

ıgure 15. Deformation sequence of an axisymmetric toroidal liquid ring, co-existing in the same space level with a droplet at the symmetry axis, and impinging onto a solid, waved surface layer on a substrate, u_0=400 m/s, Re=76229, We=33897, ε/d_0=0.33, λ/d_0=2.8.

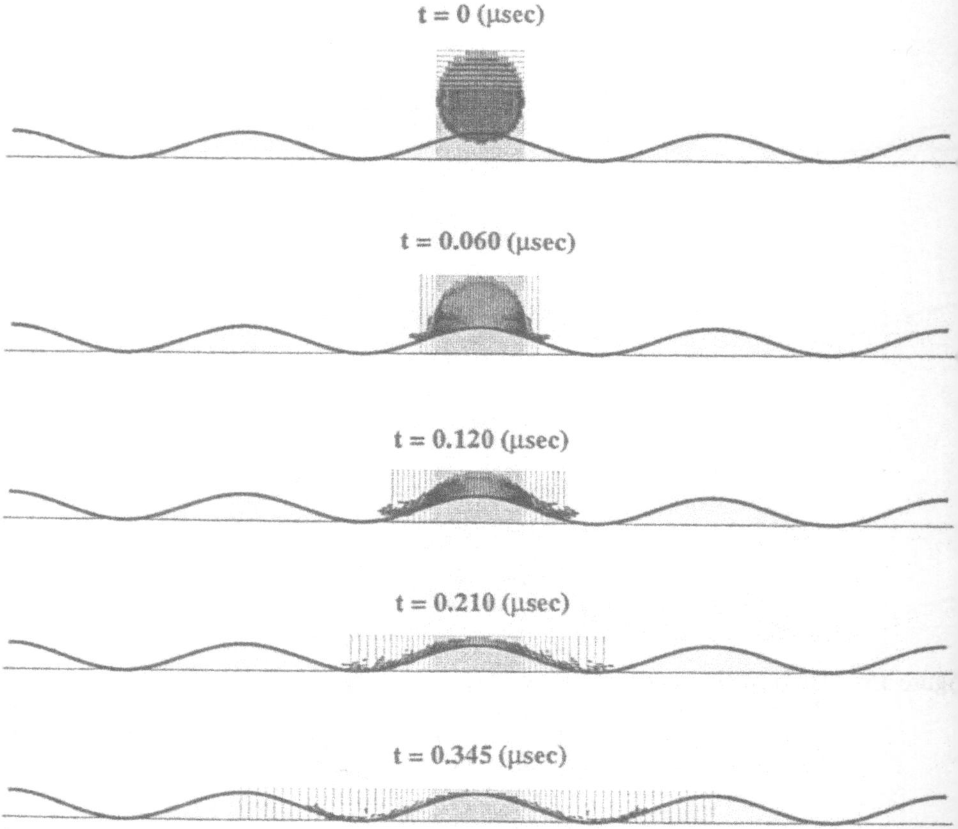

Figure 16. Deformation sequence of a single droplet impinging onto a solid, waved surface layer on a substrate, u_0=200 m/s, Re=38115, We=8474, ε/d_0=0.33, λ/d_0=2.8.

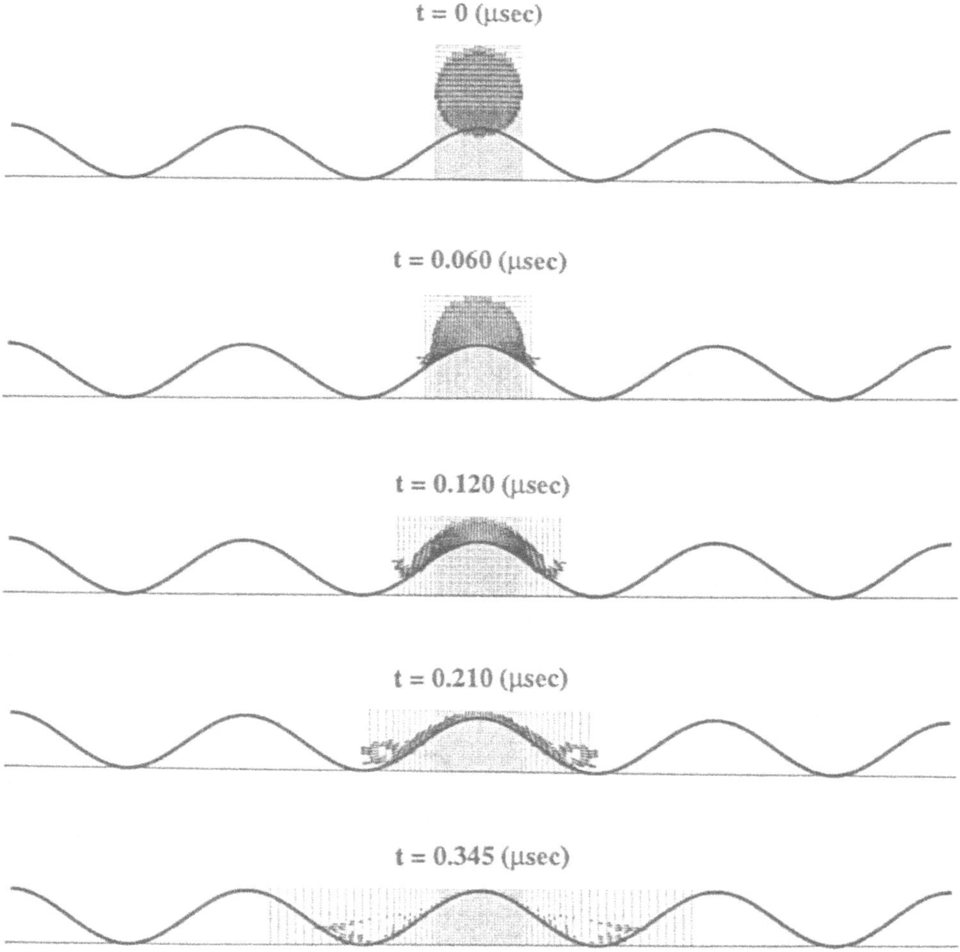

'igure 17. Deformation sequence of a single droplet impinging onto a solid, waved surface layer on a substrate, u_0=200 m/s, Re=38115, We=8474, ε/d_0=0.66, λ/d_0=2.8.

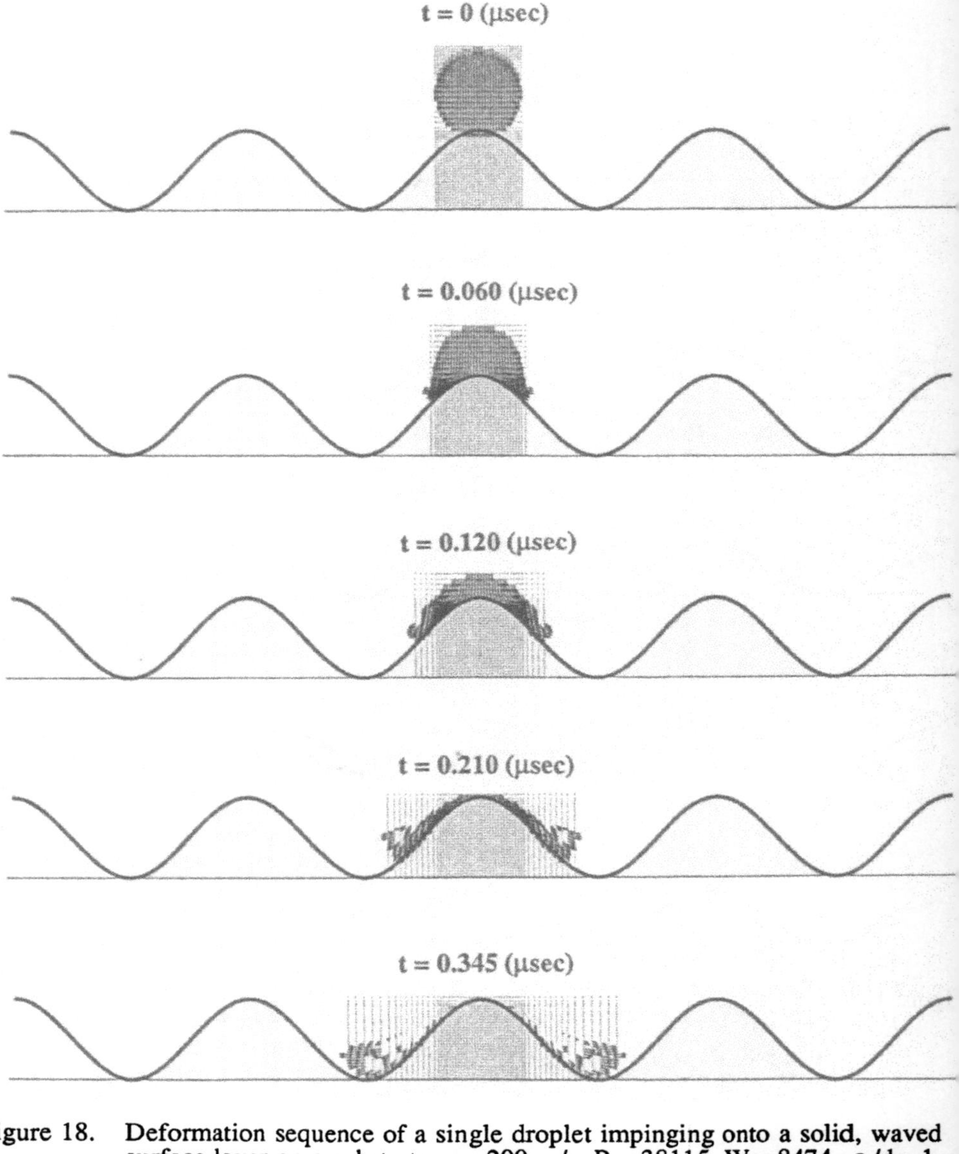

Figure 18. Deformation sequence of a single droplet impinging onto a solid, waved surface layer on a substrate, u_0=200 m/s, Re=38115, We=8474, ε/d_0=1, λ/d_0=2.8.

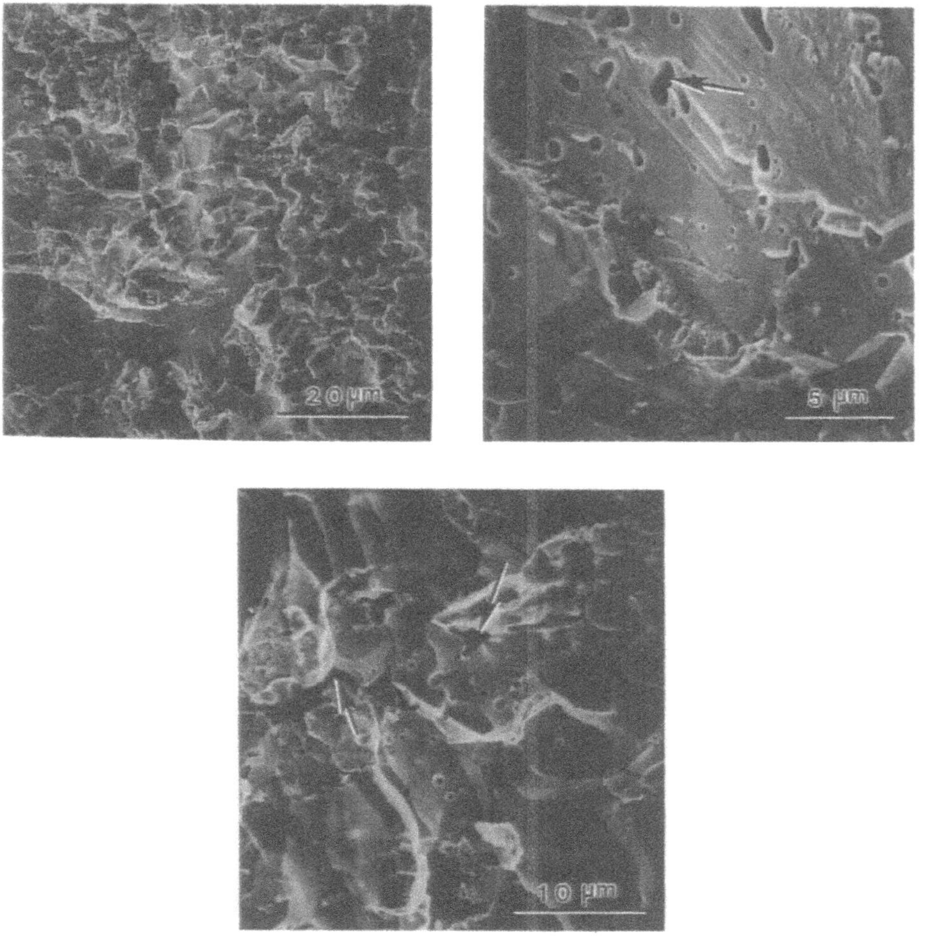

Figure 19. Morphology of fracture surface showing (a) pores present both transgranularly and intragranularly, (b) intergranular porosity, (c) transgranular porosity (from ref. [32]).

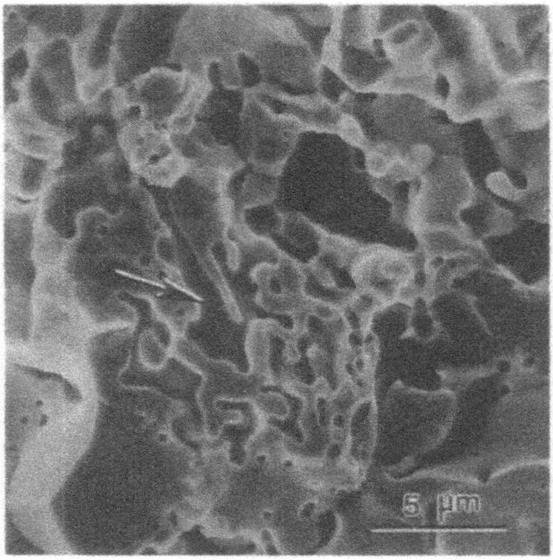

Figure 20. Morphology of fracture surface showing the porosity resulting from solidification shrinkage (from ref. [32]).

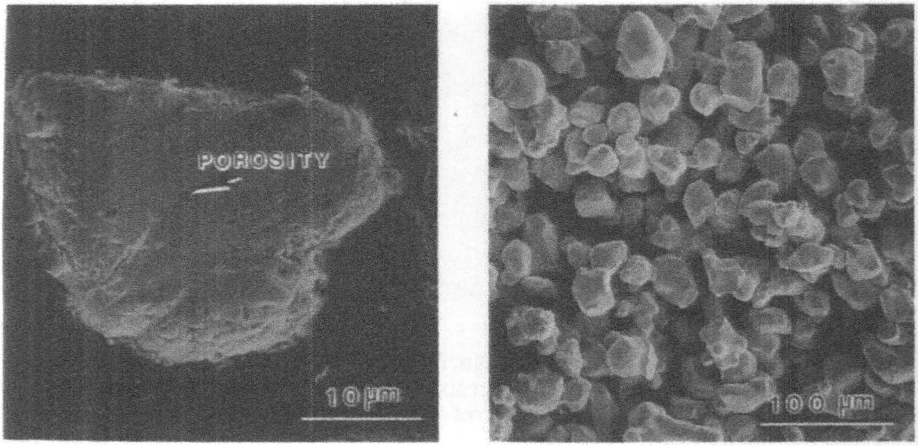

Figure 21 Morphology of W powder showing (a) pores existing in the powder, (b) few pores present on the surfaces of powders (from ref. [32]).

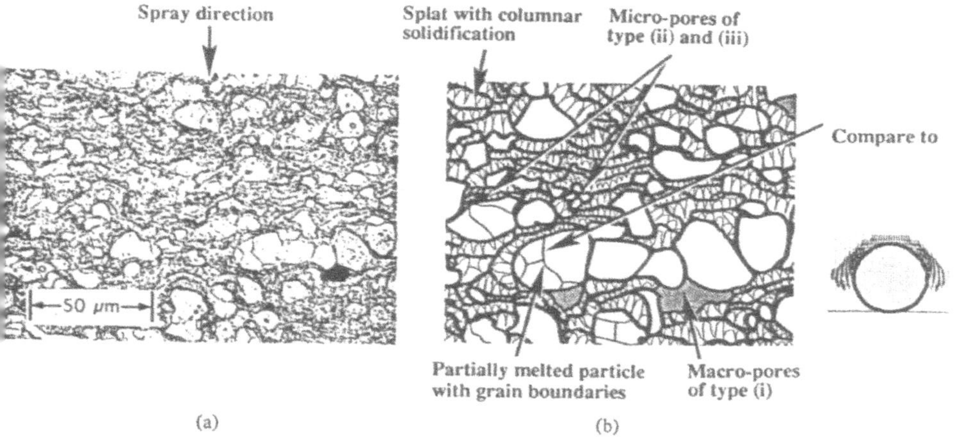

Figure 22. Microstructure of low-pressure plasma sprayed W deposited on graphite, (a) microstructure, (b) schematic of the microstructure in (a) [23].

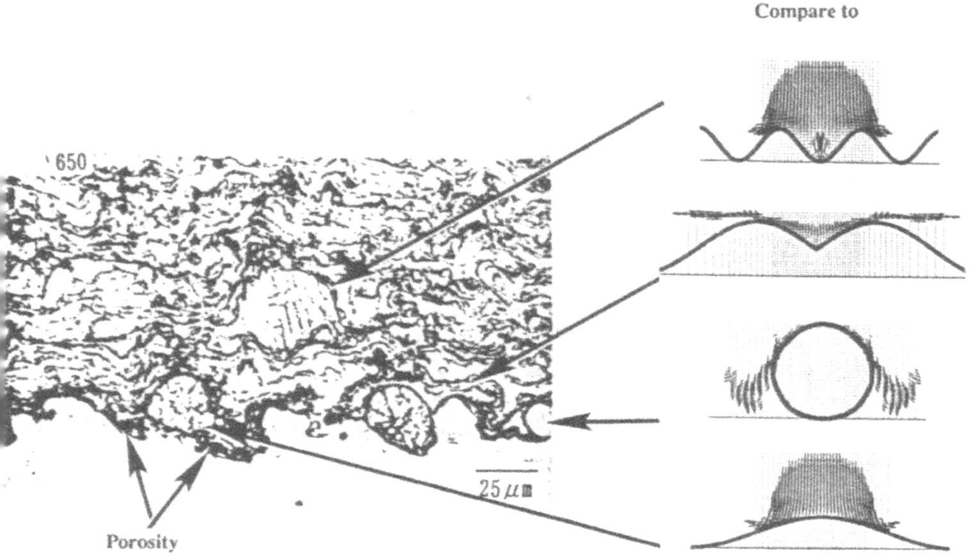

Figure 23. Microstructure of Ni-20 wt.%Cr coating sprayed at a substrate temperature of 650 °C [24].

SPRAY PROCESSING OF Ti METAL MATRIX COMPOSITES

P.S. Grant, Y.Y. Zhao, J.H. Li, M.L. Jenkins and B. Cantor
Oxford Centre for Advanced Materials and Composites
Department of Materials, Oxford University
Parks Road, Oxford OX1 3PH, U.K.

ABSTRACT. Ti monotapes reinforced with continuous SiC fibres can be manufactured by VPS with no as sprayed interfacial reaction and good fibre alignment. Plasma H_2 flow rate and plasma current have a strong effect on plasma power which determines Ti powder melting in the plasma. Increased plasma power gives higher monotape temperatures during manufacture and reduced matrix porosity. Chamber pressure and spray distance have no effect on plasma power and powder melting. In order to promote sufficient melting of Ti powder feedstock, the Ar plasma contains up to 25% H_2 which leads to wide scale formation of TiH_2 in the as sprayed Ti matrix. As sprayed monotapes have been cut, stacked and vacuum hot pressed to form a bulk Ti MMC. During VHP, the SiC fibre duplex C/TiB_2 coating reacts with the Ti matrix to form TiB needles and other reaction products. The extent of this reaction increases with increasing VHP temperature and time. In the case of Ti-6Al-4V, the presence of matrix H_2 during VHP stabilises the β phase and allows lower VHP temperatures to be used, with attendant improvements in interfacial condition.

1. Introduction

Ti alloy matrices reinforced with continuous ceramic fibres offer improved high temperature stiffness and strength in comparison with monolithic Ti alloys, while maintaining the toughness, impact and corrosion resistance of the Ti matrix [1]. However, the manufacture of fibre reinforced Ti metal matrix composite (MMC) components is complicated by material anisotropy, inhomogeneous reinforcement distribution and alignment, and chemical and physical mismatch between matrix and fibre which leads to interfacial reactions and residual stresses [1]. Spray processes are attractive for the manufacture of ceramic reinforced metal matrices because the 1-500μm sized liquid metal droplets in the spray allow intimate mixing with either a discontinuous particulate or continuous fibre ceramic reinforcing phase, and rapid droplet cooling minimises ceramic surface exposure times to potentially reactive molten metal [2-4]. This paper describes the manufacture and characterisation of SiC fibre reinforced Ti alloys manufactured by vacuum plasma spray forming and subsequent vacuum hot pressing.

M. A. Otooni (ed.), Science and Technology of Rapid Solidification and Processing, 109–121.

In vacuum plasma spraying (VPS) of MMCs, powder feedstock is injected into a hot gas plasma flame, which is usually a mixture of Ar and combinations of either H_2, He or N_2. The plasma flame accelerates, melts and projects the molten droplets onto a pre-wound fibre preform so that droplets infiltrate between fibres and solidify to form an integral MMC monotape, as shown schematically in Fig 1.

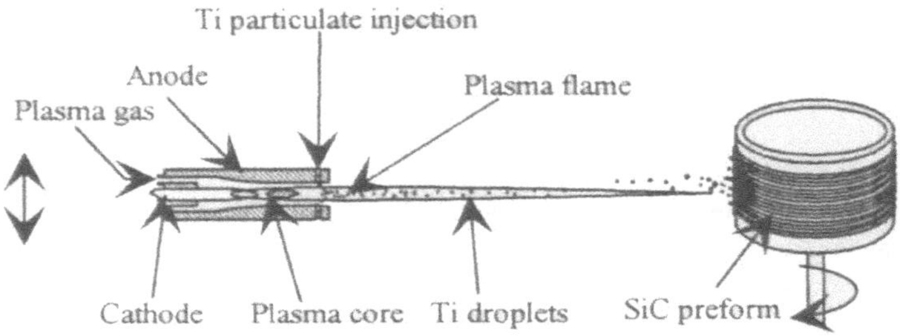

Fig 1. Schematic of the vacuum plasma spray process (VPS) for the manufacture of Ti based SiC reinforced MMC monotapes.

Spraying is performed in a sealed, low pressure, inert gas chamber so that oxygen sensitive materials may be processed. Robot manipulation of the plasma gun is frequently employed to coat large preforms evenly. The thermal condition of the metallic phase at deposition is important in determining final as sprayed MMC microstructural features such as matrix porosity, grain size, segregation behaviour and interfacial reaction. As sprayed monotapes can be cut, stacked and vacuum hot pressed (VHP) to form a bulk sample or component.

2. Experimental

Commercial purity Ti powder (VACTEX 22-00) with diameters 32-90μm and Ti-6Al-4V powder (Nuclear Metal Industry) with diameters 150-180μm were vacuum plasma sprayed in Ar atmosphere onto 100mm x 100mm flat and 20cm diameter cylindrical continuous SiC_f preforms using a Plasmatechnik A2000 VPS system under a range of process conditions. In each case, fibre preforms consisted of up to 50x100μm diameter SIGMA W cored SiC fibres with 1-3μm surface layers of C and then TiB_2, at a fibre centre to fibre centre spacing of 220μm. For flat wound substrates, the droplet spray was robotically scanned over the preform at a surface speed of 300mm/s. Cylindrically wound substrates were rotated at 60rpm and the droplet spray reciprocated along the cylinder axis at 100mm/s, as shown in Fig 1. As sprayed monotapes were sectioned for optical, scanning electron and transmission electron microscopy (SEM and TEM), image analysis and X-ray diffraction (XRD). As sprayed monotapes were also cut, stacked and vacuum hot pressed to form multilayer MMCs under a range of temperatures, times and pressures.

3. Results

For good fibre infiltration, impacting Ti droplets should be fully liquid to promote flow of matrix material around fibres, filling fibre shadows and minimising matrix porosity and monotape surface roughness. The degree of powder melting for a range of VPS processing parameters can be determined by rapidly traversing the droplet spray over the surface of a small polished substrate in a wipe test [5,6] Liquid particles in the spray deform into thin discs on impact, mushy droplets partially splat, and solid powders simply bounce off the substrate surface The degree of powder melting is assessed by observing the substrate surface under a range of VPS conditions Figs 2(a)-(c) are SEM micrographs of typical unmolten, mushy and liquid droplet impact morphologies respectively for Ti powder Fig 2(a) shows an angular, non spherical unmelted particle with width and length of ≈40 and 80μm respectively, and is characteristic of the Ti feedstock powder morphology Fig 2(b) shows a partially melted particle impact of diameter ≈350μm, and Fig 2(c) shows a fully liquid droplet impact which has deformed radially to a disc of diameter ≈90μm To identify individual droplet impacts, the relative velocity between the plasma gun and substrate must be sufficiently high to prevent impacts overlapping To assess the degree of Ti powder melting during VPS of monotapes in this study, a polished stainless steel substrate was rotated on a turntable at 60rpm, at a radius of 300mm from the turntable centre The droplet spray was robotically moved in a single pass across the spinning substrate in the the opposite direction at a surface speed of 300mm/s at a distance of 600mm from the turntable centre, to give a relative velocity of 2,200mm/s and substrate distance of 300mm

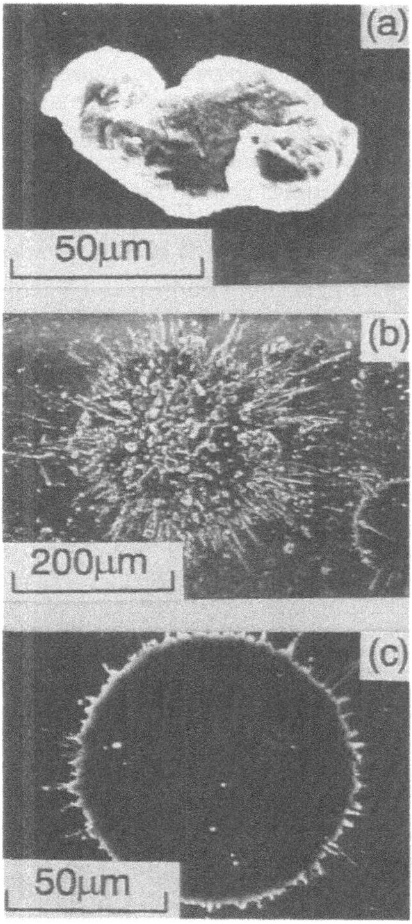

Fig 2. SEM micrographs of typical (a) unmolten, (b) mushy and (c) liquid Ti powder impacts during VPS wipe tests

Figs 3(a)-(c) show a series of wipe tests for Ti powder under a range of plasma H_2 flow rates of 0, 2 and 8 standard litres per minute (SLPM), with an Ar plasma gas flow rate of 35SLPM, Ti powder feed rate of 0 4g/s, Ar carrier gas flow rate of 1 6SLPM, spray distance of 300mm and chamber pressure of 150mbar Fig 3(a) shows a mixture of mushy droplet impacts and liquid droplet discs with diameters of 15-180μm, sprayed with an Ar only plasma Fig 3(b) shows that keeping all other parameters constant and introducing H_2 at a flow rate of 2SLPM sharply increases the total number of droplet impacts, and increases the maximum disc diameter to ≈220μm Fig 3(c) shows that further increasing the H_2 flow rate to 8SLPM, again increases the number of impacts so that droplet overlaying is now apparent The maximum disc diameter again increases to ≈375μm Further increases in the H_2 flow rate show no further increases in number of impacts or maximum disc diameter Wipe tests also showed that increasing the plasma current promoted droplet melting, and that changes in the chamber pressure had little effect The effect of changes in substrate distance were harder to identify because of the strong reduction in deposition efficiency at long substrate distances

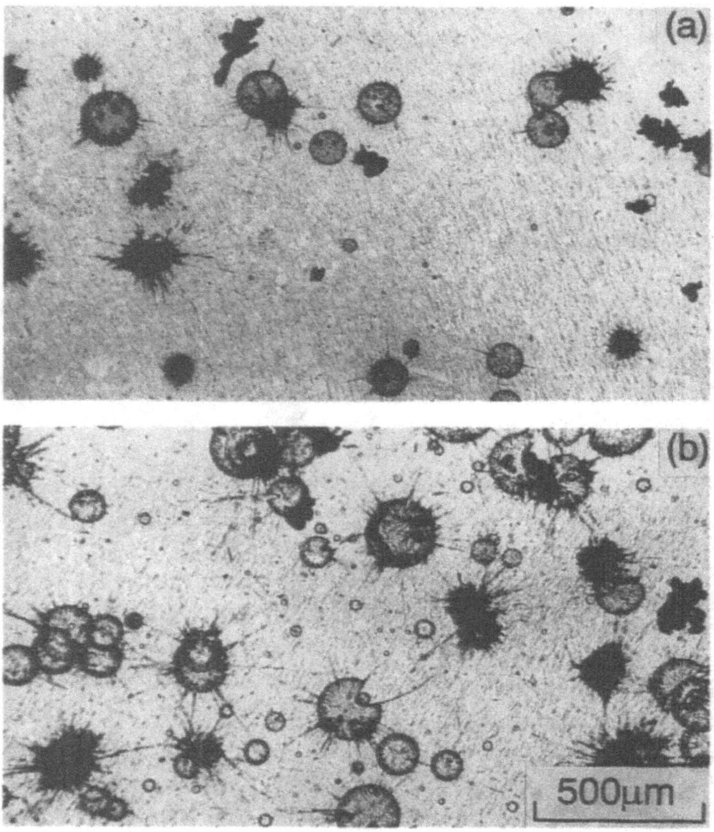

Fig 3 Wipe tests for Ti using plasma H_2 flow rates of (a) 0 and (b) 2 SLPM.

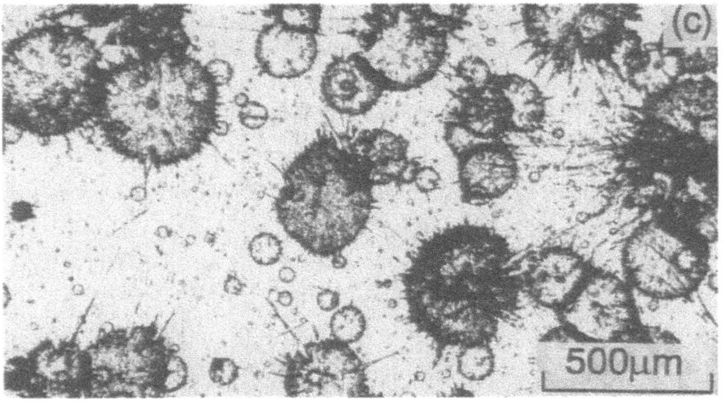

Fig 3(c) Wipe tests for Ti using a H_2 plasma gas flow rate of 8 SLPM.

Fig 4 Typical cross sectional microstructure of as sprayed Ti monotapes

Fig 4 shows a typical cross section of an as sprayed cylindrically wound Ti monotape manufactured using VPS conditions identical to those employed in Fig 3(c) The fibres are regularly spaced at 220μm intervals and show only slight fibre movement during VPS The underside of the tape is relatively smooth and the top surface rougher with a matrix porosity ≈0 6%

Figs 5(a)-(d) show a series of embedded thermocouple measurements as a function of time from flat wound monotapes during manufacture by VPS using the same process conditions as for Fig 3(c) and Fig 4, but with a range of (a) H_2 concentrations of 2, 4 and 8SLPM, (b) plasma currents of 500, 700 and 850A, (c) chamber pressures of 50, 150 and 250mbar, and (d) spray distances of 200, 300 and 400mm, respectively In Fig 5(a), H_2 flow rates of 2, 4 and 8SLPM gave corresponding plasma powers of 32 8, 38 4 and 43 5kW respectively In Fig 5(b), plasma currents of 500, 700 and 850A gave plasma powers of 26 5, 43 5 and 51 9kW respectively Changes in chamber pressure and spray distance in Figs 5(c) and (d) respectively gave no change in the plasma power Figs 5(a)-(d) show that in all cases, there is a steady increase in monotape temperature as spraying progresses, with maximum temperatures occurring at the end of deposition, after which the monotape cools Spraying times were all 5mins, except for the varying chamber pressure in Fig 5(d), where longer times were employed to offset the reduction in deposition efficiency with increasing substrate distance. Figs 5(a)-(d) show that increases in the H_2 flow rate and plasma current, and decreases in the chamber pressure and substrate distance progressively increase the maximum monotape temperature.

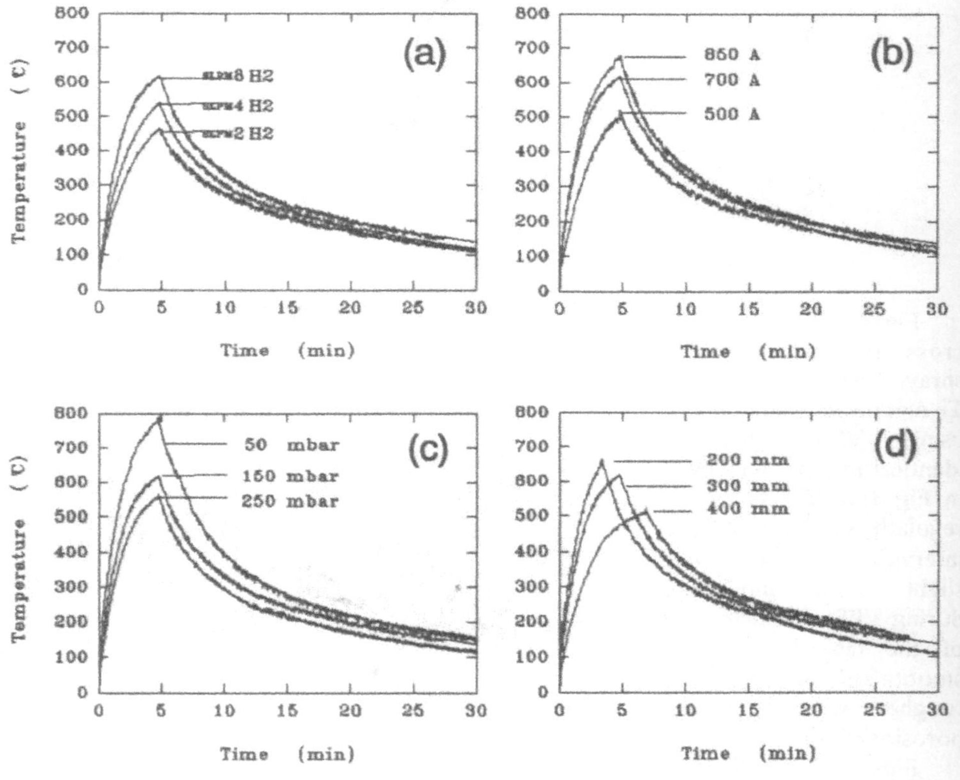

Fig 5. Embedded thermocouple measurements of monotape temperature as a function of time showing the effect of (a) plasma H_2 flow rate, (b) plasma current, (c) chamber pressure and (d) substrate distance.

Figs 6(a)-(d) show the variation of matrix porosity as a function of (a) plasma H_2 flow rate, (b) plasma current, (c) chamber pressure and (d) spray distance, corresponding to the thermal histories in Figs 5(a)-(d). The peak temperatures from Figs 5(a)-(d) are also shown. Figs 6(a) and (b) show that monotape porosity decreases from 1.2 to 0.6% as the plasma H_2 flow rate increases from 2 to 8SLPM, and from 1 to 0.4% as the plasma current increases from 500 to 850A, with corresponding increases in peak temperature from 460 to 620°C and 510 to 680°C respectively Fig 6(c) shows no systematic effect of chamber pressure on porosity Fig 6(d) shows an increase in porosity from 0 5 to 1 1% with increasing spray distance from 200 to 400mm, with a corresponding decrease in peak temperature from >700

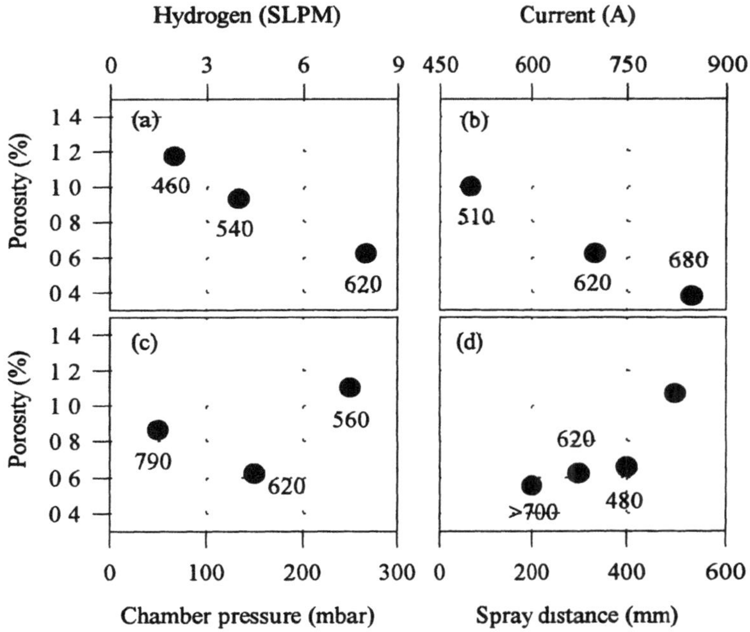

Fig 6 Ti monotape matrix porosity as a function of (a) plasma H_2 flow rate, (b) plasma current, (c) chamber pressure and (d) substrate distance, corresponding to the embedded thermocouple measurements in Fig 5

to 480°C

Fig 7 is a bright field TEM micrograph of the as sprayed SiC/C/TiB_2-Ti interface manufactured using the same VPS conditions in Figs 3(c) and 4, showing that there is no reaction between the outer TiB_2 layer and Ti matrix during manufacture Fig 7 shows some crystallisation of TiB_2 from the amorphous TiB_2 coating which may be induced by exposure to high temperatures during manufacture

Fig 8(a) is a dark field TEM micrograph from matrix regions of an as sprayed Ti-6Al-4V/SiC/C/TiB_2 monotape manufactured using the same VPS conditions as Fig 7, imaged using a $(01\overline{1})$ TiH_2 zone axis Porosity in Ti-6Al-4V monotapes was much higher (up to 10%) than for Ti monotapes because of the larger powder diameters, and therefore poorer melting Fig 8(a) shows the formation of TiH_2 within α' grains during VPS [7], as confirmed by the selected area diffraction pattern (SADP) and corresponding schematic SADP taken from the labelled TiH_2 region in Fig 8(a) and shown in Figs 8(b) and (c) respectievly Fig 8(d) is an XRD trace from matrix regions of the as sprayed Ti-6Al-4V monotape showing α' and β phases, as well as TiH_2

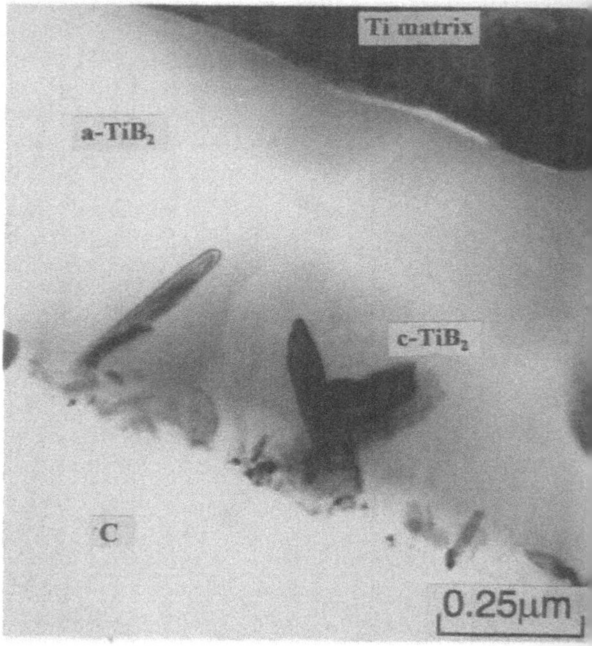

Fig 7 Bright field TEM micrograph of as sprayed Ti/TiB_2/C/SiC interface

Fig 9(a) is an optical micrograph of the typical etched microstructure of Ti monotapes after VHP at 870°C and 30MPa for 20min, showing that under these conditions the material is fully dense The as sprayed microstructure has coarsened during VHP although the original monotape splat structure is still evident Fig 9(b) is an SEM micrograph of the interfacial area of the Ti-6Al-4V monotape after VHP at 870°C and 30MPa for 120min, showing the growth of large TiB needles from the outer TiB_2 layer at the fibre/matrix interface during VHP at extended times [8,9]

Fig 10(a) is a typical bright field TEM micrograph of matrix regions of a fully dense VPS Ti-6Al-4V monotape after VHP at 800°C and 30MPa for 60min, showing that the TiH_2 in the as sprayed monotape has decomposed during VHP, and that the matrix now consists of α' and β grains only, as shown in the corresponding XRD trace in Fig 10(b)

4. Discussion

4 1 EFFECT OF VPS PARAMETERS

Figs 3(a)-(c) and 5(a) show that increasing the plasma H_2 flow rate promotes powder melting by increasing plasma power, with an optimum flow rate of 8SLPM for Ti powder over the range of conditions investigated Fig 6(a) shows that improved powder melting results in increased monotape temperatures and lower monotape porosity Wipe tests show that increasing plasma current to an optimum of 850A also promotes powder melting by

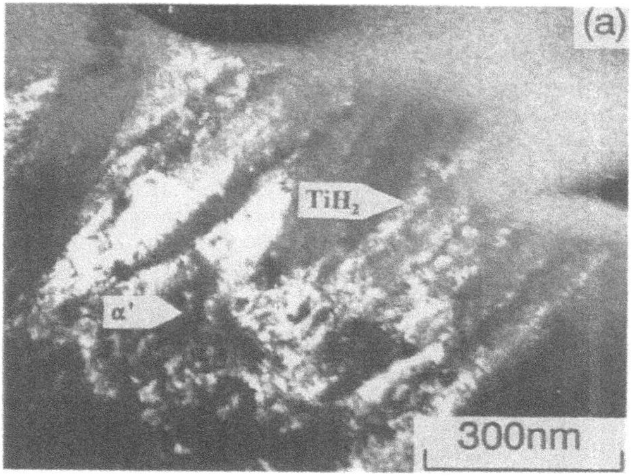

Fig 8(a) Dark field TEM micrograph of matrix regions of as sprayed Ti-6Al-4V monotapes imaged using a $(01\overline{1})$ TiH_2 zone axis from matrix region of Fig 7, (b) and (c) SADP and schematic SADP from TiH_2, (d) XRD trace from as sprayed Ti-6Al-4V.

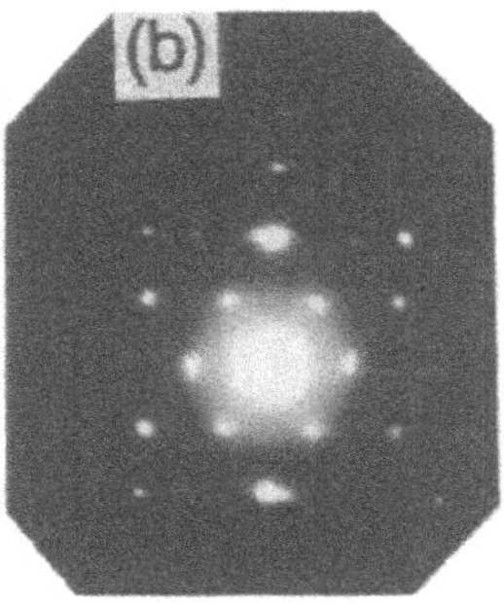

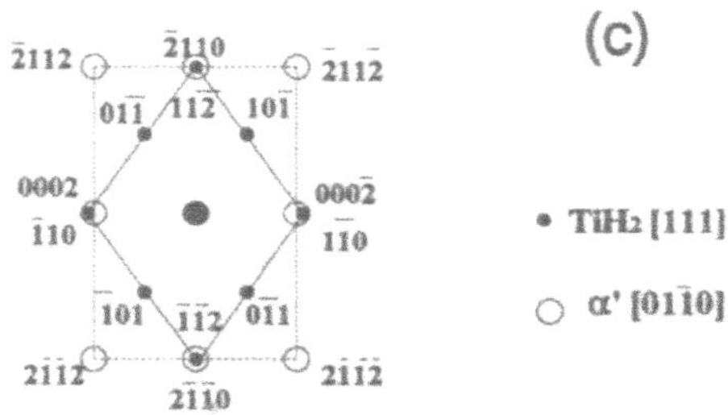

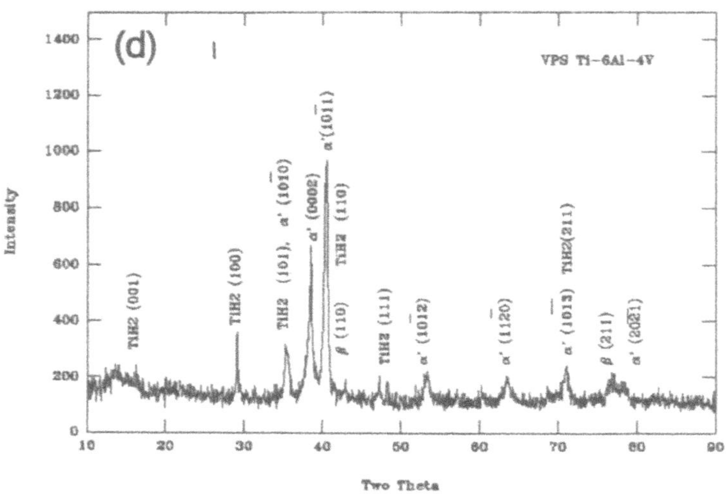

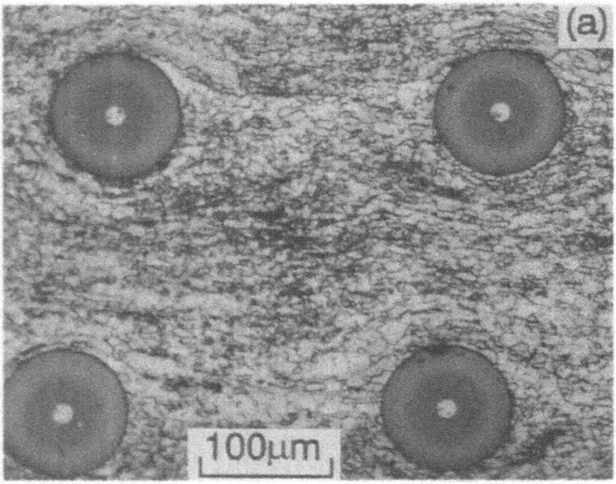

Fig 9(a) Typical cross sectional microstructure of etched Ti monotapes

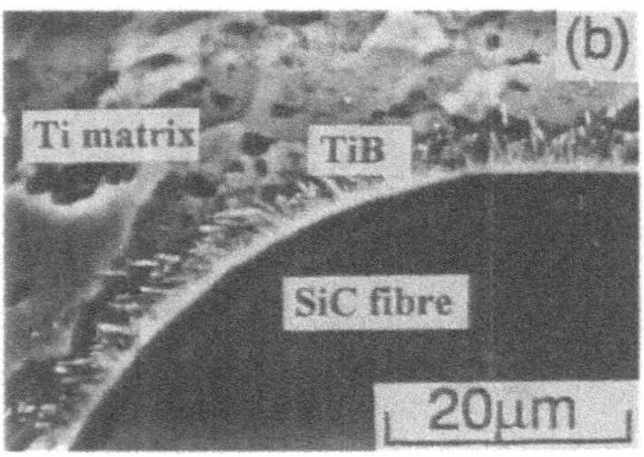

Fig 9(b) SEM micrograph of the SiC fibre/Ti matrix interface after prolonged thermal exposure during VHP showing the formation of TiB needles

Fig 10(a) Bright field TEM micrograph of matrix regions of Ti-6Al-4V monotapes after VHP

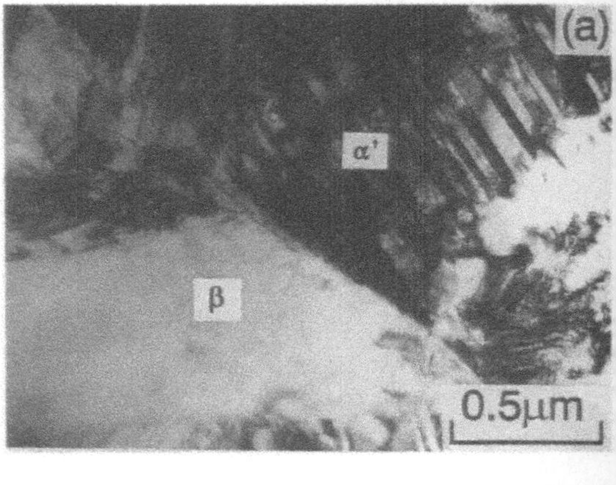

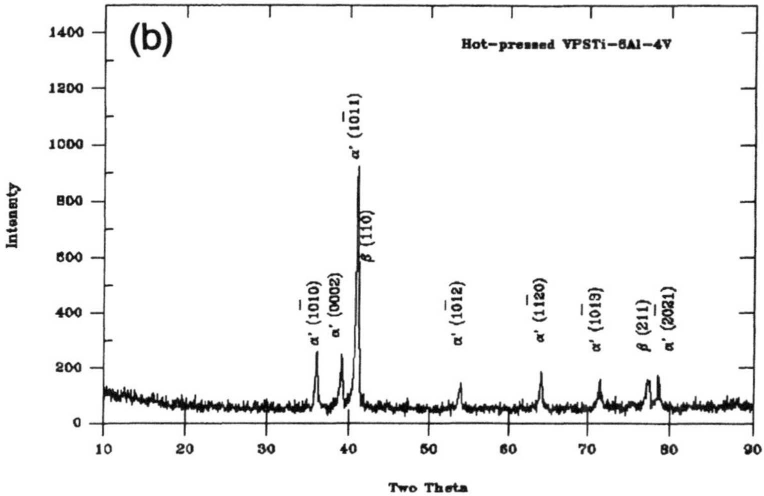

Fig 10(b). XRD trace respectively of Ti-6Al-4V monotapes after VHP.

ncreasing plasma power and similarly results in increased in monotape temperature and ower porosity. Although reducing the chamber pressure also increases monotape emperature as shown in Fig 6(c), there is no change in either plasma power or monotape orosity. In this case, monotape temperatures increase because of a longer plasma flame t low pressure, not from higher plasma powers and improved powder melting. The eneficial effect of increased flame length on powder melting is offset by an increase in lasma gas velocities and therefore lower powder plasma dwell times [10], and a eduction in heat transfer to the powder at lower pressures [11]. Similarly, reductions in pray distance do not affect plasma power, but increase monotape surface temperatures, s shown in Figs 5(d). Too short spray distances give insufficient powder plasma dwell imes for melting, whilst too long spray distances allow droplet cooling and re-olidification, as shown in Fig 6(d) [12].

Fig 11 shows that the variations of monotape porosity and temperature with H_2 flow ate and plasma current can be re-plotted on a single graph as a function of plasma power. urface temperature increases and monotape porosity decreases with increasing plasma ower, as shown in Fig 11.

.2 VPS MICROSTRUCTURE

ylindrically wound fibre preforms show good fibre alignment and infiltration during VPS s shown in Fig 4. As sprayed porosity is typically 0.4-1.2% for the Ti powder, but rises to 10% for the coarser Ti-6Al-4V powder. For all the range of VPS process conditions nvestigated, except very short spray distances, there was no evidence of as sprayed fibre/matrix interfacial reaction, although some crystallisation of the TiB results from monotape temperatures >500°C for several minutes during manufacture, as shown in Fig 7. All the monotapes show evidence of TiH_2 formation as shown in Fig 8, because of the need for plasma H_2 contents of up 25% to promote adequate powder melting.

4.3 VPS/VHP MICROSTRUCTURE

After vacuum hot pressing of Ti monotapes to full density, the formation of TiB needles occurs at the fibre-matrix interface. After prolonged consolidation times at high temperature, the duplex C/TiB_2 coating is completely consumed and a complicated mixture of reactions products are formed [8,9]. Fully dense Ti monotapes can be achieved at 870°C and 30MPa for 20mins with minimum interfacial reaction. For Ti-6Al-4V monotapes, the presence of H in the as sprayed monotape, stabilises the softer, more deformable β-bcc phase in comparison with the α'-hcp phase during VHP. The depression of the β transus in the presence of H_2 allows fully dense multilayers to be manufactured at temperatures <850°C, even from relatively porous monotapes, which are lower than those used in conventional Ti foil/SiC fibre layups [7]. Figs 10(a) and (b) show that after VHP, TiH_2 fully decomposes and H_2 is removed from the final α'/β microstructure.

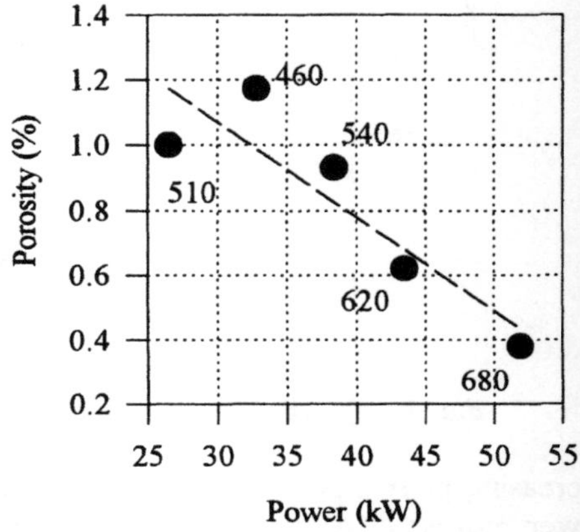

Fig 11. Porosity as a function of plasma power for Ti monotapes.

5. Conclusions

Ti monotapes reinforced with continuous SiC fibres can be manufactured by VPS with no as sprayed interfacial reaction and good fibre alignment. Plasma H_2 flow rate and plasma current have a strong effect on plasma power which determines Ti powder melting in the plasma. Increased plasma power gives higher monotape temperatures during manufacture and reduced matrix porosity. Chamber pressure and spray distance have no effect on plasma power and powder melting. In order to promote sufficient melting of Ti powder feedstock, the Ar plasma contains up to 25% H_2 which leads to wide scale formation of TiH_2 in the as sprayed Ti matrix.

As sprayed monotapes have been cut, stacked and vacuum hot pressed to form a bulk Ti MMC. During VHP, the SiC fibre duplex C/TiB_2 coating reacts with the Ti matrix to form TiB needles and other reaction products. The extent of this reaction increases with increasing VHP temperature and time. In the case of Ti-6Al-4V, the presence of matrix H_2 during VHP stabilises the β phase and allows lower VHP temperatures to be used, with attendant improvements in interfacial condition.

Acknowledgements

The authors would to thank the U.K. Engineering and Physical Science Research Council for the funding of this work. Two of us (Y.Y. Zhao and J.H. Li) would like to thank the British Council and the Chinese Education Commission for additional financial support. We would also like to thank Bob Shatwell at B.P. Metal Composites for the supply of SIGMA SiC fibre.

References

1. P.G. Partridge and C.M. Ward-Close, "Processing of advanced continuous fibre composites: Current practice and potential developments", Int. Met. Rev., **38** (1993), 1-24.
2. P.R. Smith and F.H. Froes, "Developments in titanium metal matrix composites", J. Metals, **36** (1984), 19-26.
3. P.A. Dearnley and K.A. Roberts, "Titanium matrix composites via vacuum plasma spraying", Powder Metallurgy, **34** (1991), 23-32.
4. S. Sampath and H. Herman, "Plasma spray forming metals, intermetallics and composites", J. Metals, **47** (1993), 42-49.
5. K.A. Roberts and T.W. Clyne, "A simple procedure for the characterization of spray deposition processes - the linescan test", Surface and Coatings Technology, **41** (1991), 103-115.
6. K.A.Q. O'Reilly and B. Cantor, "Spray formed TiAl", Proc. 3rd Japan Int. SAMPE Symposium, ed. T. Kishi, N. Takeda and Y. Kagawa, (SAMPE, 1993), 1576-1583.
7. J.H. Li, M.L. Jenkins and B. Cantor, "Control of interfacial reactions in titanium-based composites reinforced with SiC fibres", (Paper presented at the Int. Sym. Control of Interfaces in Metal and Cermic Composites, TMS Annual Meeting, San Fransico, 1994).
8. Z.X. Guo, B. Derby and B. Cantor, "Comparison of interfaces in Ti composites reinforced with uncoated and TiB_2/C-coated SiC fibres", J. Microscopy, **169** (1993), 279-287.
9. J.H. Li et al., to be published in Composites.
10. P. Fauchais et al., "Influence of the thermal-kinetic phenomena during particulate impact on the properties of ceramic coatings", Proc. 4th National Thermal Spray Conference (Metals Park, OH:American Society for Metaks, 1991), 205-213.
11. R.W. Smith, "Plasma spray processing...the state of the art...and future - from a surface to a materials processing technology", 2nd Plasma Technik Symposium, vol 1, ed. S. Blum-Sandmeier, H. Eschnaeur, P. Huber and A.R. Nicoll (Hafliger Druck, AG, Wettingen, Switzerland), 17-35.
12. Y.Y. Zhao, P.S. Grant and B. Cantor, "Fabrication of Ti/SiC_f metal matrix composites by vacuum plasma spraying", Journal de Physique IV (Colloque C7, supplément au Journal de Physique III, vol 3), 1993, 1685-1691.

SPRAY FORMING WITH UNIFORM DROPLETS

C.-A. CHEN, S. SAHU, J.-H. CHUN AND T. ANDO*
Laboratory for Manufacturing and Productivity,
Massachusetts Institute of Technology,
Cambridge, MA 02139 U.S.A.
**Department of Mechanical Engineering,*
Tufts University
Medford, MA 02155 U.S.A.

ABSTRACT. A Sn - 37 wt.% Pb eutectic alloy was spray formed with a uniform-droplet spray generated by a non-gas atomization process based on the controlled break up of a laminar jet. Uniform droplets were produced to have identical thermal history and velocity and were deposited onto a substrate in controlled motion to provide distinctly different thermal states of the droplets and the deposit surface. The spray formed alloy presented fine, equiaxed two-phase microstructures when partially solidified droplets were deposited and expitaxial, columnar microstructures when superheated droplets were deposited, on either a solidified or partially solidified deposit surface. The range of microstructural control possible by uniform-droplet spray forming was illustrated with a process-structure map.

1. Introduction

Spray forming combines the production of a molten alloy spray and the *in situ* deposition of the spray onto a substrate to form a rapidly solidified, near-net shape preform [1-9]. Unlike other rapid solidification processes where particulate production and consolidation are separate steps, spray forming potentially offers an integrated and economical means for advanced materials processing. Most spray forming processes developed to date, however, use a gas-atomized spray which inevitably consists of non-uniform droplets, all in different thermal and kinetic states. In addition, the spray parameters have spatial distributions and are coupled [3, 10-13], making both the analysis and control of gas-atomized spray forming processes difficult. More importantly, the coupling of the spray parameters restricts the range of applicable spray conditions and hence limits the extent of microstructural control.

These limitations can be largely removed if the spray consists only of uniform droplets. A non-gas atomization technique has been developed that can generate uniform droplets by the controlled break-up of a molten jet using imposed vibrations [14-16]. The uniform droplets in a spray have an identical thermal history, dependent only on the distance from the spray nozzle. Spray parameters such as the mass flux, droplet undercooling, liquid fraction, droplet velocity and droplet diameter can be controlled independently to maximize the range of microstructural control. This paper reports on preliminary results with a Sn-37 wt.% Pb alloy to demonstrate the potential range of microstructural control by the uniform-droplet spray forming process.

M. A. Otooni (ed.), Science and Technology of Rapid Solidification and Processing, 123–134.

2. Experiments

Fig. 1 schematically shows the uniform-droplet spray forming apparatus used in this study. The apparatus consists of a stainless steel melting crucible with a 100 μm diameter orifice at the bottom, heating elements, a piezoelectric transducer, a charging plate and a moving copper substrate placed on a computer-controlled linear lead-screw x-y table, all contained in a chamber filled with a nitrogen-2 % hydrogen gas mixture.

Four controlled spray forming experiments were performed. A Sn-37 wt.% Pb alloy prepared from 99.99 wt.% pure reagent grade Sn and Pb was used in all the spray forming experiments. In each run, about 400 g of the Sn-37 wt.% Pb alloy was melted in the crucible and kept at a constant temperature during spray forming. The melt was gas-pressure ejected from the bottom orifice while vibrations of a controlled frequency of 10 kHz were imposed on the melt through a metal rod connected to the piezoelectric transducer. The ejected laminar jet of the melt broke into uniform droplets with a diameter of 198 μm. The droplets were charged, as they were generated, by the charging plate, to prevent them from merging in their flight to the substrate where they were deposited. A charging voltage of 850 V was maintained relative to the melt throughout each run. This resulted in a field strength of 267 V/mm and a droplet charge of 7.1×10^{-12} C.

The motion of the x-y table was programmed to permit repeated deposition on a rectangular path on the substrate, either 5.0×10^{-2} m wide and 7.5×10^{-2} m long or 2.5×10^{-2} m wide and 5.0×10^{-2} m long. The mode of spray deposition was therefore not continuous but periodic. The programming was done such that one of the longer edges of the rectangular path experienced a constant substrate speed of 3.8×10^{-2} m/s to assure an identical instantaneous deposition rate along this edge in all four runs. This translates into adding 1.42 kg/m^2 to the prior deposit, or growing the deposit by 170 μm, in each deposition cycle. The substrate moving cycle was adjusted to be either 2.3 or 3.8 s to provide a desired overall deposition rate. The substrate surface temperature was kept at 433 K prior to each experiment.

The four controlled spray forming experiments represent distinctly different spray forming conditions of (1) mushy droplet spray vs. mushy deposit surface, (2) mushy droplet spray vs. solid deposit surface, (3) liquid droplet spray vs. solid deposit surface and (4) liquid droplet spray vs. mushy deposit surface. Table I summarizes the thermal states of the uniform droplets and deposit surfaces used in the four spray forming experiments. The thermal states of the deposit surface refer to those at the start of a new deposition cycle at a given point on the deposit surface and may differ from the instantaneous thermal states of the deposit surface during its passage through the uniform spray.

To produce the four conditions (1)-(4), the melt superheat, melt ejection velocity, flight distance, substrate temperature and substrate motion were set as shown in Table II by calculating the thermal states of uniform droplets assuming equilibrium solidification. Since the assumption of equilibrium solidification, of both the uniform droplets and the deposit, used to determine these spray forming conditions is crude at best, exact values of liquid fractions in the mushy droplets in experiments (1) and (2) and in the deposit surfaces in experiments (1) and (4) are not known. Nonetheless, observation of intercepted droplets and the deposit surfaces during these experiments indicates that the desired mushy states were maintained under the spray and substrate conditions used.

The spray formed materials were examined for their microstructures with a JEOL JSM6100 scanning electron microscope (SEM) at an acceleration voltage of 10 kV.

Only the portions of spray formed materials that were produced at the constant substrate velocity of 3.8 m/s were examined. SEM micrographs were taken on a vertical cross section of the deposits at different distances from the bottom surface that faced the substrate during the spray forming experiments.

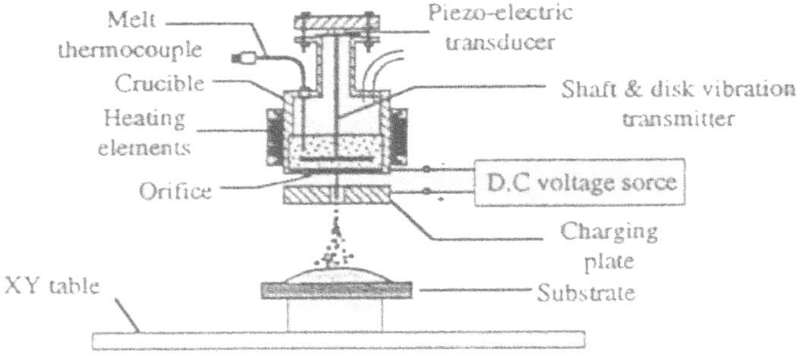

Fig. 1 Schematic illustration of the uniform spray forming apparatus.

Table I Spray forming conditions used in this investigation. Thermal states refer to conditions at the beginning of each deposition cycle. (f_l : liquid fraction)

Experiment	Spray	Deposit surface
1	Partially solidified ($f_l = 0.8$)	Partially solidified ($f_l = 0.2$)
2	Partially solidified ($f_l = 0.8$)	Solidified
3	Fully molten (superheated at 462 K)	Solidified
4	Fully molten (superheated at 462 K)	Partially solidified ($f_l = 0.4$)

Table II Spray and substrate parameters used in this investigation

Parameters	Experiments			
	(1)	(2)	(3)	(4)
Melt temperature (K)	486	486	566	566
Jet velocity (m/s)	5.2	5.2	5.2	5.2
Flight distance (m)	0.3	0.3	0.3	0.3
Substrate temperature (K)	443	443	443	443
Substrate velocity (m/s)	0.038	0.038	0.038	0.038
Deposition cycle (s)	2.3	3.8	3.8	2.3

3. Results and Discussion

3.1 DEPOSIT APPEARANCE

The rectangular substrate motion produced spray deposits in a rectangular frame shape as schematically shown in Fig. 2. The deposits showed smooth top and bottom surfaces and a bell shaped cross section about 20 mm high in the middle section. The mass of the deposit was essentially the same as the mass of the melt produced in the crucible in all four experiments, indicating a nearly 100 % spray yield.

Fig. 2 Schematic illustration of the spray deposit. The shaded portion was spray formed under the controlled conditions shown in Tables I and II.

3.2 MICROSTRUCTURES

Microstructures were studied by SEM using the shaded portions shown in Fig. 2 of the deposits that represented the controlled spray forming conditions given in Tables I and II. Fig. 2 also shows the positions on the cross section of the spray deposits where SEM photomicrographs were taken.

3.2.1 Mushy droplets vs. mushy deposit surface Experiment (1), in which mushy droplets were deposited onto a mushy deposit surface in each deposition cycle, produced a fine, non-lamellar two phase microstructure in the entire deposit. A deposition cycle of 2.3 s was used. Fig. 3 shows SEM micrographs taken at the bottom and mid portions of the spray deposit (areas 1 and 2 in Fig. 2). The matrix phase is Sn and the discrete second phase is Pb, although the exact compositions of these phases are not known. The two-phase microstructure presents an equiaxed morphology although in areas away from the bottom, some low degree of directionality normal to the substrate is noticed.

Splat boundaries, where they are discernible, are delineated with a thin layer of the Pb-rich phase and preserve deformed droplet shapes in the bottom region (Fig. 3 (a)) but are flatter and roughly parallel to the substrate surface at greater distances from the bottom (Fig. 3 (b)) reflecting the high constant liquid fraction in the droplets used and lower rates of heat extraction in later stages of the experiment. The portions of splat boundaries delineated with a Pb-rich layer attest that these portions of the deposit surface were solid when the next droplets arrived. Conversely, the absence of the Pb-rich layer is indicative of a molten splat surface where liquid intermixing between successive splats occurred. It follows that most of the splat boundaries produced between the final layer of splats in a deposition cycle and the first splat layer in the next cycle are more visible than those between splats within a deposition cycle.

The solid phase(s) that were present in the droplets were deformed and probably fragmented into smaller pieces which then grew into the observed equiaxed, microstructures. The degree to which an equiaxed morphology is produced in the splats depends on the form and extent of prior solidification in the incoming droplets and the local solidification time in the splats. High liquid fractions in the droplets and slow local solidification in the splats promote columnar solidification in the splats. The exact form and extent of the prior droplet solidification depend on prior droplet undercooling [17, 18] and are not known at present. Work is underway to determine the prior undercooling of uniform droplets. The somewhat directional microstructure noticed in the mid portion (Fig. 3 (b)) suggests that this region completed solidification less rapidly than regions closer to the substrate.

The above results confirm that depositing mushy droplets onto a mushy deposit surface does produce the so-called equiaxed microstructure in the spray deposit as suggested by many workers [3, 10, 19-21]. Spray and substrate parameters used in gas-atomized spray forming are therefore adjusted to maintain both the spray and the deposit surface in a mushy state. It is important to note, however, that a gas-atomized spray contains droplets in different sizes, which may be fully molten, mushy or even fully solidified depending on their thermal history, and therefore is 'mushy' only as a whole spray. It has been suggested that fine solid particles in a gas-atomized spray are important as they may act as randomly oriented 'nuclei' that grow to produce an equiaxed microstructure in a spray formed alloy [10, 20]. This experiment with uniform mushy droplets, however, has shown that a fine, equiaxed microstructure can result in the absence of such artificially introduced nuclei.

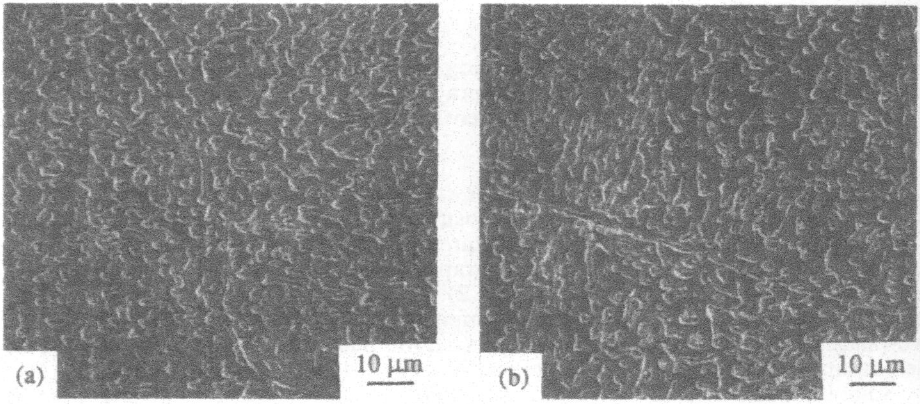

Fig. 3 SEM micrographs of the spray deposit produced with mushy uniform droplets and a mushy deposit surface. (a) near the bottom surface (area 1 in Fig. 2); (b) middle portion (area 2 in Fig. 2).

The spray deposit contained no noticeable pores. This can be attributed to the use of uniform droplets at a high uniform liquid fraction. In spray forming, liquid is required to fill the interstices between splats. A constant liquid fraction in uniform droplets assures that there is always liquid at any point on the deposit surface where a splat interstice is formed. A sufficiently high droplet fraction promotes a flat splat geometry, as seen in Fig. 3 (b), which does not produce large splat interstices. Such conditions do apply to uniform-droplet spray forming, making the porosity level very low.

3.2.2 Mushy droplets vs. solid deposit surface Experiment (2) is a modification of experiment (1) in that a longer deposition cycle of 3.8 s was used to complete the solidification of a deposited layer before the next cycle of deposition began. The mushy droplets were identical to those in experiment (1). Fig. 4 shows an SEM micrograph taken at a mid portion (area 2 in Fig. 2) of the resultant deposit. As expected from the longer deposition cycle used, many splat boundaries are delineated with Pb, indicating completeness of solidification, and are roughly parallel to the substrate with an average spacing between them of about 40 μm. Given the droplet diameter of 198 μm and assuming a flattening ratio of 5 : 1, each of these splats was most likely produced from one droplet. This indicates that many of these splats completed their solidification before the next ones arrived at the high instantaneous deposition rate that occurred in the spray. The microstructures within the splats consist of equiaxed grains of the Sn-rich and Pb-rich phases and are basically the same as those produced in experiment (1). This deposit also showed flatter splats at greater distances from the bottom. No noticeable porosity was found.

3.2.3 Liquid droplets vs. solid deposit surface Experiment (3), in which fully molten, superheated droplets were deposited onto a solidified deposit surface in each cycle, produced columnar microstructures in the spray deposit except in regions near the bottom where a fine, equiaxed microstructure was observed. Virtually no porosity was noted. Fig. 5 (a) shows the equiaxed microstructure near the bottom (area 1 in Fig. 2). Both the Sn-rich and Pb-rich phases are evenly distributed in random orientation, making the microstructure truly equiaxed. No splat boundaries are visible despite the

Fig. 4 SEM micrograph of the spray deposit produced with mushy uniform droplets and a solidified deposit surface. Middle portion (area 2 in Fig. 2).

solidified deposit surface maintained throughout this experiment. The absence of Pb films at splat boundaries in this region is indicative of a high solidification rate.

Fig. 5 (b) shows a columnar microstructure at a mid portion (area 2 in Fig. 2) of the deposit. The splats are flat, showing a thickness of 10 - 40 μm, and nearly parallel to the substrate surface, reflecting the use of superheated droplets which readily spread upon deposition. They also indicate that the individual splats completed solidification very quickly, during a passage of the spray, before the next ones arrived. Such a high local solidification rate would result in these very thin splats. The splat boundaries are not delineated with the Pb-rich phase but are still noticeable in areas where morphological changes took place. The columnar microstructure consists of lamellar (regular) and non-lamellar (irregular) regions. The ratio of non-lamellar to lamellar regions in the columnar microstructures was found to increase with decreasing the distance from the substrate. The columnar microstructure just above the bottom equiaxed region was almost entirely non-lamellar and showed a dendritic morphology, Fig. 5(c). Near the top surface of the deposit, about 20 mm from the bottom, the columnar microstructure was entirely lamellar eutectic (Fig. 5 (d)).

The lamellar eutectic regions consist of fine eutectic cells, about 10-20 μm in diameter. Many adjacent cells having similar orientations make up a columnar eutectic grain, as indicated by two such eutectic grains in Fig. 5 (d). Observation at a lower magnification (Fig. 5 (e) and (f)) reveals that the transverse diameter of these eutectic grains increased with increasing the distance from the substrate.

The non-lamellar Sn grains in the columnar microstructure is contiguous through many splats. Some of the lamellar eutectic cells also grew through the splats. The growth of the Pb lamellae, however, is disturbed at the splat boundaries, as indicated by the non-uniform interlamellar spacings and by the frequent replacement of the regular lamellar morphology with the non-lamellar irregular morphology. The Sn phase in the non-lamellar regions that replaced the lamellar eutectic, however, grew directly

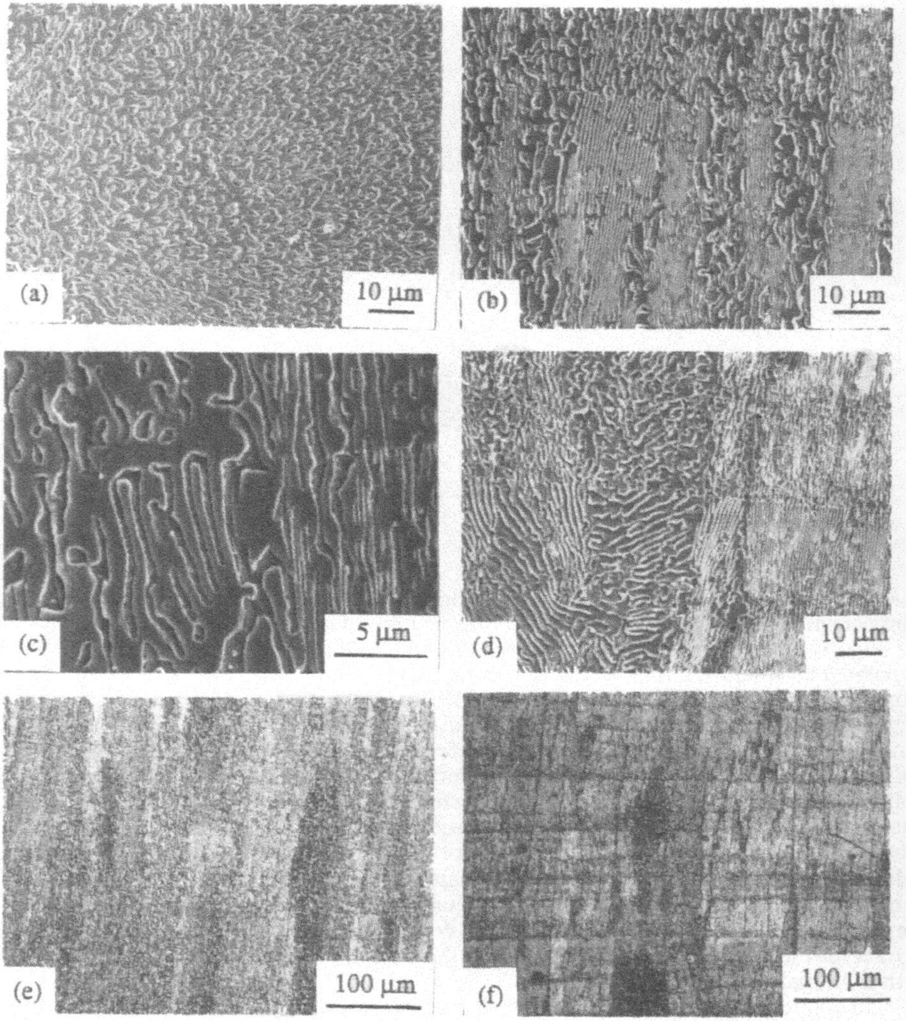

Fig. 5 SEM micrographs of the spray deposit produced with superheated droplets and a solidified deposit surface. (a) equiaxed two-phase microstructure near the bottom surface (area 1 in Fig. 2); (b) columnar microstructure at the middle portion (area 2 in Fig. 2) showing both non-lamellar and lamellar eutectic growth; (c) columnar microstructure just above the equiaxed region showing dendritic growth; (d) columnar microstructure at the top portion (area 3 in Fig. 2) showing entirely lamellar eutectic growth; (e) low magnification view of the columnar microstructure at the mid portion (area 2 in Fig. 2); (f) low magnification view of the columnar microstructure at the top portion (area 3 in Fig. 2).

from the Sn lamellae in the lamellar eutectic without nucleation. Thus, the growth of the columnar microstructure, whether it is non-lamellar or lamellar eutectic, is characterized as epitaxial, at least with respect to the growth of the Sn phase.

Such epitaxial growth should be promoted when incoming droplets contain no solid phase and when the nucleation of a solid phase in the liquid splat is more difficult than the epitaxial growth of a solid phase in the prior splat into the liquid splat. Epitaxial growth in spray forming was first reported by Singer and Evans [3] in a steel deposit produced by gas-atomized spray forming. However, in their work, epitaxial growth was confined in very small isolated volumes near the deposit peripheries where the local deposition rate was low. Larger epitaxial regions could not be produced because of the presence of solid and semisolid particles in the gas-atomized spray. In the present experiment, the above requirements for epitaxial growth are strictly met, permitting epitaxial growth throughout the deposit except in the initial stage.

The preference of dendritic growth to eutectic growth in columnar regions near the substrate suggests that the melt, and therefore the solid/liquid interface, in the solidifying splats were substantially undercooled in the initial stages. Undercooling often shifts the coupled zone for eutectic growth to off-eutectic compositions, permitting dendritic growth in a eutectic melt [22, 23]. Since the uniform droplets were superheated just before impinging on the deposit surface, the assumed undercooling is considered to have resulted from the rapid cooling of the droplets (splats) by the substrate which was kept at 443 K. The transition from lamellar to non-lamellar morphology at splat boundaries also indicates abruptly increased growth rates due to the quench effect of the prior solidified deposit surface,.

3.2.4 Liquid droplets vs. mushy deposit surface In experiment (4), the deposition cycle was decreased, from 3.8 s used in experiment (3), to 2.3 s to maintain a mushy deposit surface while depositing fully molten, superheated droplets. Despite the retention of liquid at the deposit surface during the experiment, the resultant deposit showed

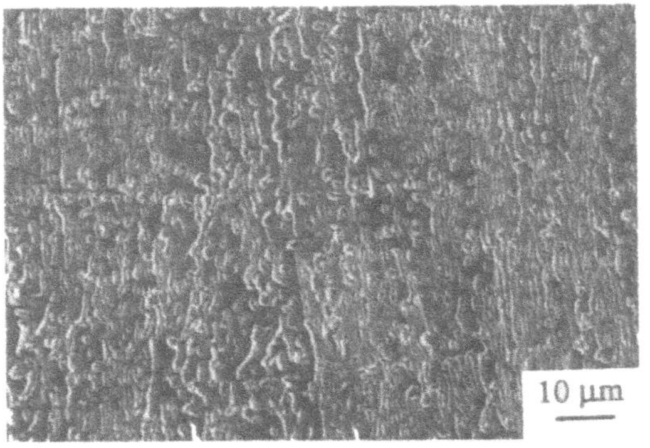

Fig. 6 SEM micrograph of the spray deposit produced with superheated uniform droplets and a mushy deposit surface. Middle portion (area 2 in Fig. 2). Columnar microstructures similar to those of the deposit produced with superheated droplets and a solidified deposit surface were found.

basically the same microstructures as those in the deposit produced in experiment (3): an equiaxed two-phase microstructure near the substrate and epitaxial columnar structures at greater distances from the substrate. Fig. 6 shows the epitaxial columnar structure at a mid portion (area 2 in Fig. 2) of the deposit. Both lamellar eutectic and non-lamellar morphologies are observed in the columnar part of the deposit, with an increased amount of lamellar eutectic with increasing the distance from the substrate. The transition from lamellar to non-lamellar growth also occurred at some splat boundaries.

The observation of epitaxial growth in this deposit suggests that the impact associated with the impingement of molten droplets onto the mushy deposit surface at a low velocity of about 5 m/s was not sufficient to cause any substantial disturbance to the growth of the solid in the prior splat and thus did not stop the epitaxial growth of the solid into the next liquid layer. The relatively low liquid fraction of the deposit surface also helped the growing solid maintain its solidification front unaffected by the impact. Maintaining these conditions for epitaxial growth is not possible in gas-atomized spray forming, in which a mixture of solid, semisolid and molten droplets impacts the deposit surface at much greater velocities exceeding 100 m/s [12].

3.3 PROCESS-STRUCTURE MAP

Fig. 7 schematically shows a general process-microstructure map for spray forming. In this figure, the liquid fractions of the droplets and the deposit surface are shown as independent process variables so that the entire process window can be described. Generally, the combination of hot droplets and a hot deposit surface produces coarse solidification structures in the deposited material (upper right corner), whereas cold droplets and a cold deposit surface produce rapidly solidified but porous structures (lower left corner). An optimum range is thus found between these two extremes. The resultant optimal microstructures are characterized by fine, equiaxed grains.

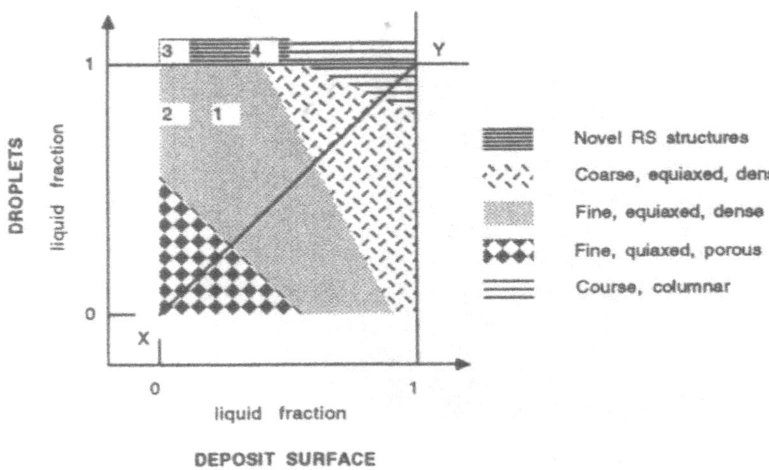

Fig. 7 Process-structure map for spray forming. The four spray forming conditions (1) - (4) are indicated assuming the equilibrium solidification of the droplets and deposit.

In actual spray forming processes, coupling and spatial distributions of the process parameters often restrict the applicable spray forming conditions. Even with the uniform-droplet spray forming process, the conditions below line XY are not applicable because the liquid fraction of the deposit surface should not exceed the droplet liquid fraction unless a secondary heat source is used to heat the deposit surface. Apart from this limitation, in uniform-droplet spray forming, any condition above line XY can be maintained at all points on the deposit surface. The resultant microstructure should reflect the single spray forming condition and therefore is expected to be uniform. In gas-atomized spray forming, a range of local conditions, including some of those below line XY, occur simultaneously. Therefore, the resultant deposit would present a range of microstructures (and frequently porosity), even produced under the most optimum spray and substrate conditions.

All of the four experimental conditions (1) - (4) used in the present study are near the upper left corner of the map where the splats are expected to solidify rapidly as evidenced by the observed non-lamellar solidification in the eutectic alloy. The conditions near this corner are important since they lead to rapidly solidified novel microstructures as demonstrated in this study. The epitaxial crystal growth reported in this paper is possible only with fully molten droplets. Spray forming with fully molten droplets may be done either with superheated droplets as done in this study or with undercooled droplets. Amorphous deposits are also considered to result with undercooled droplets at a sufficiently low, controlled deposition rate.

4. Conclusions

Distinctly different microstructures of fine, equiaxed microstructures and fine, columnar structures were produced in a Sn-37 wt.% Pb alloy by uniform-droplet spray forming with controlled thermal states of the uniform droplets and the deposit surface.

The fine, equiaxed microstructures consisted of Sn and Pb phases in non-lamellar, microcrystalline morphologies, and were formed in deposits produced with partially solidified uniform droplets deposited on either a solidified or partially solidified deposit surface.

The columnar microstructures are characterized by epitaxial growth through hundreds of splats and were formed in deposits produced with fully molten, superheated uniform droplets deposited on either a solidified or partially solidified deposit surface. The epitaxial columnar structures presented lamellar and non-lamellar morphologies with progressively increased amounts of lamellar eutectic in later stages of spray forming.

The fine, equiaxed two-phase microstructures resulted from the deformation and fragmentation of the solid phase(s) in the droplets caused by the impact and their subsequent growth and coarsening in the splats. The latter mechanism alone can produce an equiaxed microstructure in a spray formed material.

The use of uniform droplets permits spray forming with independently controlled spray and substrate parameters allowing for ultimate levels of microstructural control in spray formed materials. A schematic process-structure map was used to illustrate the range of structural control attainable by uniform-droplet spray forming.

Acknowledgment
The authors are grateful to the National Science Foundation for the support of this work under the grant No. DMII-9301375.

References

1. A. R. E. Singer, Met. Mater., 4, (1970), p. 246.
2. A. R. E. Singer, J. Inst. Met., 100, (1972), p. 185.
3. A. R. E. Singer, and R. W. Evans, Met. Technol., 10, (1983), p. 61.
4. A. G. Leatham, R. G. Brooks, and M. Yaman, in Modern Developments in Powder Metallurgy, vol. 15, (eds., E.N. Aqua and C.I. Whitman), APMI, Princeton, NJ, (1985), p. 157.
5. N. J. Grant, in Casting of Near Net Shape Products, (Y. Shirai, J. E. Battles, R. S. Carbonara and C. E. Mobley, eds.), TMS, Warrendale, PA, (1988), p. 203.
6. E. J. Lavernia and N. J. Grant, Int. J. Powder Metall., 2(2) (1986), p. 93.
7. Y. Ikawa, K. Kumagai, H. Hamabe, Y. Ozaki and S. Yamauchi, J. Jap. Inst. Met., 30(6), (1991), p. 548
8. I. Ucok, T. Ando and N. J. Grant, Int. J. Rapid Solidification, 27(3), (1991), p. 237.
9. S. Matsuo, T. Ando, M.C. Zody and N.J. Grant, in Advances in Powder Metall., vol. 5, (eds., L.F. Pease III and R.J. Sansoucy), APMI, NJ, (1991), p.161.
10. P. Mathur, D. Apelian and A. Lawley, Acta Metall., 37 (2), (1989), p. 429.
11. E. Guiterres-Miravete, E. J. Lavernia, G. M. Trapaga, J. Szekely and N. J. Grant, Metall. Trans., 20A, (1989), p. 71.
12. B. P. Bewlay and B. Cantor, Mater. Sci. Eng., A118, (1989), p. 207.
13. B. P. Bewlay and B. Cantor, Metall.,Trans., 20B, (1990), p. 899.
14. C. H. Passow, MS. Thesis, Dept. of Mechanical Engineering, Massachusetts Institute of Technology, (1990).
15. C. H. Passow, J.-H. Chun and T. Ando, Metall. Trans., 24A, (1993), p. 1187.
16. T. Ando, J.-H. Chun and S. Sahu, Proc. 93 PM World Congress, Part II., July 12-15, Kyoto Japan, Jap. Soc. of Powder and Powder Metall., (1993), p. 971.
17. M. G. Chu, Y. Shiohara, M. C. Flemings, Metall. Trans., 15A, (1984), p. 1303.
18. K. P. Cooper, J. E. Anderson and J. H. Perepezko, Proc. 4th Int. Conf. on Rapidly Quenched metals, Sendai, (1991), p. 107.
19. S. Annavarapu, D. Apelian and A. Lawley, Metall. Trans., 19A, (1988), p. 3077.
20. E. J. Lavernia, Int. J. Rapid Solidification, 5, (1989), p. 47.
21 X. Liang J. C. Earthman and E. J. Lavernia, Acta Metall. Mater. v 40, (1992), p. 3003.
22. W. Kurz and D. J. Fisher, Fundamentals of Solidification, Trans Tech Publications, Adermannsdorf, Switzerland, (1986).
23. W. Boettinger, in Rapidly Solidified Amorphous and Crystalline Alloys (eds., B. H. Kear, B. C. Giessen), Elsevier, New York, (1982).

MICROSTRUCTURE AND MECHANICAL PROPERTIES OF ALLOYS AND METAL MATRIX COMPOSITES (MMCs) PROCESSED BY THE SPRAY FORMING TECHNIQUE

J.S.ZHANG, W.J.ZHANG and G.L.CHEN
University of Science and Technology Beijing
Beijing 100083
P.R.China

ABSTRACT. In this paper, the recent research activities in the field of spray forming performed at University of Science and Technology Beijing have been briefly reviewed. We have established a spray forming facility recently, and the microstructure and mechanical properties of several alloys (including Ni_3Al–base alloys and high speed steels) and particulate reinforced metal matrix composites (MMCs) (including aluminium alloy 2618+SiCp and maraging steel+Al_2O_3) have been evaluated. The results of characterization of the microstructure and mechanical properties of the representing spray deposited materials (Ni_3Al–base alloys and Al 2618+SiCp MMC) are given here.

1. Introduction

Spray forming (Osprey Process or Liquid Dynamic Compaction(LDC)) is a unique integrated processing technique combining various advantages of powder metallurgy and rapid solidification[1–3]. During spray forming process, a molten stream of metal is gas atomized into a droplet spray which is subsequently propelled away from the atomization region by the fast flowing atomizing gas, to be interrupted by a suitably placed substrate so that a large deposited billet is formed[4]. The major characteristics of this processing technique include: high density, low oxygen content, unique microstructure with rapid solidification characteristics, improved properties, high flexibility in production, high productivity, short production route and so on. Because of the above outstanding features, this technique is highly promising for the processing of such materials that normally suffer from segregation problems and difficult or impossible to be prepared by the conventional casting technique or too costly to be produced by the powder metallurgy process. In recent years, great efforts have been made to develop the spray forming technique into practical industrial applications. In this paper, the recent research activities in this field at University of Science and Technology Beijing have been briefly summarized.

2. Experimental

To make a general evaluation of the microstructure and mechanical property features of the spray deposited materials, two advanced materials including Ni_3Al–base alloys and aluminium alloy 2618+12vol%SiCp metal matrix composite(MMC) have been selected for evaluation. These materials were processed and provided by the Osprey Metals Ltd. The chemical compositions of the Ni_3Al–base alloys are listed in Table 1. The compositions of the matrix alloy Al 2618 is Al–0.18Si–2.3Cu–1.6Mg–1.1Fe–1.0Ni–0.07Ti (in wt%) and about 12vol%SiC particles were incorporated during spray deposition.

M. A. Otooni (ed.), Science and Technology of Rapid Solidification and Processing, 135–146.

The as–deposited and heat treated microstructure of the materials were examined in detail by optical microscope, scanning electronic microscope and imaging analyzer.

The tensile properties of the materials were measured with an Instron 1185 test machine. Stress–rupture experiments were also done with the Ni_3Al–base material. The Ni_3Al alloys were evaluated in as–deposited form. While the aluminium alloy MMC was examined after hot extrusion and either peak aged (200℃ / 20hr) or under aged (200 ℃ / 6hr) after solution treatment at 530℃ for 2hr and water quenched.

3. Results And Discussion

3.1 MICROSTRUCTURE OF SPRAY DEPOSITED ALLOYS AND MMCs

3.1.1 *Ni_3Al–Base Alloys.* Figure 1 shows the typical optical microgragh of as–deposited, nitrogen atomized Ni_3Al–base alloys without and with Cr addition. It is revealed that a fine, equiaxed grain structure is obtained by the Osprey process. No abnormal structures such as prior particle boundaries, layered structure and deformed grains were observed. Cr addition changed the grain structure obviously. In Cr–free alloys (A–1 and A–2), a recrystallized equiaxed grain structure is observed with a grain size typically between 30 and 70μm (Fig.1(a)). The grain size of the alloy with Zr addition is slightly lowered. It suggests that the recrystallization took place after the solidification because of the high temperature annealing effect during deposition[5] . In the Cr–containing alloys (A–3 and A–4) , quite different microstructural features were observed(Fig.1(b)). Although equiaxed grains (size between 20 and 50μm) were still observed, the grain exhibits a spherical shape (or a daisy like flower pattern) with no evidence of recrystallization. The flower pattern structure is quite similar to the γ / γ' eutectic often observed in superalloys. SEM–EDS analysis shows that the grey phase in the "flower" is the ordered γ' phase with higher Al and lower Cr contents, while the white network phase is the disordered γ phase with lower Al and higher Cr contents. According to the phase diagram[6], A–3 alloy is located in the γ+γ' two phase region so that the γ / γ' network structure is believed to be resulted from the Cr microsegregation during deposition solidification. The core region structure of the flower pattern keeps the rapidly solidified structure formed during droplet flight, while the outer region structure is resulted from the solidification of the molten network of low melting point phase surrounding the spherical primary phase[7]. Unlike in the case of nickle–base superalloys, on average, no obvious elemental segregation between core and outer regions of a γ / γ' eutectic pool could be found, which indicates that the advantage of homogeneous element distribution of rapidly solidified spray droplets has been kept in the final preform.

Table 2 shows the results of density measurements of the Ni_3Al–base alloys. The theoretical densities of the alloys were also calculated based on a lattice structural model. From the Table, it can be seen that a very low porosity level(0.35~0.93), thus a very high relative density (99.07~99.65%) can be achieved in the Osprey deposited Ni_3Al–base alloys atomized by nitrogen, which is typical for the spray deposited preforms[7–10]. Study on the porosity size, density and distribution of the Osprey deposited Ni_3Al–base alloys shows a result identical to the density measurement and calculation. The distribution of porosity in the preforms is relatively uniform and the density of porosity is very low (less than 1%). Typical morphology and distribution of porosity are shown in Fig.1(a). Porosity size in the preforms is very small, typically less than 5 μm, except for a rather small fraction of larger porosity (about 20μm). In the Cr–free alloys, most of the porosities are located at the recrystallized grain boundaries (Fig.1(a)), which indicates that most of the micro–porosities which are mainly located at the pre–solidified particle boundaries during deposition are inherited and presented at newly recrystallized grain boundaries.

The levels of the impurities including B, N_2, H_2 and O_2 in the Ni_3Al–base master alloys and the Osprey deposited preforms are given in Table 3. Usually, the nitrogen content will be increased in the preforms processed by nitrogen atomization[8,11]. While it can be seen that the nitrogen level in the Ni_3Al–base alloy preforms obviously depends on the alloy composition. In the alloy without Zr and Cr additions (e.g., A–1), the nitrogen level is very close to that of the master alloy. While in alloys A–2 and A–3, the nitrogen contents increased about 20 and 100ppm above the master alloys, respectively. But, no nitride was observed by optical and electron microscopes. From this result, it is clearly proved that the nitrogen content of the Osprey deposited Ni_3Al–base alloys is not explicitly connected to the nitrogen atomization, but obviously related to the alloy composition. The low level of oxygen achieved in the spray deposited preforms is one of the major characteristics of the technique. It is also the case for the Ni_3Al–base deposited preforms (Table 3). The oxygen content is rather low (less than 76ppm) compared to the materials processed by powder metallurgy (about 300~1000ppm), and only increased a little compared to the casting master alloys. The hydrogen levels in the preforms are also kept rather low ($<$5ppm). Oxygen and hydrogen are commonly known as the elements responsible for environmental embrittlement of materials. They could be very important for the mechanical properties of the Ni_3Al–base alloys used in an intermediate temperature range around 600 to 900 ℃. The increase in oxygen and hydrogen contents will lower the ductility and fatigue resistance of the Ni_3Al–base alloys[12]. From this point of view, the spray deposited materials with much reduced oxygen and hydrogen contents will show better mechanical behavior in the environmental sensitive temperature range.

3.1.2 *Aluminium 2618 Alloy+12vol%SiCp MMC.* Figure 2 shows the microstructure of extruded Al 2618+SiCp MMC. The SiC particles distribute quite homogeneously throughout the matrix. The particles are only slightly aligned along the extrusion direction. The alignment of the SiC particles is inherited from slight enrichment of the particles in the interdendritic regions or grain boundaries during spray co–deposition because of the rejection force acting on the particles by the moving solid / liquid interface[13–16]. In the MMCs processed by the conventional casting procedure, the reinforce particles are usually pushed in front of the solid / liquid interface and finally settled at interdendritic regions or grain boundaries. Such a distribution of the reinforce particles is apparently unfavorable to the properties of MMCs. The aggregation of the reinforce particles during processing can be largely reduced by spray forming process because of the change of the interaction mechanisms between particles and solid / liquid interface [17–19]. Through a careful selection of the processing parameters, it is possible to incorporate the reinforce particles into inner part of the grains. Therefor, the particles could be homogeneously distributed throughout the matrix.

Two types of SiC particle morphology can be observed (Fig.2(b)), i.e., the bar shape and the sharp edged shape particles with an average size about 10μm. In the locations where SiC particles aggregated, small porosities ($<2\mu$m) can be formed. The content of SiC particles has been determined by imaging analysis and chemical analysis of the carbon content in the MMC. The chemical composition of carbon (1.24wt%) can be transferred into a SiCp volume fraction of 12.56% which is quite similar to the value determined by imaging analysis (12.9vol%). The theoretical density (2.823g / cm^3) of the MMC can be calculated according to the densities and fractions of the SiC and Al 2618 alloy. The density of the extruded MMC has been measured to be 2.806g / cm^3. Therefore, the relative density and porosity level of the MMC evaluated can be determined as 99.4% and 0.6%, respectively, which indicates that high density MMCs can be obtained by spray forming and hot extrusion (the relative density of as–deposited MMC is about 97.5~99%).

The grain size of the extruded MMC is about 1~ 8μm with no obvious recrystallization. It is one of the outstanding features of the spray co–deposited MMCs that the grain sizes of the MMCs can be reduced markedly compared to the same matrix alloys spray deposited under the same conditions[20–22]. The refine of the grain size of MMCs can not be attributed to a single mechanism. It is a rather complicated effect combining the solidification, heat transfer and solid state cooling effects[20].

There are some phases (spherical or bar shape) precipitated either within grains or at grain boundaries (Fig.2(b)). SEM–EDS analysis showed that the spherical phase (about 1μm in size) is $FeNiAl_9$, while the bar phase (about 2μm in length and 0.2~0.5μm in width) is the S phase (Al_2CuMg). Compared to the conventional casting and wrought aluminium alloys[23], the microstructure of MMCs have been greatly refined, the precipitates also been refined and more homogeneously distributed throughout the matrix.

The microstructure of the heat treated Al 2618+SiCp MMC is shown in Fig.3. Al 2618 is an Al–Cu–Mg–Fe–Ni type heat resistant alloy which is located in the α–Al+S (Al_2CuMg) two phase region. The additions of Fe and Ni cause the precipitation of $FeNiAl_9$ phase with extraordinary structural stability at high temperature. Together with the enough quantity of the precipitated strengthening phase Al_2CuMg, the matrix alloy 2618 shows an excellent high temperature behavior. During solution treatment, the S phase has been resolved completely, while the $FeNiAl_9$ phase only changed slightly. In the aging process, phase transformation takes place in the matrix according to the following sequence: saturated α–solusion ➤G.P. zone ➤partially coherent S′ phase ➤non–coherent S phase. The microstructure of peak aged and under aged specimens are quite different (Fig.3(b) and (c)), because they are located in the different stages of the aging process. In the under aged specimen, only rather small precipitates (G.P. zones and S phases) can be observed. After prolonged the aging to the peak aged stage, the S phase in the matrix grows obviously. The difference of the microstructure of the specimens aged to different stages will results in different mechanical properties, which will be described in the following section.

3.2 MECHANICAL PROPERTIES OF OSPREY DEPOSITED Ni_3Al–BASE ALLOYS AND Al 2618+12vol%SiCp MMC

3.2.1 *The Ni_3Al–Base Alloys.* The temperature dependence of yield strength and elongation of the Osprey deposited Ni_3Al–base alloys is shown in Fig.5, in which the data of IC50 and IC218[24] (with compositions near A–2 and A–3 alloys ,respectively) are also included for comparison. The Osprey deposited Ni_3Al–base alloys also show the common anomalous temperature dependence of yield strength with the peak strength and the corresponding temperature of 358MPa(450℃), 732MPa(600℃), 813MPa(750℃) and 630MPa(600℃) for A–1, A–2, A–3 and A–4 alloys respectively. The additions of Cr and Zr can increase both the peak strength and the corresponding temperature. Another important features is that the high temperature ductility of the deposited Ni_3Al–base alloys is also improved by the Cr addition. For example, the high temperature elongation of A–3 alloy is at least greater than 4% and its strength is increased to a level higher than that of the IC218 wrought alloy (about 680MPa at 600℃), which can be related to the refined microstructure and limited elemental segregation during spray forming. A–4 alloy shows a relatively lower high temperature strength, which could be resulted from the lowered Al content in the alloy as well as somewhat higher porosity levels as observed on the fracture surface. A–2 alloy shows a similar high temperature strength to IC50 alloy, and both of them show poor ductility at high temperature. A–1 alloy shows the lowest strength and ductility, which can also be related to the higher porosity level in the alloy (Table 2). In fact, on fracture surface, the solid particles formed during droplet flight can be observed occasionally (Fig.5(a)), which may indicate the lowered bonding between the solidified particles in certain regions of the alloy. Therefore, the processing parameters should be further optimized for A–1 as well as for A–4 alloys. Figure 5 compares the fracture surface morphology of the A–1 and A–3 alloys, in which obvious difference can be noticed. In A–1 alloy (similar to A–2), intragranular fracture is the major mechanism for cracking, which is one of the reasons for the lowered ductility. While in A–3 alloy, transgranular ductile dimple fracture is dominant (Fig.5(b)). At 750℃ , many small nodules can be found on the fracture surface of A–3 alloy (Fig.5(c)) , which may be considered as the result of dynamic recrystallization[25,26]. At this temperature, the lowest elongation of A–3 alloy has been got (about 4%). Another factor concerning the lowered ductility of Ni_3Al–base alloy in this temperature range may be the dynamic oxidation effect proposed by C.T.Liu et al[27,28].

The stress–rupture properties of the representing A–4 alloy are compared with the conventional superalloy Waspaloy in Fig.7. It can be seen that in the intermediate temperature range, the rupture strength of the Osprey deposited Ni_3Al–base alloy is near to that of Waspaloy. While in the high temperature range, the rupture strength is lowered. This may be explained according to the fracture mechanism involved in different temperature ranges. Fracture surface analysis (Fig.7) shows that in the intermediate temperature range, the transgranular crack growth is the dominant fracture mechanism (Fig.7(a)). While in the high temperature range, fracture will governed by the intragranular crack growth (Fig.7(b)). As mentioned above, grain size in the spray deposited materials is reduced. Therefore, their high temperature creep resistance may be reduced accordingly. This result is in accordance with the general tendency of the effect of grain size on high temperature creep resistance of metals and alloys. An important conclusion could be drawn from this result, i. e., if the spray deposited materials are to be used in the high temperature environment where creep resistance is required, their grain structures should be optimized accordingly or the grain boundaries should be strengthened further.

3.2.2 *Aluminium 2618+12vol%SiCp MMC.* The results of tensile experiments on the Osprey co–deposited Al 2618+12vol%SiCp MMC in temperature ranging from 25 to 370 ℃ have been summarized in Fig.9 combined with the data of the matrix Al 2618 alloy[29]. It is commonly observed that the proportional limit, yield strength, ultimate strength as well as elastic modulus can be improved by incorporating the reinforce particles such as SiC or Al_2O_3 into Al alloy matrix. The selection of the matrix alloy composition and heat treatment is also important for optimizing the properties of MMCs. The SiC particle reinforces the Al matrix at high temperature markedly. For example, from the present result, the service temperature capability of Al 2618 alloy can be increased as high as 100 ℃ over the matrix 2618 alloy by incorporating 12vol%SiCp, which is an outstanding improvement.

It can be seen that heat treatment also affects the mechanical properties of Al

2618+ SiCp obviously. For the under aged(UA) and peak aged(PA) specimens, although their ductility is kept almost the same, the strength of the UA alloy is much higher than that of the PA alloy, especially in the temperature range of 100~ 300 ℃. This may be related to the change of strengthening phase precipitation kinetics by the SiCp incorporation. In such cases, the aging process is usually accelerated because of the large quantity of dislocations introduced by the internal stress caused by the difference in the thermal expansion coefficients between the ceramic particles and matrix metals during cooling[30–33]. Dislocations are known to be the sites for heterogeneous nucleation and paths for increased atomic transport during growth, both of which may account for the observed accelerated aging behavior.

Figure 9 shows the typical fracture surface morphology of the UA and PA alloys fractured at 260℃, in which different features can be found for the two alloys. In the UA alloy, finer dimples and less SiC particles were observed. At lower temperatures, debonding of the interface between reinforce particles and matrix is a major source for crack nucleation. Therefore, more SiC particles can be found on the fracture surface. As temperature increases, matrix softening may occur, which will reduce the ratio of debonding. Therefore, less SiC particles appears on the fracture surfaces and the ductility of the alloys is also increased remarkably (Fig.9).

Figure 10 shows the room temperature elastic modulus of the Al 2618+12vol%SiCp MMC as well as the matrix Al 2618 alloy. The results are similar to those obtained by Ledbetter et al[34]. It should be noticed that unlike in the case of tensile strength, elastic modulus for the UA alloy is lower than that of the PA alloy. It may be related to the fact that more S phase precipitated in the PA alloy, which may be equivalent to the increase of the effective SiCp content in the alloy.

4. Conclusions

The study of the mirostructure of spray deposited Ni_3Al–base alloys with the additions of Zr and Cr has revealed that high quality preforms can be obtained with rather low impurity levels (e.g., the N, H and O contents are less than 120, 3.6 and 76ppm, respectively), high relative density (>99.65%) and fine macrosegregation–free equiaxed grains (ranging from 20 to 70μm). Except for the outstanding microstructural features mentioned above, the spray co–deposited Al 2618+SiCp MMC shows the rather homogeneously distributed reinforcement particles throughout the matrix, which is beneficial to improve the property. Very limited interface reaction between reinforce particles and matrix can be observed, which indicates that spray forming is suitable for the processing of MMCs.

The spray deposited Ni_3Al–base alloy with Zr and Cr additions shows the excellent mechanical properties at room and high temperatures. The peak strength and the corresponding temperature as well as the ductility at intermediate temperature of the deposited Ni_3Al–base alloy have all been improved to the levels superior to the conventional casting and wrought alloys with the similar compositions. The deposited alloy also shows promising stress–rupture properties in the intermediate temperature range. The proportional limit, yield and ultimate strength as well as elastic modulus of the Al 2618+12vol%SiCp MMC are improved remarkably compared to the control alloy, especially in the intermediate temperature range. The corresponding service temperature of the MMC has been increased about 100℃, which is a very encouraging result. The incorporation of the reinforce particles changes the aging kinetics of the matrix alloy obviously. Although the ductility of the MMC is reduced, it is still in an acceptable range for practical applications.

Acknowledgement. The authors are very grateful for the Osprey deposited materials provided by the Osprey Metals Ltd.

References

[1] A.G.Leatham, P.G.Brooks, J.S.Coombs and A.J.W.Ogilvy, Proc. First Inter. Conf. on Spray Forming, Osprey Metals Ltd., Neath, U.K., (1990) Paper4.
[2] A.G.Leatham, A.J.W.Ogilvy and L.Elias, Powder Metallurgy: Aerospace, Defense and Demanding Applications, ed. F.H.Froes, Metal Powder Industries Federation, Princeton, N.J., (1993) p.165.
[3] N.J.Grant, Metall. Trans. A, 23A(1992)1083.
[4] A.R.E.Singer, Metals and Mater., 4(1970)246.
[5] X.Liang and E.J.Lavernia, Mater. Sci. Eng., A161(1993)221.
[6] A.Taylor and R.W.Floyd, J. Inst. Met., 81(1952–53)451.
[7] A.G.Leatham, A.J.W.Ogilvy, P.E.Kopp and O.Metelmann, Modern Development in Powder Metallurgy, vol.19, eds. P.U.Gummeson and D.A.Gustafson, Princeton, N.J., (1988) p.475.
[8] R.H.Bricknell, Metall. Trans. A, 17A(1986)583.
[9] J.L.Estrada and J.Duszczyk, J. Mater. Sci., 25(1990)1383.
[10] P.Mathur, S.Annavarapu, D.Apelian and A.Lawley, J. Metals, 41(1989)23.
[11] H.C.Fiedler, T.F.Sawyer, R.W.Kopp and A.G.Leatham, J. Metals, 38(1987)28.
[12] J.L.Estrada, J.Diszczyk and B.M.Korevaar, J. Mater. Sci., B, 25(1989)1381.
[13] P.K.Rohatgi, R.Asthana and S.Das, Inter. Met. Rev., 31(1986)115.
[14] K.C.Russell, J.A.Cornie and S.Y.Oh, Interfaces in Metal Matrix Composites, eds. A.K.Dhingra and S.G.Fishman, New Orleans, USA, TMS / AIME, (1986) p.61.
[15] D.M.Stefanescu and B.K.Ohindaw, Metals Handbook, vol.15, ASM International, (1988) p.142.
[16] A.Mortensen and I.Jin, Inter. Met. Rev., 37(1992)101.
[17] M.Gupta, F.Moharmed and E.Lavernia, Metals and Ceramic Matrix Composites: Processing, Modelling and Mechanical Behavior, eds. R.B.Bhagat, A.H.Clauer, P.Kumar and A.M.Ritter, The Minerals, Metals and Materials Society, (1990) p.91.
[18] E.J.Lavernia, Sample Quart., jan.,(1991)2.
[19] Y.Wu and E.J.Lavernia, Metall. Trans. A, 23A(1992)2923.
[20] M.Gupta, F.Mohamed and E.J.Lavernia, Metall. Trans. A, 23A(1992)831.
[21] J.White, I.G.Palmer, I.R.Hughes and S.A.Court, Aluminium–Lithium Alloys V, vol.3, Williamburg, Virginia, (1989) p.163
[22] M.Gupta, F.A.Mohamed and E.J.Lavernia, Inter. Sympo. on Advances in Processing and Characterization of Ceramic and Metal Matrix Composites, Halifax, Canada, (1989) p.236.
[23] Altars of Metallogragh of Wrought Aluminium Alloys. Publication Corporation of Metallurgical Industry (in Chinese), (1976) p.176.
[24] V.K.Sikka, High Temperature Ordered Intermetallic Alloys III, Mater. Res. Soc. Symp. Proc., vol.133, (1989) p.487.
[25] M.Takeyyama and C.T.Liu, Acta Metall., 36(1988)1241.
[26] E.M.Schulson, Metall. Trans. A, 9A(1978)527.
[27] C.T.Liu and V.Sikka, J. Metals, 38(1986)19.

[28] C.A.Hippsley and J.H.Devan, Acta Metall., 37(1989)1485.
[29] ASM Metals Handbook, 9th edition, vol.9.
[30] L.M.Friend and S.D.Luxton, J. Mater. Sci., 23(1988)2173.
[31] R.J.Arsenault and N.Shi, Mater. Sci. Eng., 54(1981)175.
[32] I.Dutta and D.L.Bourell, Mater. Sci. Eng., A112(1989)67.
[33] T.Christman and S.Suresh, Acta Metall., 36(1988)1691.
[34] H.L.Ledbetter and S.K.Patta, Mater. Sci. Eng., 67(1984)25.

Table 1 Chemical compositions of the Ni_3Al–base alloys used in this study (at%).

Alloy	Al	Ni	Zr	Cr
A–1	24.37	75.63	–	–
A–2	24.19	75.37	0.44	–
A–3	18.21	73.81	0.51	7.47
A–4	15.76	76.11	0.61	7.52

Table 2 Densities of the Osprey deposited Ni_3Al–base alloys.

alloy	theoretical* (g/cm^3)	measured (g/cm^3)	porosity (%)	relative density (%)
A–1	7.470	7.401	0.93	99.07
A–2	7.475	7.469	0.35	99.65
A–3	7.726	7.648	0.54	99.46

* theoretical density (ρ_{100}) = $4M / (N_A \cdot V_C)$
$M = \Sigma Mi \times (at\%)i$, Mi—atomic weight of the ith element;
N_A = Avogadro constant;
$V_C = a_0^3$, a_0—lattice parameter measured by X–ray diffraction.

Table 3 Impurity levels in the Ni_3Al–base master alloys and Osprey deposited preforms (ppm).

alloy	form	B	N	H	O
A–1	master alloy	330	4	1.7	25
	preform	370	7	4.8	33
A–2	master alloy	340	3	1.2	14
	preform	350	23	2.7	76
A–3	master alloy	320	18	1.0	55
	preform	290	120	3.4	63

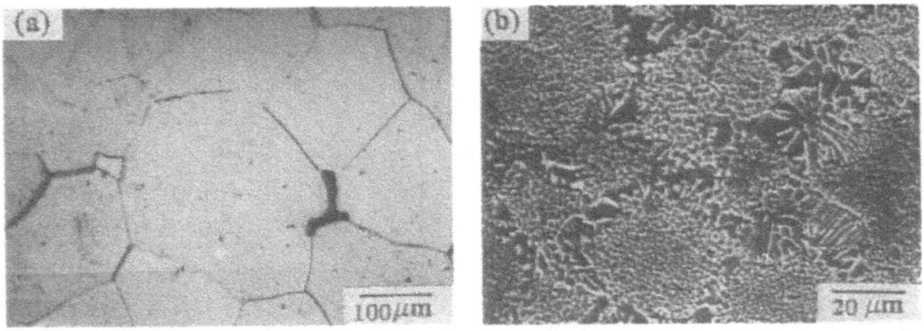

Fig.1 Typical optical microgragh*s* of as–deposited Ni_3Al–base alloys.
(a)Without Cr (A–2) and (b) with Cr (A–3).

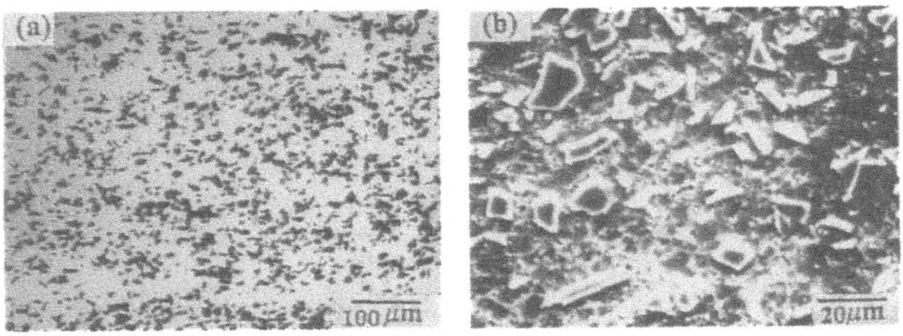

Fig.2 Typical microstructure of Al 2618+SiCp MMC (as–extruded).
(a) Low magnificent and (b) high magnificent.

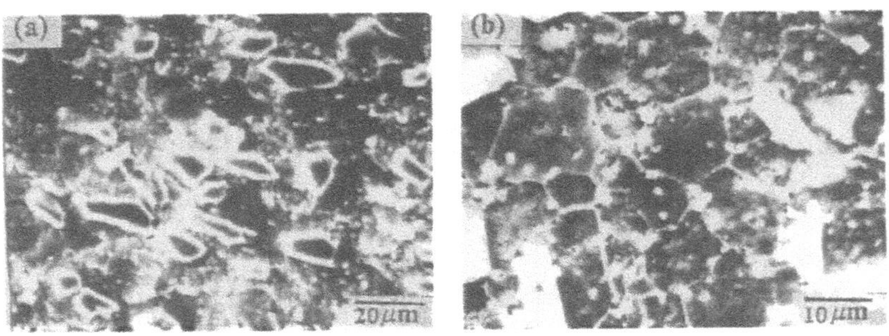

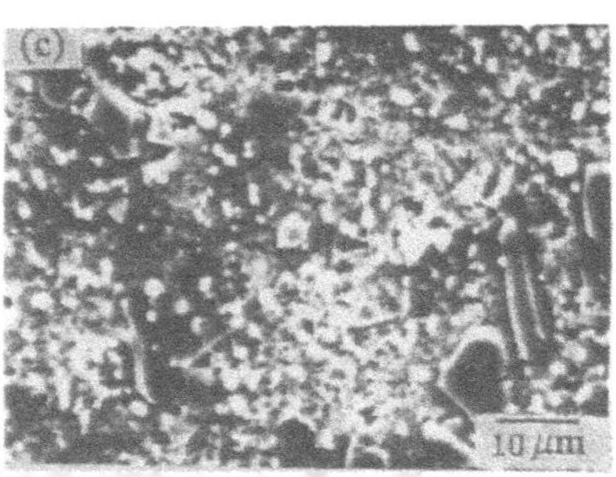

Fig.3
Microstructure of Al 2618+SiCp MMC after heat treatment.
(a) Solution treated, (b) peak aged and (c) under aged.

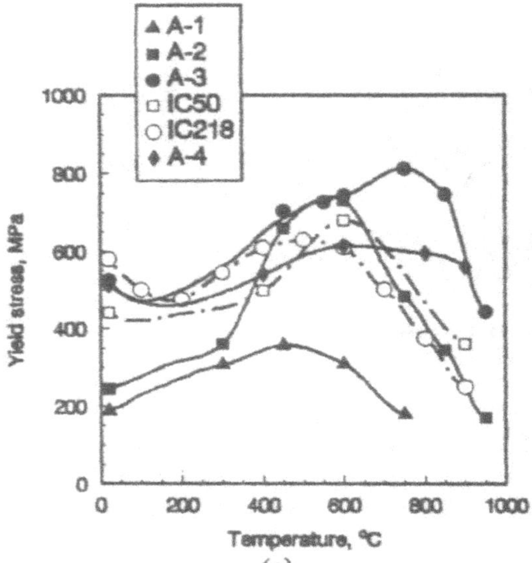

(a)

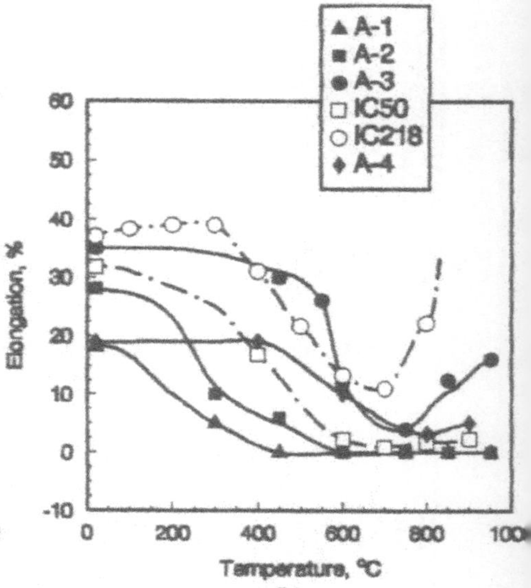

(b)

Fig.4 Temperature dependance of yield strength and elongation of the Osprey deposited Ni$_3$Al–b alloys.(a) Yield strength($\sigma_{0.2}$) and (b) elongation(δ).

(a)

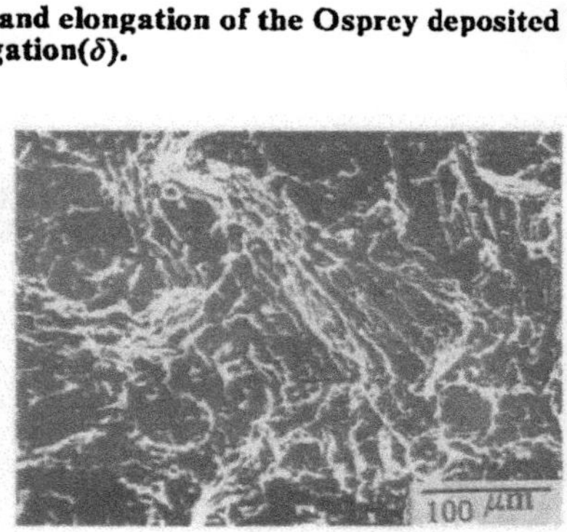

(b)

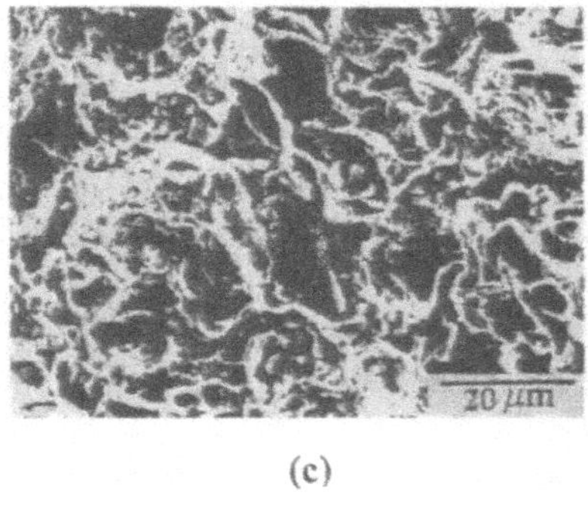

(c)

Fig.5 Typical fracture surface morphology of the A–1 and A–3 alloys. (a) A–1, at 450℃, (b) A–3, at 450℃ and (c) A–3, at 750℃.

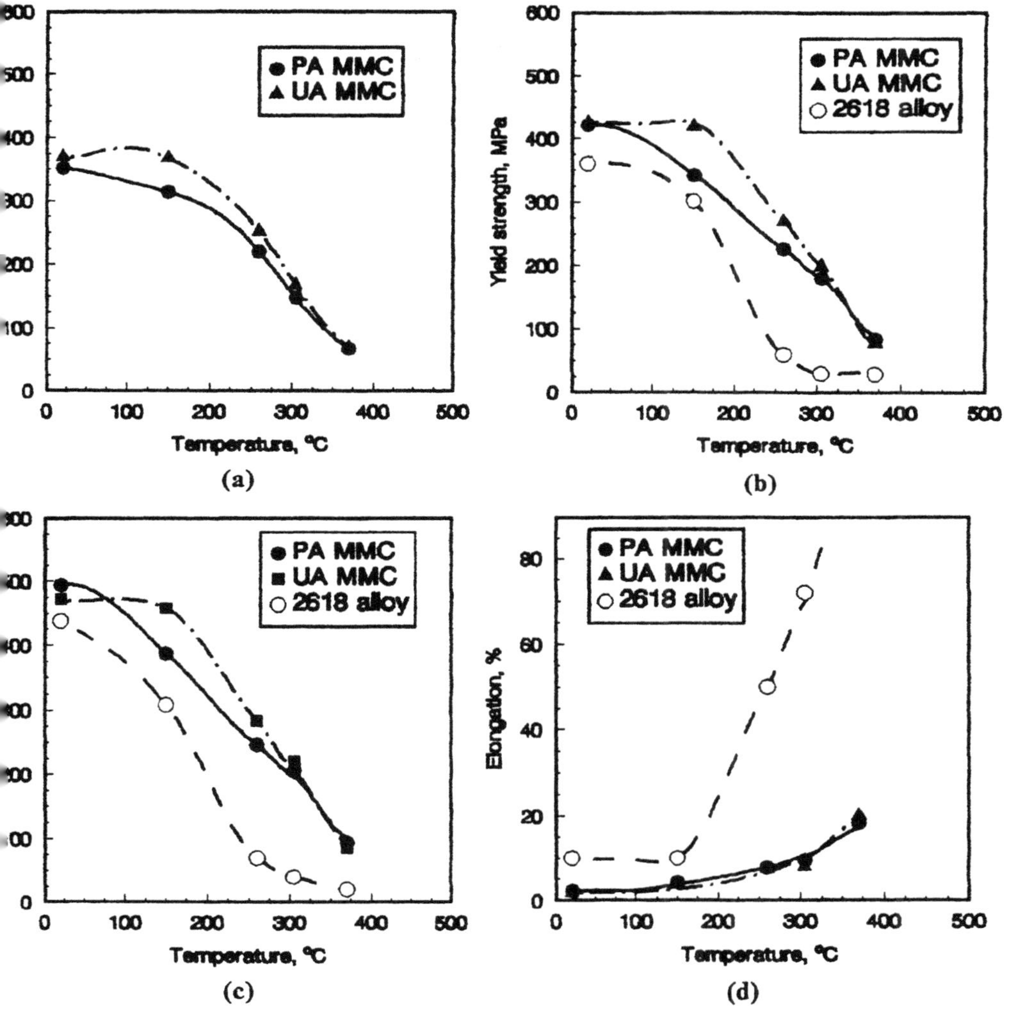

Fig.6 Tensile properties of Al 2618 and Al 2618+12vol%SiCp MMC in the temperature range 25～370℃.(a) Proportional limit, (b) yield strength, (c) ultimate strength and (d) elongation.

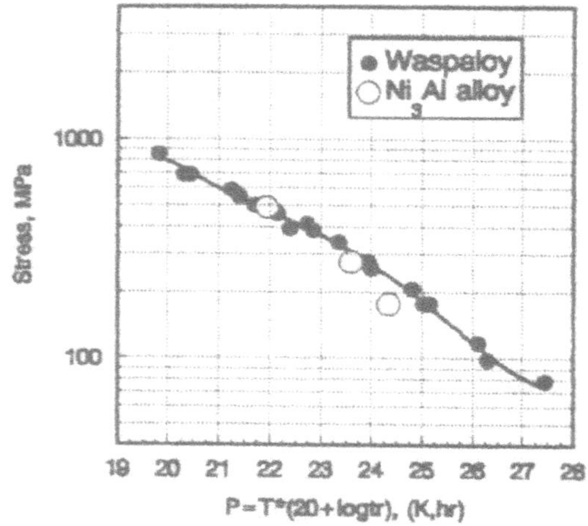

Fig.7 Comparison of the stress rupture life of A–4 alloy with the conventional superalloy Waspaloy on the basis of Larson–Miller curve.

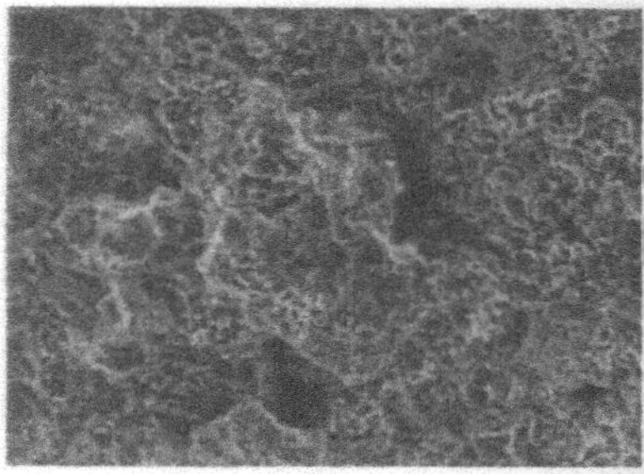

(a)

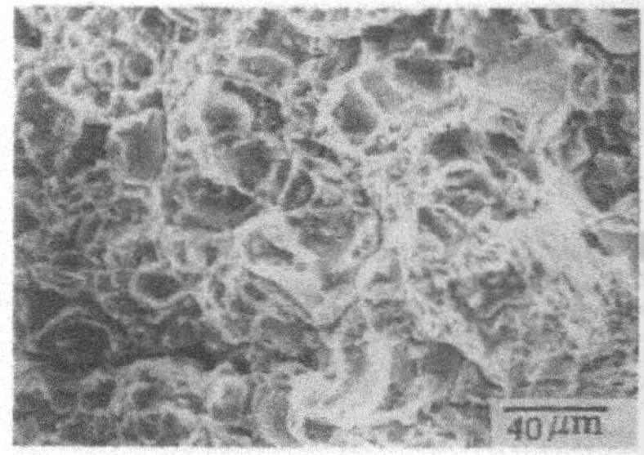

(b)

Fig.8 Fracture surface morphology of stress ruptured A–4 alloy. (a) 730℃ / 490 MPa and (b) 870℃ / 176MPa.

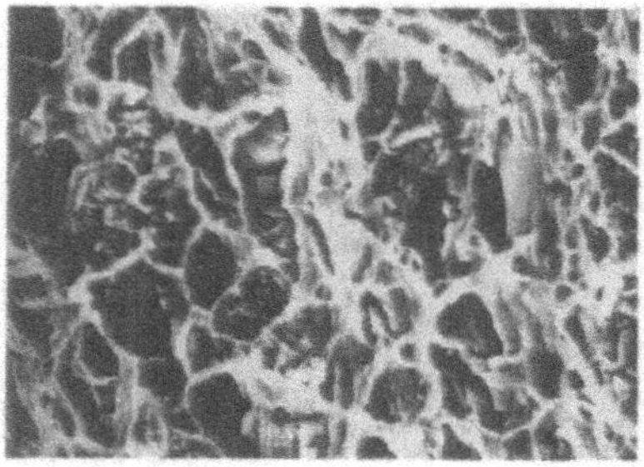

(a)

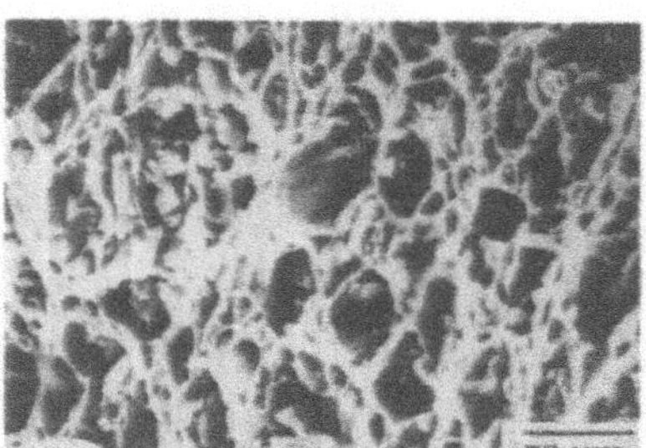

(b)

Fig.9 Typical fracture surface morphology of Al 2618+12vol%SiCp MMC fractured at 260℃.
(a) Under aged and (b) peak aged.

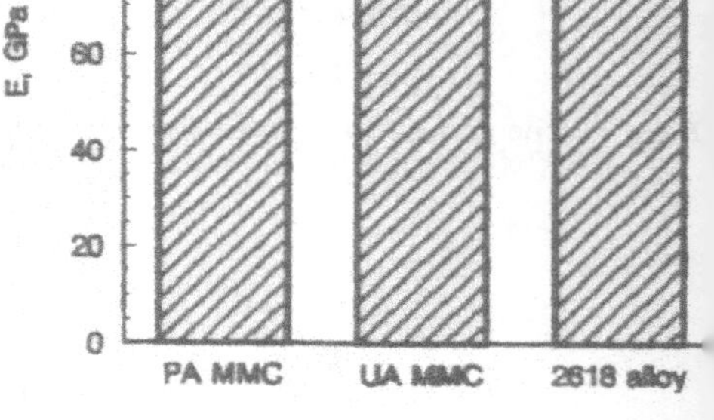

Fig.10 Comparison of the room temperatu elastic modulus of Al 2618+12vol%SiC MMC and Al 2618 alloy.

SURFACE CRYSTALLIZATION AND FIELD INDUCED ANISOTROPY OF COBALT-RICH AMORPHOUS ALLOYS

C. K. Kim, R. C. O'Handley, Department of Materials Science and Engineering, Massachusetts Institute of Technology
Cambridge, MA 02139
W. K. Ho, Sensormatic Electronics Corp. Deerfield Beach, FL 33442

Abstract

A comparative microstructural study of field annealed Co-rich amorphous alloys (CoFeBSi) was carried out using transmission electron microscopy (TEM) and scanning transmission electron microscopy (STEM) with thin sections parallel to the substrate.

Our results reveal new details of the process of surface crystallization in these metastable materials. Surface crystallization occurs for annealing in air well below the bulk crystallization temperature and involves a sequence of three steps: 1) formation of an amorphous borosilicate surface oxide layer, 2) depletion of glass stabilizing elements (boron and silicon) from the underlying amorphous metal substrate, and 3) primary crystallization of the destabilized, metal-enriched sub-surface layer to an fcc or hcp cobalt phase.

Also we have found evidence for a new microscopic mechanism for field annealing of Co-rich amorphous alloys. Striking differences in the microstructural morphology were observed for different Co-rich amorphous alloy compositions. The degree of response to field annealing correlates with the structure and faulting of the surface crystallized layer. Field induced anisotropy is stronger when the crystallized surface is dominated by faulted fcc Co-rich crystallites. It is weak when the surface crystallites have the hcp structure and are less faulted.

Introduction

It is widely recognized that surface crystallization of amorphous alloys involves different mechanisms, occurs at lower temperatures and often results in different crystallization products relative to bulk crystallization (Koster, 1988). It is not unusual for the crystallization products at the surface to differ from those observed after bulk crystallization. Alloys that transform by eutectic or polymorphous crystallization in bulk form, often transform by primary crystallization near the surface (Koster, 1988, Koster and Herold, 1981).

Field annealing of polycrystalline or multiphase magnetic materials can lead to anisotropic microstructures which impose a magnetostatic anisotropy on the material (Chikazumi and Graham, 1969). In single and polycrystals and in amorphous materials the phenomenon of field induced anisotropy is often explained in terms of atomic pair ordering with a preferred direction imposed by the direction of magnetization (Slonczewski 1963, Fukunaga 1981, Fujimori, 1983). There is abundant indirect evidence for pair ordering (Fujimori 1983, Mizoguchi et al 1978, Luborsky and Walter,

M. A. Otooni (ed.), Science and Technology of Rapid Solidification and Processing, 147–155.

1977) based on the compositional dependence of induced anisotropy: K_u goes as $x^2(1-x)^2$ for $A_x B_{1-x}$. We can define the deviation of the ith bond direction or easy axis from its orientation in an isotropic distribution as $d\theta_i$. The net anisotropy is the average of all $d\theta_i$s normalized by 4π : $d\Omega/4\pi$, where $d\Omega = 1/N \Sigma_{i=1} d\theta_i$. A directed pair anisotropy of order $d\Omega/4\pi \approx 10^{-3}$ is probably adequate to account for the strength of induced anisotropy; such an anisotropic pair ordering is very difficult to observe directly by scattering experiments. Suzuki et al (1987) have detected stress-induced bond orientational ordering in a metallic glass.

Field induced anisotropy in Fe-Co-Ni rich crystalline alloys of face centered cubic (fcc) structure (Perminvars) correlates strongly with the formation of oxygen impurity faults on {111} planes. This observation is rarely discussed in texts on magnetic materials (Chikazumi and Graham 1969, Cullity 1972, Chikazumi 1965, Chin and Wernick 1980, Slonczewski 1963) but it was well documented in 1959 (Heidenreich et al., 1959, Nesbitt et al., 1959). This model applies also to the Permalloys (Nesbitt et al 1960) and to Ni_3Fe (Nesbitt et al., 1965). These faults begin with thin plates as Guinier-Preston zones which condense to highly ordered {111} faults with time. They are readily observed as bands of contrast in direct TEM images or as streaking along <111> directions in diffraction spots arising from {111} planes. When the oxygen content of an alloy is below 10 ppm there are no faults and no field-induced anisotropy. There has so far been no evidence that the faults form with any preferred orientation among the various {111} planes. We were surprised to find a similar correlation between the strength of field induced anisotropy in amorphous alloys and the degree of faulting in the fcc surface crystallized layer.

Experiment

Very low magnetostriction Co rich amorphous ribbons with a cross section of 20 μm x 3mm were cut to a length of 4cm. The amorphous ribbons were made by planar flow casting at the Metglas Products Division of Allied-Signal Corporation, Parsippany, NJ. Composition was verified by chemical analysis with Auger electron spectroscopy and electron probe microanalysis (EPMA). Samples were then annealed under various time (3min.-7hrs.), and temperature (250°C-440°C) and atmosphere in the presence of a field that saturated them along their length. Maximum annealing temperature was set well below the crystallization temperature, 520°C. The field annealing induced anisotropy was measured by hysteresis loop tracing.

In order to obtain direct microstructural evidence of the early stages of crystallization and of field annealing in amorphous cobalt rich alloys, we used transmission electron microscopy (JEOL 200 CX and Akashi 002B operating at 200 keV). TEM possesses three unique capabilities suitable for this work: 1) high structural spatial resolution (< 0.3nm), 2) high-resolution chemical analysis probe (< 2 nm) and 3) the ability to image defects not seen by many other methods.
STEM observation was made using a VG HB5 FEG-STEM with LINK LZ-5 X-ray detector.

Several series of electron-transparent plan-view TEM specimens (thin dimension parallel to the oxide/ribbon interface) were prepared using a single or double gun Argon milling method. TEM studies on such samples allow a three dimensional reconstruction of structural information across the annealed ribbons. Bright and dark field diffraction contrast imaging were used for the real space structural imaging; selected area electron diffraction was used for the reciprocal space information. Electron microprobe analysis with STEM was used to obtain local compositional information.

Results and Discussion

Surface crystallization of Co-rich amorphous alloys

This paper emanated from some findings we have made in a TEM study of the micromechanisms of field annealing cobalt-rich amorphous alloys.
Amorphous alloy with a nominal composition of $Co_{74} Fe_6 B_{18} Si_2$ has a crystallization temperature of approximately 520° C. However, heating in air or flowing nitrogen at 200° to 450° C leads to the formation of a surface oxide. Our TEM investigations show the oxide at the surface of the ribbon to be amorphous, non-magnetic borosilicate (B_2O_3 + SiO_2) of order 1 to 10nm in thickness. This layer shows no evidence of rocksalt, spinel, corundum or other transition metal oxides. Some transition metal impurities may be present in the borosilicate glass. The formation of the surface oxide depletes a thin layer of the underlying amorphous alloy of its glass-stabilizing elements (B and Si). This has been confirmed by in-situ microanalysis. This metalloid depletion leads to a decrease in the crystallization temperature T_x of the amorphous alloy near the surface.
Consequently, after sufficient time (e.g. 15 min at 350° C, well below the bulk T_x) a crystalline, magnetic, metallic layer (10 to 100 nm in thickness) of predominantly fcc or hcp cobalt (depending on starting alloy composition) forms beneath the oxidized surface.

The striking microstructural feature we observed during annealing of some Co rich amorphous ribbons was the crystallization of the amorphous substrates to a face-centered cubic cobalt phase containing a high density of faults. The faulting was clearly demonstrated both in a real and reciprocal space. Fig. 1 shows the faulted fcc grains (30 nm diameter) near the surface of annealed (380° C, 60 min) cobalt-rich ribbons. These ribbons still appear to be amorphous by x-ray diffraction. Fig. 2 shows for the same alloy the streaked diffraction maxima typical of faulted crystals. The detailed sequence of the surface crystallization of these amorphous ribbons can be summarized as follows. First, the fully amorphous ribbon developed regions of short-range order over a scale of one to five nanometers. These regions of short-range order show up in high resolution TEM images as small patches of lattice fringes in a random speckled background typical of fully amorphous metallic alloys. Some of these regions mature into highly faulted crystallites of a size less than 50 nm. In the aggressively annealed samples (e.g. 380°C, 90 min, or 410°C, 10 min.), the ribbon cross section was covered with highly faulted crystallites which appeared to form on the {111} planes of face centered cubic (fcc) Co. The possible source of these faults are thin sheets of oxygen atoms which interrupt the normal stacking of planes of fcc Co along <111> directions. The major source of oxygen for these faults is the atmosphere during annealing. However, another source of oxygen such as dissolved oxygen in the material is necessary since vacuum annealed material and material annealed in Ar atmosphere also form oxygen faulted Co crystallites but to a lesser degree. In case of Ar annealed material, five min. of ion milling of the "air-side" (shiny side of the ribbon) revealed the layer just beneath the surface which exhibited increased amount of faults compared with the surface. This implies that oxygen may be supplied from inside of the material, which was apparently the case in Perminvars (Heidenreich et al and Nesbitt et al 1959).

We have also discerned for the first time the effects of the relative concentration of metalloids B and Si on the oxidation, crystallization and the consequent microstructural and magnetic properties of field annealed Co-rich ribbons. Samples studied are listed in Table 1.

Table 1 : Composition of some of the alloys studied.

Material Designation		Amorphous Alloy Composition (at%)
S-0	:	$Co_{74.26}Fe_{4.74}B_{21}$
S-2	:	$Co_{74.26}Fe_{4.74}Si_{2.1}B_{18.9}$
S-8	:	$Co_{74.26}Fe_{4.74}Si_{8.4}B_{12.6}$
S-13	:	$Co_{74.26}Fe_{4.74}Si_{12.6}B_{8.4}$

These materials were annealed under the same conditions (380°C, 60 min.). Dramatic microstructural differences were observed between samples of different metalloid composition. The surface microstructure of annealed S-0 and S-2 ribbons consisted of faulted nanocrystalline Co with a face centered cubic structure (Fig. 3, upper left). That of S-13 ribbon was composed of unfaulted nanocrystalline Co with a close packed hexagonal structure (Fig. 3, upper right). A mixture of hexagonal and cubic Co crystallites was observed in S-8 ribbons. These differences in microstructures appear to result from the difference in the nature of the surface oxide formed during annealing. Electron diffraction patterns obtained from these oxides (which are Fourier transforms of local spatial atomic correlations) indicated that they retain amorphous structures, but with different packing densities. In the case of S-0 and S-2 ribbons, the oxide is largely B_2O_3 type and its diffraction pattern indicates only weak spatial correlation of constituent atoms, i.e., it is highly disordered and porous (Fig. 3, lower left). In the S-13 ribbons, the oxide is predominantly SiO_2 with a compact (or protective) structure showing more short range order in its electron diffraction pattern (Fig. 3, lower right). The oxide formed on the S-8 ribbon appears to possess an intermediate state of a packing.

It is well known that SiO_2 is a protective oxide ,i.e., a barrier to further diffusion. Also, oxygen appears to stabilize the fcc phase over the hcp phase in our cobalt-rich alloys. Therefore, the difficulty of oxygen penetration to the S-13 ribbon resulted in a close packed hexagonal Co structure (which is the stable structure of pure cobalt at room temperature). The oxide structural results are summarized in Table 2.

Table 2 : Boron and silicon content of selected amorphous alloys and the nature of the surface oxide observed to form upon annealing.

Sample	**B content**	**Si content**	**Dominant Oxide**	**Oxide Structure**
S-0	21	0	B_2O_3	porous
S-2	19	2	B_2O_4	porous
S-8	13	8	mixed	mixed
S-13	8	13	SiO_2	dense, diffusion barrier

S-0 and S-2 ribbons which oxidize to the porous B_2O_3 phase contain larger oxygen concentrations in the crystallized sub-surface layer. We postulate the following sequence of steps for phase stabilization. Oxygen enters the high-temperature-stable fcc phase during annealing. Upon cooling, the oxygen stabilizes the fcc structure against transformation to the hcp phase and condenses as faults in fcc {111} planes.

Microstructure and field induced anisotropy

Response to field annealing (H_{offset}) strongly correlates with the presence and amounts of oxygen faults in fcc surface crystallites. Fig. 4 shows increased field induced anisotropy with increasing annealing time and temperature. The initiation of nucleation of faults was observed in a 15 min. annealed sample while the surface of 105 min. annealed sample was covered with heavily faulted fcc Co crystallites (Fig. 4-a). In contrast to surface crystallization, bulk crystallization of the ribbon which was observed at 440°C after 10 min. rendered the offset negligible (Fig. 4-b).

It is tempting to speculate that these faults may form on those {111} planes which bear a particular orientation relative to the direction of the magnetization, however, there is no evidence of this yet. A {111} oxygen fault plane is locally coordinated like a {111} oxygen plane in $CoOFe_2O_3$ spinel ferrite. This local spinel-like planar defect has stronger magnetic anisotropy energy (for cobalt ferrite, $K_1 \approx 2.5 \times 10^5$ J/m^3) than the parent alloy (for Perminvar, $K_1 \approx 10^3$ J/m^3, Permalloy $K_1 \approx 10^2$ J/m^3 or amorphous Co-rich alloys $K_1 \approx 10$ J/m^3). The details of how the oxygen faults select field-induced anisotropy axis are unclear. It is believed to be related to the fact that they suppress long-range ordering and so may favor local ordering in the field direction (Nesbitt et al, 1965).

This new mechanism which we have found to operate in cobalt-rich amorphous oxides appears to be similar to that identified by Heidenreich and Nesbitt in 1959 to be responsible for field annealing in crystalline Co-Ni-Fe (Permendur and Permalloy) alloys and field-induced ordering in Ni_3Fe by Nesbitt et al (1965).

Summary

Striking differences in the microstructural morphology were revealed for different glass former ratios B/Si. For high B/Si ratios the surface crystallites are predominantly fcc Co and show a high density of oxygen faults. For low B/Si ratios the surface crystallites are predominantly hcp Co and almost free of faults. Response to field annealing is proportional to the B/Si ratio and correlates with the presence of oxygen faults in surface crystallites.

Acknowledgment

The financial support by the Sensormatic Electronics Corporation is gratefully acknowledged.

List of figures

Fig. 1 TEM bright field image of the $Co_{74.26}Fe_{4.74}Si_{2.1}B_{18.9}$ (identified as S-2 material in the paper) annealed in air at 380° C for 60 min., showing faulted nanocrystalline Co with FCC structure.

Fig. 2 Electron diffraction pattern taken in the $Co_{74.26}Fe_{4.74}Si_{2.1}B_{18.9}$ (S-2 material) embracing a large number of faulted nanocrystallites presented in Fig. 1. Pattern consists of FCC polycrystalline diffraction ring and multiple streaks. These streaks result from the Fourier transformation of the thin plate-like faults (several atomic layer thick) in the crystal.

Fig. 3 (upper left) TEM bright field image of the $Co_{74.26}Fe_{4.74}B_{21}$ (S-O material) annealed in the air at 380° C for 60 min., showing highly faulted nanocrystalline Co with FCC structure.

(lower left) Electron diffraction pattern taken from the surface oxide formed on the $Co_{74.26}Fe_{4.74}B_{21}$ (S-O material) during annealing, showing an amorphous structure with a very weak correlation of atoms.

(upper right) TEM bright field image of the $Co_{74.26}Fe_{4.74}Si_{12.6}B_{8.4}$ (S-13 material) annealed in air at 380° C for 60 min., showing fault free nanocrystalline Co with HCP structure.

(lower right) Electron diffraction pattern taken from the surface oxide formed on the $Co_{74.26}Fe_{4.74}Si_{12.6}B_{8.4}$ (S-13 material) during annealing, showing an amorphous structure with a nearly close packing correlation of atoms.

Bibliography

Chikazumi, S. (1965), Physics of Magnetism, John Wiley.

Chikazumi, S. and Graham Jr., C.D. (1969) in Magnetism and Metallurgy, Berkowitz, A.E. and Kneller, E. (eds.), Academic Press, p. 577.

Cullity, B. (1972), Introduction to Magnetic Materials, Addison Wesley.

Chin, G.Y. and J.H. Wernick, in Ferromagnetic Materials II, ed. E.P. Wohlfarth (North Holland, Amsterdam, 1980), p. 55.

Datta, A., DeCristafaro, N.J., and Davis, L.A. (1982), in T. Masumoto and K. Suzuki (eds.), Proc. 4th Int. Conf. on Rapidly Quenched Metals, Sendai, August 1981, Japan Institute of Metals, Sendai, p. 1007.

Fujimori, H. (1983) in F.E. Luborsky (ed.), Amorphous Metallic Alloys, Butterworths, p. 300.

Fujimori, H., Morita, H., Obi, Y., and Ohta, S. (1977), Amorphous Magnetism II, Eds. Levy, R.A. and Hasegawa, R., Plenum Press, p. 393.

Fukunaga, T. and Suzuki, K. (1981), Sci. Rep. Res. Insts. Tohoku Univ., A29, 153.

Heidenreich, R.D., Nesbitt, E.A., and Burbank, R.D. (1959), J. Appl. Phys. 30, 995; E.A. Nesbitt and R.D. Heidenreich (1959), ibid. 30, 1000.

Kisfaludi, G., Schay, Z., Guczi, L., Konczos, G., Lovas, L., and Kovacs, P. (1987), Appl. Surf. Sci. 28, 111.

Koster, U. (1988), Mater. Sci. and Eng. 97, 233.

Koster, U. and U. Herold (1982), in Glassy Metals I, ed. H.J. Guntherodt and H. Beck (Springer Verlag, Berlin) p. 225.

Luborsky, F.E. and Walter, J.L. (1977), IEEE Trans. Magn. MAG-13, 953.

Miyakazi, T. and Takahashi, M. (1978), Jap. J. Appl. Phys. 17, 1755.

Nesbitt, E.A., R.D. Heidenreich and H.J. Williams, J. Appl. Phys. 31, 228S (1960).

Nesbitt, E.A., B.W. Batterman, L.D. Fullerton, and A.J. Williams (1965), J. Appl. Phys. 36, 1235.

Prinz, G. (1990) Science 250, 1092.

Slonczewski, J.C. (1963), Magnetic Annealing, in Magnetism (G.T. Rado and H. Suhl, eds.) Vol. I, Chapter 5, Academic Press.

Suzuki, Y., J. Haimovitch and T. Egami (1987), Phys. Rev. B 35, 2162.

Wagner, H.-G., Ackermann, M., Goa, R. and Gonser, U. (1985), in S. Steeb and H. Warlimont (eds.), Rapidly Quenched Metals, North-Holland, Amsterdam, p. 247.

Fig. 1

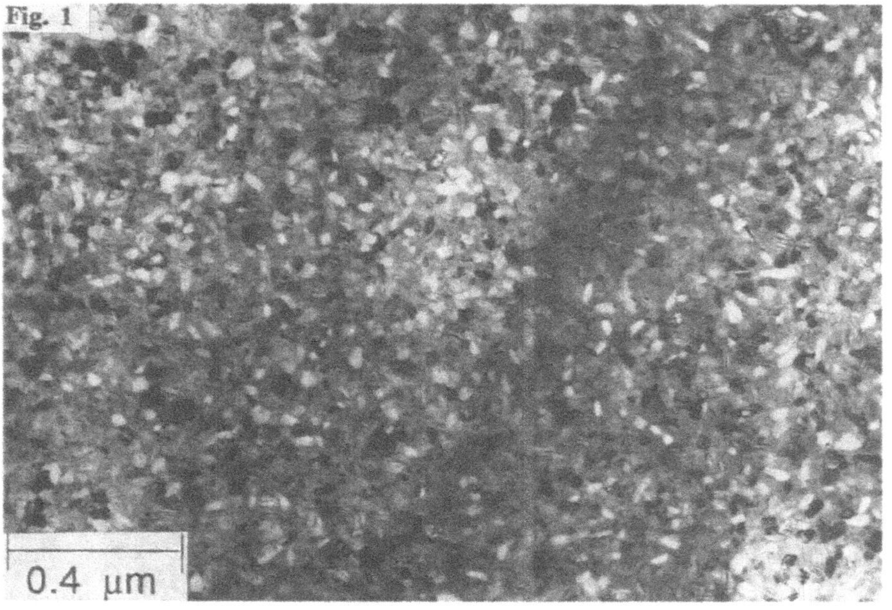

Fig. 2

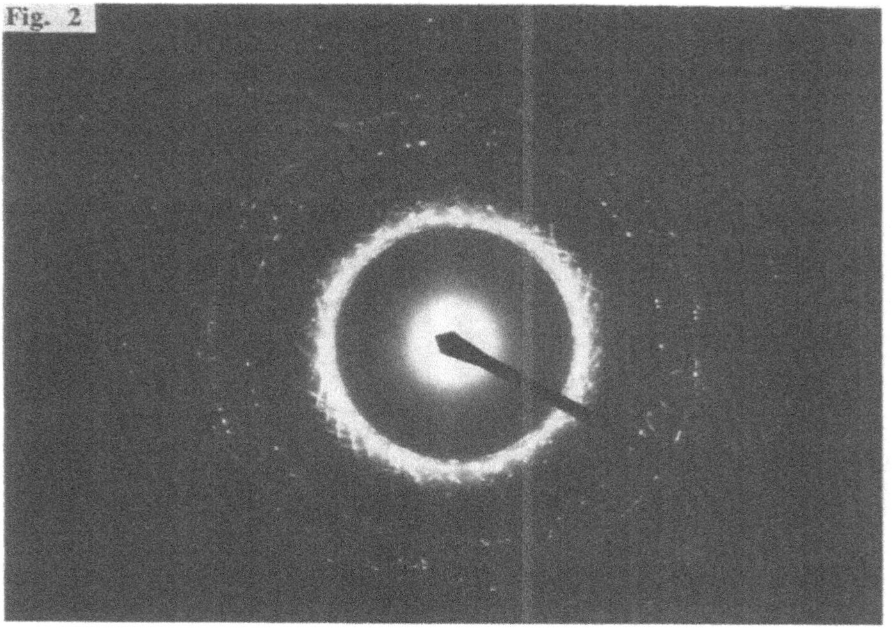

Fig. 3

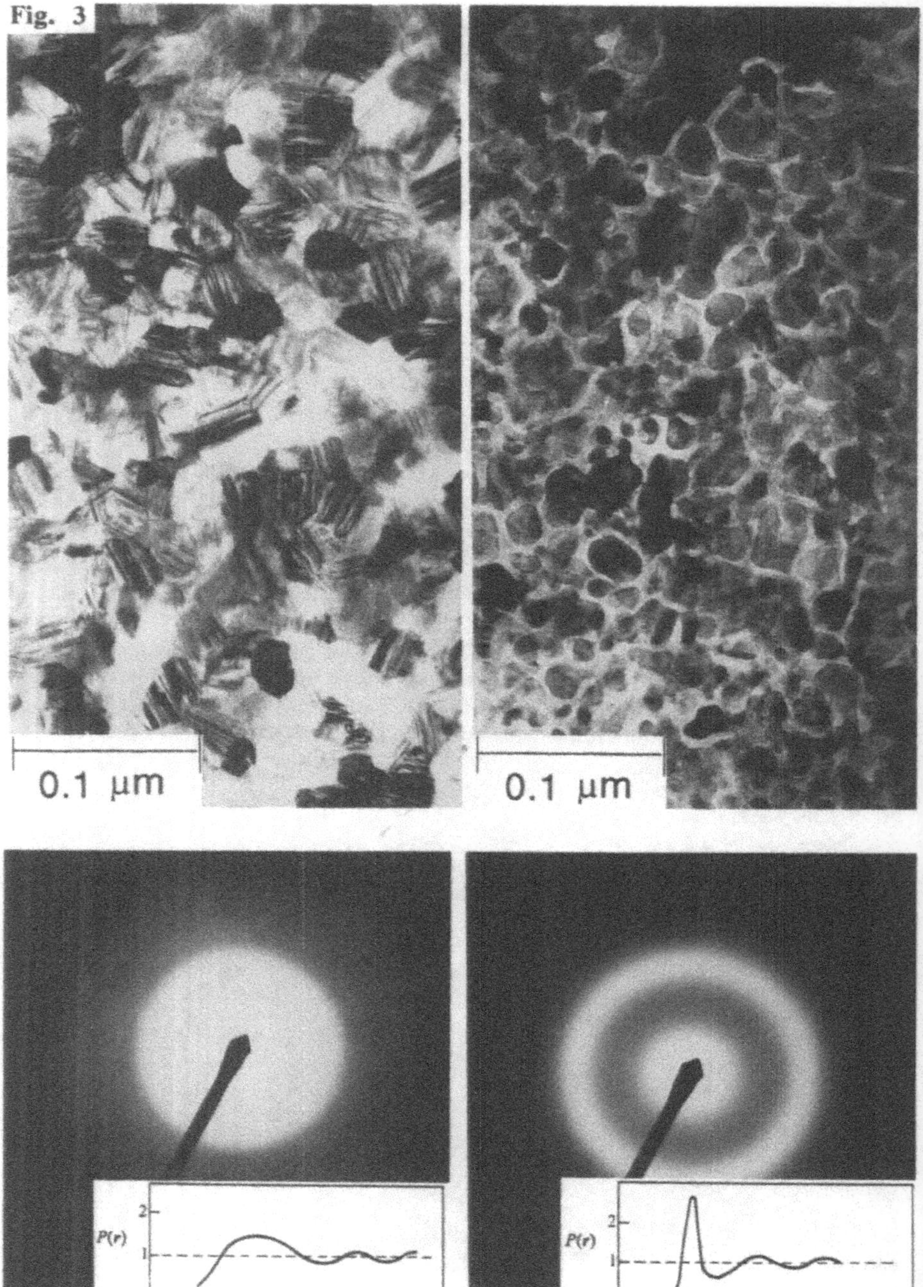

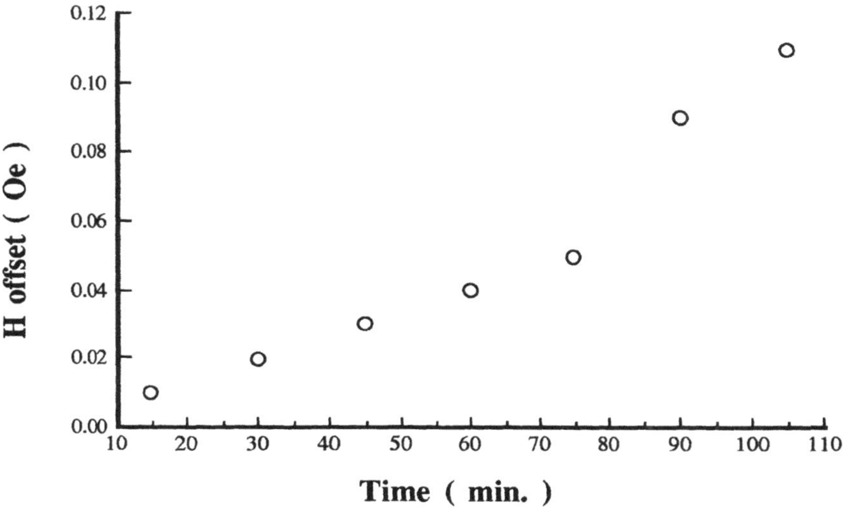

Fig. 4-a Field annealing response of $Co_{74.26}Fe_{4.74}Si_{2.1}B_{18.9}$ amorphous ribbon annealed at 380°C for various duration of time.

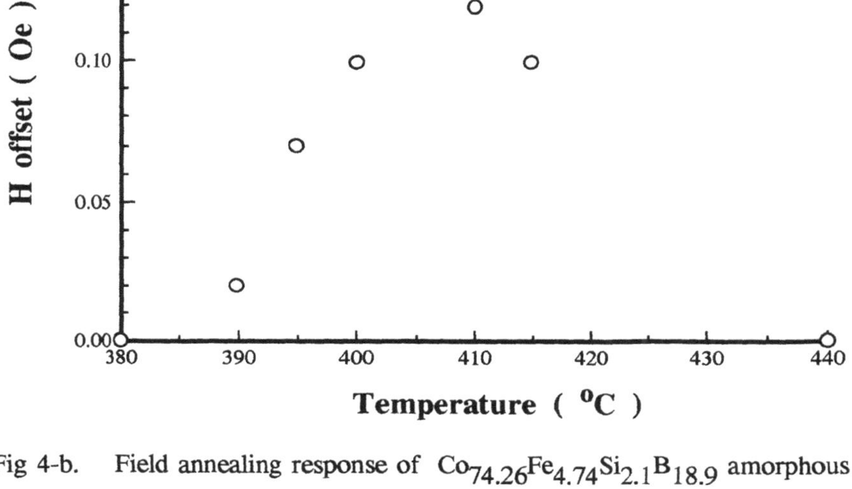

Fig 4-b. Field annealing response of $Co_{74.26}Fe_{4.74}Si_{2.1}B_{18.9}$ amorphous ribbon annealed for 10 min. at various temperature.

A PERSPECTIVE ON THE HIGH RESISTANCE TO COARSENING AND DURABILITY OF SILICIDE DISPERSIONS IN RSPM Al-Fe-V-Si ALLOYS.

D.M.J. WILKES and H. JONES

Department of Engineering Materials,
University of Sheffield,
Mappin Street,
Sheffield, S1 4DU
U.K.

Abstract

Aluminium alloys for elevated temperature aerospace applications need a fine distribution of dispersoids which are stable to transformation and coarsening if they are to compete with Ti.

The α-silicide phase in rapidly solidified Al-Fe-V-Si alloys appears to meet these requirements and the present study is involved with developing an understanding of the mechanisms behind this stability.

While the phase is stable at the highest expected service temperatures, there is a transformation temperature well below the melting point of the alloy. No transformation of the α phase was observed after 6 months at 425°C but 100h at 540°C resulted in significant formation of $Al_{13}Fe_4$ throughout the material.

The low coarsening rate reported for the α phase in these alloys was corroborated by measurements at 425°C, but it is higher than predicted by theory because of enhanced diffusion through the substructure. This is supported by the activation energy for coarsening, determined to be 195±15 $kJmol^{-1}$, which is significantly less than would be expected for volume diffusion or interface controlled coarsening.

The relatively poor performance of other Al-Fe base alloys is attributed to their mixed, non-spherical phases which are not resistant to transformation. Measurement of coarsening in these alloys is difficult and is dominated by large particles at grain boundaries.

M. A. Otooni (ed.), Science and Technology of Rapid Solidification and Processing, 157–171.

A PERSPECTIVE ON THE HIGH RESISTANCE TO COARSENING AND DURABILTY OF SILICIDE DISPERSIONS IN RSPM Al-Fe-V-Si ALLOYS.

D.M.J. WILKES and H. JONES

Department of Engineering Materials,

University of Sheffield,

Mappin Street,

Sheffield, S1 4DU

U.K.

1. Introduction

For many years one of the major areas of research in the aluminium industry has been the development of Al-base alloys for elevated temperature aerospace applications. This is driven by the significant cost and weight savings to be gained by the replacement of Ti alloys with thermally stable Al alloys [1].

The main applications of interest are wingskins and fins for existing military and proposed civilian hypersonic aircraft, and in components such as landing wheels which experience high temperatures even in conventional aircraft [2]. In these circumstances the specific strength and stiffness of the alloys must be high (i.e. approaching those of Ti) but these properties must also be maintained and remain stable for long periods at high temperature.

In conventional aluminium alloys the stability of mechanical properties is limited by the effect of temperature on the strengthening precipitates. In some cases there will be transformation to particles with less strengthening effect (e.g. δ'-Al_3Li $\rightarrow$ δ-AlLi in Al-Li alloys [3] ; and $\theta'' \rightarrow \theta'$ $\rightarrow \theta$ in Al-Cu alloys [4]) and there will also be coarsening of the existing particles.

Coarsening is the process where larger particles grow at the expense of smaller ones so that the surface energy/unit volume of the particle distribution is decreased. In this way the average particle size increases and the strengthening contribution decreases.

The kinetics of the process are governed by the solubility and diffusivity of the solute elements which make up the particles, and the interfacial energy with the matrix of the particles themselves. The problem of stability in aluminium alloys has, therefore, been tackled by using additions of transition metal (TM) and rare earth (RE) elements, which typically have very low diffusivities and solubilities in aluminium, and, in some cases, by exploiting precipitates with low interfacial energy such as Al_3Zr [5].

This has meant that rapid solidification has been employed to incorporate high levels of solute while avoiding coarse, embrittling, primary phases and to give fine homogenous distributions of the desired strengthening phase [6]. One drawback of this is that the phases formed on solidification can be unstable and so, while perhaps being resistant to coarsening, are not resistant to transformation [7].

Many combinations of TM and RE additions have been investigated since the early work of Jones on rapidly solidified Al-Fe alloys [8], the most potential being shown by the Al-Cr-Zr, Al-Zr-V, Al-Fe-Ce, and Al-Fe-V-Si systems. Of these, Al-Fe-V-Si , developed by Allied-Signal, has emerged as the system most likely to compete with titanium [1].

This paper will outline the characteristics and properties of these Al-Fe-V-Si alloys and the strengthening silicide phase and attempt to establish the mechanisms and reasons behind the stability of this material.

2. Al-Fe-V-Si alloys

2.1. INTRODUCTION

The early work of the Allied group focused on the Al-Fe-V (-Zr) system [9] until silicon was introduced into one alloy, resulting in superior stability [10]. It was found that this alloy was strengthened by a silicide phase which was stable in the presence of V and some other TM [11] elements and the resulting work involved many different compositions, varying the amounts and ratios of Fe, TM, and Si. Early studies by Allied showed that Mo gave the strongest alloys with the finest precipitate distribution [11, 12] but V was favoured as the stabilising addition, presumably because it gave the best combination of properties.

In the Al-Fe-V-Si system certain Fe:V and V:Si ratios appeared to give the most stable alloys [7] and within these boundaries (i.e. in a certain phase field) three alloys with different volume fractions of second phase were selected for further development. Al-6.5Fe-0.6V-1.3Si (FVS611) , Al-8.5Fe-1.3V-1.7Si (8009) , Al-11.7Fe-1.2V-2.4Si (FVS1212) have silicide volume fractions of 0.16, 0.24 and 0.43 respectively, with correspondingly higher strength and stiffness and lower ductility (see Table 1) . 8009 offers a combination of high strength and toughness, relatively low density and high formability [13].

Table 1 : Mechanical properties of Al-Fe-V-Si alloys. [14, 15]

Alloy	UTS (MPa)	YS (MPa)	Young's Modulus (GPa)	Elongation (%)
FVS611	352	310	-	17
8009	460	414	88	13
FVS1212	560	530	97	7

Very high cooling rates during solidification are necessary to produce these materials because Fe has an extremely low solid solubility in Al [16] and a strong tendency to form coarse, embrittling intermetallics such as $Al_{13}Fe_4$ and Al_6Fe [17]. To achieve such high cooling rates on an industrial scale, Allied have developed a commercially viable production route involving planar flow casting ($\dot{T} = 10^6$ to 10^7 K/s), comminution and compaction [18].

2.2. 8009

The structure of 8009 in the as consolidated form (Figure 1) consists of a homogeneous distribution of spherical silicide particles about 50nm in diameter, the majority of which sit on dislocation boundaries of sub-grains which are 200 to 500nm across [11]. The stability of this microstructure (Figures 1 to 4), is carried through to the room temperature properties of 8009 which are unaffected by exposure for up to 1000h at 425°C and for 100h at up to 475°C [16].

At temperatures above ≈220°C sub-grain strengthening dominates the properties of the alloy as dislocations can move around particles by thermally activated mechanisms [10, 11].

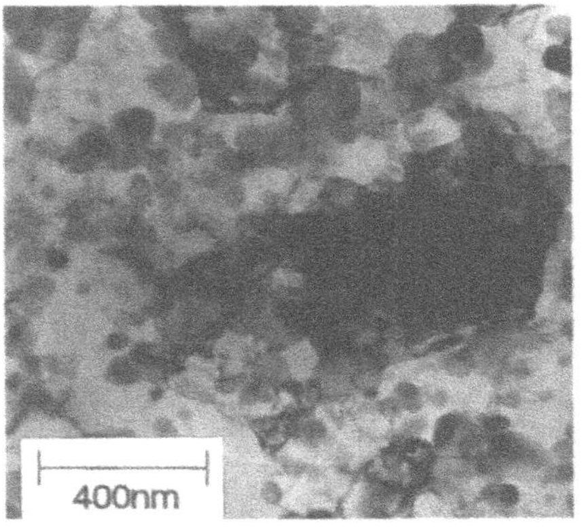

Figure 1: TEM micrograph of 8009 plate, as consolidated.

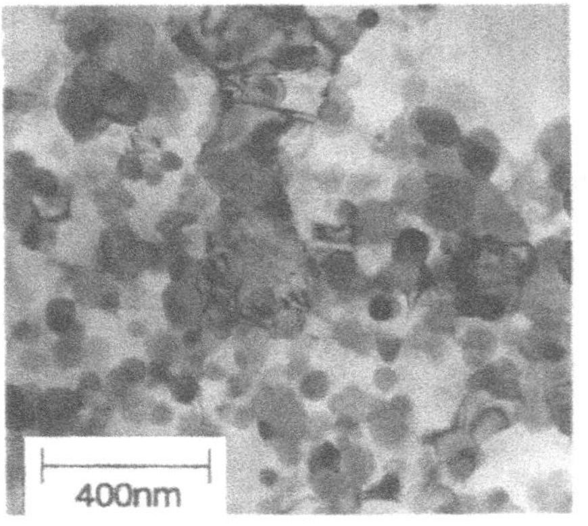

Figure 2: 8009 plate after 200h at 425°C.

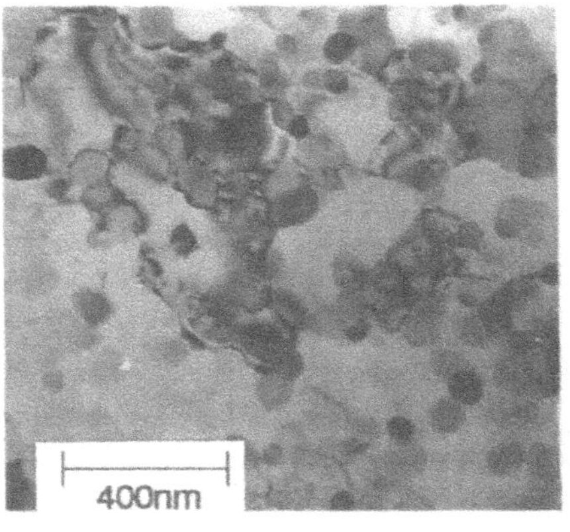

Figure 3: 8009 plate after 1000h at 425°C.

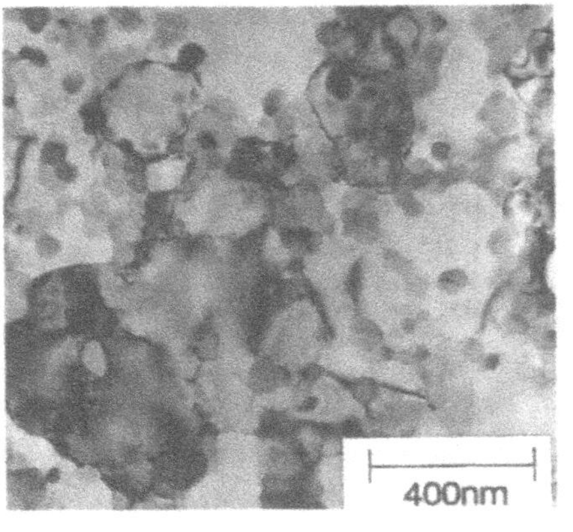

Figure 4: 8009 plate after 4000h.

The alloy retains a high percentage of its strength at elevated temperatures as the sub-grain boundaries are pinned by the silicide particles [19] The sub-grain contribution to room temperature strength is also not degraded by prolonged exposure, giving the alloy stability and elevated temperature strength unmatched by other competing RSPM Al alloys (Figure 5 and 6)

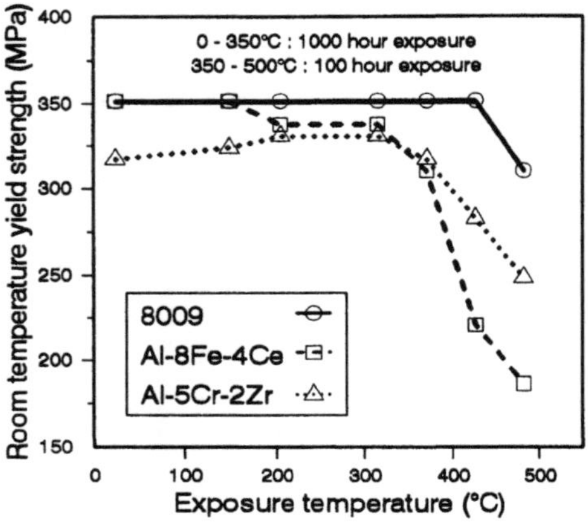

Figure 5 : Room Temperature strength of 8009 and competing alloys.[40]

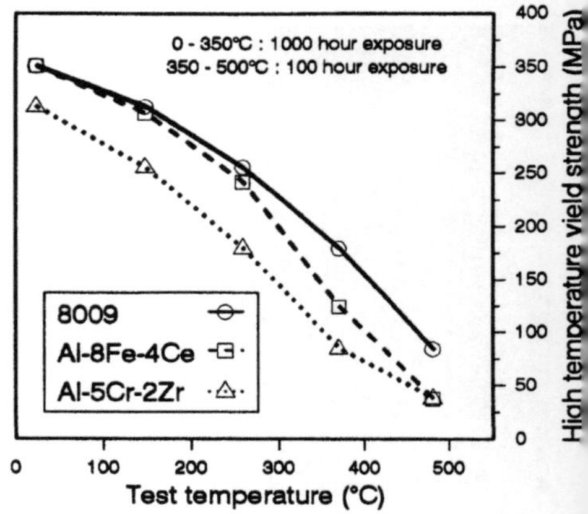

Figure 6 : High Temperature strength of 8009 and competing alloys.[40]

3. The Silicide Phase

The silicide strengthening phase in these rapidly solidified Al-Fe-V-Si alloys is the complex cubic α-$Al_{13}(Fe,V)_3Si$ phase (138 atoms per unit cell ; a≈12.6Å) based on the metastable α-$Al_{12}Fe_3Si$ [20]. In ternary alloys this phase would normally revert to the equilibrium α'- Al_8Fe_2Si and /or $Al_{13}Fe$ during heat treatment [11] but is stabilised by the addition of the bcc transition metal elements V, Mo, Cr, Mn, and W.

The alpha phase has been thoroughly investigated as it is often seen in commercial purity Al alloys which have been solidified at cooling rates exceeding $10Ks^{-1}$, in which Fe, Si, Cr, Mn, and V are common impurities [21, 22]. Two different types of alpha have been identified in these dilute alloys and have also been observed in the more concentrated Al-Fe-V-Si alloys. The first, $Al_{13}(Fe,V)_3Si$, is based on the bcc α-$Al_{12}Fe_3Si$ phase and has two interlocking sub-lattices (Al and Si on one and Fe and V on the other) giving overall bcc symmetry. The second, $Al_{12}(Fe,V)_3Si_2$, is isostructural with the primitive cubic, equilibrium α-$Al_{15}Mn_3Si_2$ phase. The structure of the second phase is very similar to the first with a few extra Al + Si lattice positions and only a single primitive lattice [7, 20]. The phase that forms depends on the chemistry of the alloy, specifically the Fe:V, Fe:Si and V:Si ratios which effectively form the phase field boundaries and which are carried through to the particles [23]. The differences between these two phases and the effects on the properties of the alloy were discounted when compared with more significant factors such as volume fraction and particle size distribution [7].

The stability of the bcc phase is achieved by the substitution of 2 to 4 Fe atoms by V in the metastable $Al_{12}Fe_3Si$ phase [20] but the reasons behind this stability are unclear. With more vanadium the structure takes on the characteristics of the equilibrium α-$Al_{15}Mn_3Si_2$ phase and as such is expected to be stable. It appears, then, that the bcc phase is a stable intermediate between the metastable α-AlFeSi and equilibrium α-AlMnSi phases. A similar effect is seen in binary Al-Fe alloys where metastable Al_6Fe becomes the stable phase upon small additions of Mn as the stability is shifted towards the equilibrium Al_6Mn [24].

The level of the stability of the α-AlFeVSi is, however, debatable. The Allied group claim it is stable up to the melting point of the alloy [20, 25] while other observers have seen transformation from α to $Al_{13}Fe_4$ at temperatures as low 480°C [26]. In the present study 6 months at 425°C did not result in any detectable transformation (Figure 4) but 4 days at 540°C was clearly sufficient for extensive formation of $Al_{13}Fe_4$ (Figure 7).

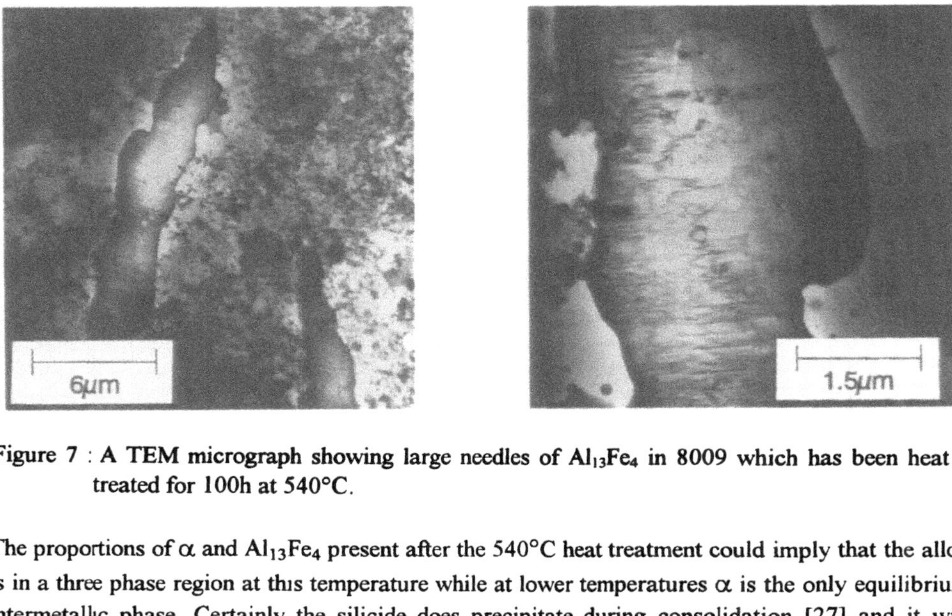

Figure 7 : A TEM micrograph showing large needles of $Al_{13}Fe_4$ in 8009 which has been heat treated for 100h at 540°C.

The proportions of α and $Al_{13}Fe_4$ present after the 540°C heat treatment could imply that the alloy s in a three phase region at this temperature while at lower temperatures α is the only equilibrium ntermetallic phase. Certainly the silicide does precipitate during consolidation [27] and it will urvive at the highest expected service temperatures (i.e. up to 450°C) but it appears that there is a hreshold temperature above which transformation to $Al_{13}Fe_4$ will occur.

. Coarsening

.1. THEORY

he theoretical model of coarsening developed by Lifshitz, Slyozov and Wagner (LSW) [28, 29] hows that if bulk diffusion of solute is the rate controlling factor, the relationship between the verage particle radius, $\bar{r}$, and time, t, will be :

$$\bar{r}^3 - \bar{r}_o^3 = kt \qquad \text{(equation 1)}$$

The coarsening rate, k, is given by :

$$k = \frac{8DC_\infty V_m \sigma}{9RT} \qquad \text{(equation 2)}$$

where D = volume diffusion coefficient at temperature T ($m^2 s^{-1}$)

C_∞ = equilibrium solid solubility at temperature T (atomic fraction)

V_m = molar volume ($m^3 mol^{-1}$)

σ = interfacial energy (Jm^{-2})

R = gas constant ($JK^{-1} mol^{-1}$)

(Note the self consistency of the units when the equation is quoted in this form, giving $m^3 s^{-1}$ as the units of k).

As both D and C_∞ have an Arrhenius dependence on temperature the LSW equation can be rearranged to give :

$$kT = k_o \exp\left(\frac{-Q_k}{RT}\right) \qquad \text{(equation 3)}$$

where Q_k , the activation energy for coarsening, is the sum of the activation energies for diffusion and solubility.

Plots of lnkT against 1/T can then be used to evaluate the coarsening activation energy and to compare values of k measured at different temperatures [30, 31].

4.1.1. Modifications to the theory. The LSW theory assumes bulk diffusion controlled coarsening in a system of discrete spheres with a volume fraction approaching zero and a steady state distribution, but many investigations have extended the theory to more realistic situations.

If the volume fraction of particles is significantly greater than zero then the coarsening rate is expected to increase and various modifications have been proposed [32] to account for this : the most widely used of these, perhaps, is that of Brailsford and Wynblatt [33].

Where the coarsening rate is not controlled by bulk diffusion, the relationship between r and t becomes non cubic : e.g. $r^2 \propto t$ for interface controlled coarsening ; $r^4 \propto t$ for grain boundary diffusion control. For each relationship there is a different equation for k [34]. Activation energies obtained from plots of lnkT vs 1/T will also vary as the mechanism changes. In the case of r^2 , Q_k will be a combination of activation energies for solubility and interface mobility, while for r^4 it will be a combination of diffusivity and solubility at grain boundaries or sub-grain boundaries.

4.2. PREDICTION OF RATE EQUATION FOR Al-Fe-V-Si

If the values put into the LSW equation are correct and the equation itself is correctly defined then it will indicate how closely the real system approximates to the theoretical case, and which factors have the greatest influence.

The factors in the rate equation to be defined and quantified are D, C_∞, V_m, and σ. To simplify the analysis, it is assumed that the diffusivity and solubility of Fe in Al are the dominant factors in the coarsening process, with no influence from the other solutes.

4.2.1. Diffusivity. There are large uncertainties associated with the measurement of D but there is close correlation between the values given for D_o and Q_D for volume diffusion of Fe in Al by [35, 36]. Taking the average of these gives :

$$D_o \approx 0.011\ m^2s^{-1} \qquad \text{and} \qquad Q_D \approx 190 kJmol^{-1}$$

4.2.2. Solubility. The value of C_∞ is also determined by an Arrhenius relationship and Lendvai [37] gives a pre-exponential factor of ≈ 224 and $Q_C \approx 106 kJmol^{-1}$ for the equilibrium solubility of Fe in Al, with C_∞ as an atomic fraction.

4.2.3. Molar volume. It can be argued that this refers to either the volume per mole of Fe atoms ($\approx 5\times10^{-6}\ m^3\ mol^{-1}$) , the volume per mole of atoms of α ($\approx 10^{-5}\ m^3 mol^{-1}$) , or the volume of α that contains one mole of Fe atoms ($\approx 5\times10^{-5}\ m^3 mol^{-1}$). The intermediate value of $10^{-5}\ m^3$ seems the most appropriate from the derivation and gives the smallest uncertainty.

4.2.4. Particle/matrix interfacial energy. A rough value of $0.25 Jm^{-2}$ is taken for σ, which is a likely value for large, incoherent particles [34, 38].

4.2.5. Predicted coarsening rate. Combining these factors gives k in this system as :

$$k = \frac{6\times10^{-7}}{T}\exp\left(\frac{-295{,}700}{RT}\right) \qquad \text{(equation 4)}$$

i.e. $k_o \approx 6\times10^{-7}\ m^2s^{-1}$ and $Q_k \approx 296\ kJmol^{-1}$

4.3. ALLIED'S RESULTS

Allied reported the coarsening rate ($d(r^3)/dt$) for the α phase in Al-Fe-V-Si as $\approx 5\times10^{-30}\ m^3s^{-1}$ at 425°C [23] although this value had already been corrected for the effect of volume fraction using the Brailsford and Wynblatt correction (i.e. the measured rate had been divided by about 3) [33]. No measurements were given but the coarsening measurements in the present study (see Figure 8) indicate that this value is valid, although difficult to determine accurately.

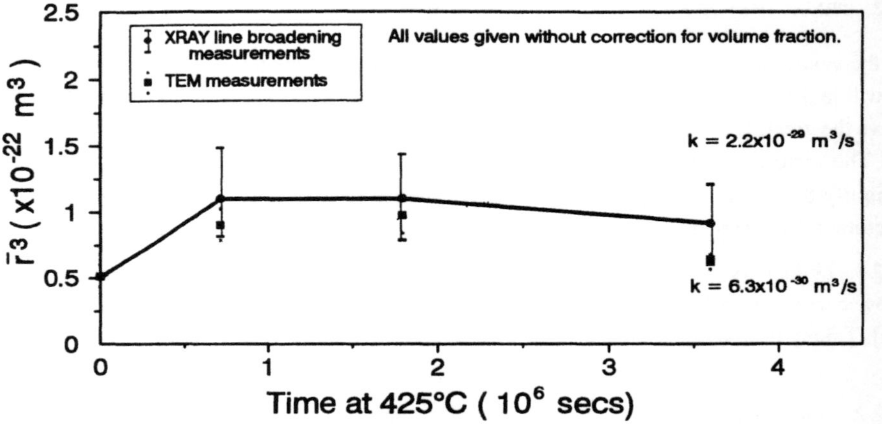

Figure 8 : Graph of $\bar{r}^3$ vs Time at 425°C for 8009 plate.

Using equation 4, the predicted rate for volume diffusion controlled coarsening in this system is as $\approx 6\times10^{-32}$ m^3s^{-1} at 425°C, which is significantly lower than the experimental value reported by Allied. This would suggest that there is some mechanism enhancing the coarsening rate relative to the bulk diffusion case. Pontikakos and Jones [38] have shown that in Al-Fe, at these relatively low temperatures, the coarsening rate is not controlled by volume diffusion and in 8009 where a high proportion of the particles are on sub-grain boundaries it is likely that diffusion along dislocation pipes will be dominant.

Figure 8 shows, however, that coarsening in this system is so slow at 425°C that it is impossible to determine in 1000h with a useful degree of accuracy. By using data obtained at higher temperatures a more complete picture is built up. The plot of lnkT against 1/T in Figure 9 was produced using coarsening data reported for a number of different alloys at different temperatures and gives $k_o \approx 1\times10^{-12}$ m^3s^{-1} and $Q_k \approx 195\pm15$ $kJmol^{-1}$.

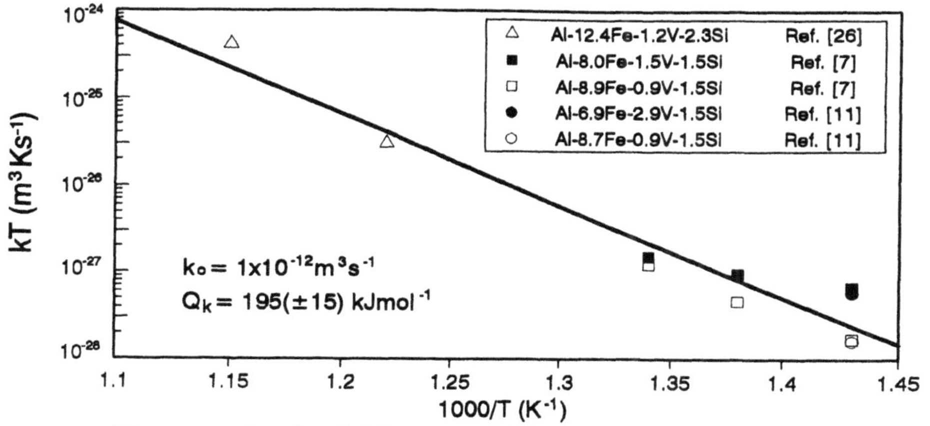

Figure 9 : ln(kT) vs 1/T for Al-Fe-V-Si alloys.

The value of Q_k is much lower than would be expected by adding the activation energies for volume diffusion and solid solubility (which would give $Q_k \approx 300kJmol^{-1}$) again suggesting that coarsening is actually enhanced by the substructure. The lower value of k_o would also be expected if sub-grain or grain boundary diffusion was operating as D_o would be lower [39].

The activation energy in this system is still high when compared with Q_k for Al-Li ($\approx$115kJmol^{-1}) and Al-Cu ($\approx$140kJmol^{-1}) [31].

5. Comparing Coarsening Rates in Different Systems

A true guide to the stability of the alloy, regardless of coarsening mechanism, is the rate of increase of r (where r is the effective strengthening dimension of the particles). An increase in r directly effects the alloy strength but this increase will not be linear with time which makes comparison of results difficult.

Whether the true relationship is $r^2 \propto t$ or $r^4 \propto t$, the coarsening measurements will, within experimental errors, give a straight line for $r^3 \propto t$ [40] from which a gradient can be determined. This will enable one to say, with absolute confidence, which alloy has the lowest rate of increase of r^3, but comparison with equation 2 will not yield accurate values for the factors in the coarsening equation unless $r^3 - r_o^3$ is strictly proportional to t. Conversely, predictions of the alloy stability to the end of its lifetime will have a greater error the more deviation there is from $r^3 - r_o^3 \propto t$.

Note that the Arrhenius form of the coarsening equation (equation 3) will apply for all coarsening mechanisms and, while the magnitude of k_o and k will change, the gradient of lnkT against 1/T should be a valid measure of the activation energy for coarsening, whether $r^2 - r_o^2$, $r^3 - r_o^3$, or $r^4 - r_o^4 \propto t$.

5.1. COMPARISON OF AL-FE-V-SI WITH OTHER AL-FE ALLOYS

The resistance to coarsening of α-$Al_{13}(Fe,V)_3Si$ in Al-Fe-V-Si can be explained by the extremely low diffusivity and solubility of Fe in Al and coarsening is probably accelerated by the fine substructure in the alloys. The coarsening rate is, however, several orders of magnitude less than for particles in other Al-Fe alloys [38, 41] and, assuming that all other factors were equal, may suggest that the interfacial energy for α is a lot lower than assumed [16]. A number of hypotheses have been put forward to explain how this phase could have a very low interfacial energy including orientation relationships [22] and stepped interfaces [7], but there is no direct evidence and most investigations show randomly oriented particles [42].

It should be remembered that, regardless of the coarsening arguments, the stability of the properties of the alloy is the important consideration. 8009 is strengthened by particles which are resistant to coarsening because they contain Fe, but the other characteristics of the alloy are more significant : a narrow, fine distribution of spherical particles of a single phase that does not readily transform.

Other Al-Fe alloys are generally produced by routes with lower solidification rates and are not as homogeneous, with broader size distributions of non-spherical particles. These alloys suffer from metastable solidification structures which undergo some transformation during heat treatment, resulting in mixed phases and mixed morphologies : e.g. $Al_6Fe \rightarrow Al_{13}Fe_4$ in binary Al-Fe alloys [38]; $Al_{13}Fe_4 \rightarrow Al_{10}Fe_2Ce$ in Al-Fe-Ce alloys [41]. Coarsening is then dominated by the dwell time of particles on grain boundaries which are mobile during heat treatment [38]. All of these factors make coarsening measurements difficult to make and interpret, with the resulting measured coarsening rate dominated by larger particles which are, or have been, on grain boundaries.

8009 is a much more homogeneous material in which coarsening occurs at the same rate throughout the alloy. Diffusion is enhanced by low angle dislocation boundaries, but these allow less movement of solute than grain boundaries [39]. This system more closely approaches the very low coarsening rates dictated by the low diffusivity and solubility of Fe in Al.

6. Conclusions

- The fine substructure in 8009 is stabilised by coarsening resistant silicide particles, giving the alloy excellent stability and high temperature strength.

- The silicide phase is stable at the highest expected service temperatures (<450°C) but there is a temperature, well below the melting point of the alloy, above which large needles of $Al_{13}Fe_4$ will form.
- The theoretical rate for volume diffusion controlled coarsening, predicted by the LSW theory, for the α-silicide phase in Al-Fe-V-Si is $\approx 6\times10^{-32}$ m^3s^{-1} at 425°C.
- The measured coarsening rate agrees with previously reported rates but is higher than the predicted rate, indicating that coarsening is enhanced by some other mechanism.
- The activation energy for coarsening is determined to be $\approx$195(±15)kJmol^{-1} , with $k_o \approx 1\times10^{-12}$ m^3s^{-1} , both much lower than would be expected if volume diffusion of Fe in Al was controlling coarsening.
- It is likely that the significant density of dislocation boundaries in this material enhances diffusion and, therefore, the coarsening rate.
- Comparison of $d(r^3)/dt$ coarsening rates for different dispersoids is a valid way of assessing their stability, regardless of the mechanism, but care must be taken when comparing these values with each other and with theory.
- Competing Al-Fe alloys suffer from metastability, mixed, non-spherical phases and large particles on grain boundaries.

7. References

1. Eckvall JC, Rainen RA, Chellman DJ : 1990, *J. Aircraft*, 27, 836.

2. Hyatt MV, Axter SE : 1991, RASELM '91 (edited by K Hirano), 273-280.

3. Makin PL, Ralph B : 1984, *J. Mater. Sci.*, 19, 3835.

4. Boyd JD, Nicholson RB : 1971, *Acta Metall.*, 19, 1379.

5. Angers LM, Konitzer DG, Murray JL, Trucker WG : 1988, in Dispersion Strengthened Al-Alloys (edited by YW Kim, WM Griffith), 355-369. TMS.

6. Couper MJ, Singer RE : 1986, in High Strength PM Al-Alloys II (edited by GJ Hildeman, MJ Koczak), 199-211.

7. Skinner DJ : 1988, in Dispersion Strengthened Al-alloys (edited by YW Kim, WM Griffith), 181-197. TMS.

8. Jones H : 1969/70, *Mater. Sci. and Eng.*, 5, 1.

9. Das SK, Okazaki K, Adam CM : 1984, in High Temperature Alloys-Theory and Design, (edited by JO Steigler). Met. Soc. AIME.

10. Lavernia EJ, Ayers JD, Srivatsan TS : 1992, *Int. Mater. Rev.*, 37, 1.

11. Skinner DJ, Bye RL, Raybould D, Brown AM, Zedalis MS : 1986, in Enhanced Properties

in Structural Metals by RS, 291-295. ASM.

12. Zedalis MS, Peltier JM, Gilman PS : 1989, in Light Weight Alloys for Aerospace Applications (edited by EW Lee, EH Chia, NJ Kim), 323-334. TMS.

13. Gilman PS : 1990, *Metals and Materials*, Aug, 504.

14. Gilman PS, Das SK, Raybould D, Zedalis MS, Peltier JM : 1988, in 2nd Int SAMPE Met. and Met. Processing Conf., 91-101.

15. Gilman PS, Zedalis MS, Peltier JM, Das SK : 1989, in Aircraft Design Systems and Operation, 1-7. AIAA.

16. Gilman PS, Das SK : 1988, in PM Aerospace Materials. Metal Powder Report Publ. Services.

17. Jha SC, Sanders TH : 1986, in High strength PM Al-alloys II (edited by GJ Hildeman, MJ Koczak).

18. Das SK, Bye RL, Gilman PS : 1991, *Mater. Sci. and Eng.*, 134, 1103.

19. Mitra S : 1992, *Scripta Metall.*, 27, 521.

20. Rodriguez MA, Skinner DJ : 1990, *J. Mater. Sci. Lett.*, 9, 1292.

21. Hodgson P, Parker BA : 1988, *J. Mater. Sci.*, 16, 1343.

22. Liu P, Dunlop GL, Arnberg L : 1988, *Mater. Sci. Eng.*, 98, 437.

23. Skinner DJ, Bye RL, Raybould D, Brown AM : 1986, *Scripta Metall.*, 20, 867.

24 Bäckerud L : 1968, *Jerkont. Ann.*, 152, 109.

25. Das SK, Davis LA : 1988, *Mater. Sci. and Eng.*, 98, 1-12

26. Lee JC, Lee S, Lee DY, Kim NJ : 1991, *Met. Trans A*, 22, 853.

27. Frank RE, Hawk JA : 1989, *Scripta Metall.* 23, 113.

28. Lifshitz LM, Slyozov VV : 1961, *J. Phys. Chem. Solids*, 19, 35.

29. Wagner C : 1961, *Z. Electrochem.*, 65, 581.

30. Fine ME, Weertman JR : 1984, Annual Technical Report, NW University.

31. Rylands LM, Wilkes DMJ, Rainforth WM, Jones H : 1994, *J. Mater. Sci.*, 29, 1895-1900.

32. Ardell AJ : 1988, in Phase Transformations '87 (edited by GW Lorimer), 485-494. Institute of Metals.

33. Brailsford AD, Winblatt P : 1979, *Acta Metall.*, 27, 179.

34. Martin JW, Doherty RD : 1976, Stability of Microstructure in Metallic Systems, Ch.4. Cambridge University Press.

35. Beke DL, Kedves FJ : 1985, *Crystal Res. Technol.*, 20, 73.

36. Alexander WB, Slifkin LM : 1970, *Phys. Rev. B*, 1, 3274.

37. Lendvai J : 1987, in Age Hardenable Al Alloys (edited by I Kovacs, J Lendvai, 13/14, 101-120. Mater. Sci. Forum.

38. Pontikakos I, Jones H : 1982, *Met. Sci.*, 16, 27.

39. Gjostein NA : 1973, in Diffusion, Ch.9. ASM.

40. Jain SC, Hughes AE : 1978, *J. Mater. Sci.*, 13, 1611-1631.

41. Lee DY, Zupon DE : 1988, in Dispersion Strengthened Al-Alloys (edited by YW Kim, WM Griffith), 265-281. TMS.

42. Cotton JD, Kaufman MJ : 1991, *Met. Trans. A*, 22, 927.

SHORT-DROP-TUBE EXPERIMENTS ON AlCuFe QUASICRYSTAL-FORMING ALLOYS

C. Voltz, B. Vinet and J. Blétry
Commissariat à l'Energie Atomique
DTA/CEREM/DEM/ Section d'Etudes de la Solidification
CEN, 17 rue des Martyrs, 38054 Grenoble Cedex 9, France

ABSTRACT. AlCuFe quasicrystal-forming alloys have been studied by short drop-tube processing. Structural and microprobe analyses of the elaborated droplets, before and after annealing treatment, reveal that the true icosahedral phase is the low-temperature (annealed) phase with composition (64/23/13), while the high-temperature (as-quenched) phase is a crystalline approximant with different composition (67/16/17) and lower density.

1. Introduction.

Since the first study of the ternary phase diagram AlCuFe by Bradley and Goldschmidt (1), and the discovery of quasicrystalline phases around the atomic composition $Al_{65}Cu_{20}Fe_{15}$ by Tsai, Inoue and Matsumoto (2), a number of experiments, as well as theoretical developments, have dealt with this complex system. However, the final structure of such polyphase alloys closely depends on both solidification path and post-annealing treatment, explaining why the understanding of the phase diagram is somewhat obscure, or even comes up against contradictions. For instance, several authors claim that the low-temperature (or annealed) form corresponds to the stable icosahedral phase (2-5), while others ensure that this low-temperature form is only a crystalline approximant of the high-temperature (or as-quenched) phase (6-7).

To clarify this fundamental aspect, experiments on AlCuFe alloys ranging around the quasicrystalline atomic composition have been realized by letting droplets freely undercool and solidify through a short drop tube. This type of facility is widely known to promote out of equilibrium solidifications (8). The stability domain of both quasicrystalline and approximant phases is discussed taking into consideration annealing treatments of the droplets.

2. Experimental procedure.

Six different compositions of AlCuFe alloys, (62/25.5/12.5), (63/25/12), (63/23.5/13.5), (64/23/13), (65/20/15) and (66.5/21/12.5), have been prepared from the same constituents, Al (4N), Cu (5N) and electrolytic Fe (3N). Each ingot, that weights 15 g, is elaborated under Ar atmosphere using the cold

M. A. Otooni (ed.), Science and Technology of Rapid Solidification and Processing, 173–178.

crucible technique with electromagnetic levitation. It is removed to the top of the 2.3-m high drop tube facility within a BN crucible in a graphite holder (9). The entire tube is evacuated to a pressure of 10^{-7} hPa. Remelting under vacuum is carried out by r.f inductive heating and the melt temperature is recorded with a two-colour pyrometer. Before run, the crucible is closed from the outside by means of a pneumatic device. Then the facility is back filled with He at 1000 hPa for faster cooling by convection. Ejection of the melt at $T \simeq 1350$ K, is realized with Ar by a differential top pressure of 500 hPa through a nozzle of diameter 0.27 mm. The generation of the droplets relies upon the phenomenon of instability and break-up of a jet flow. Since their initial velocity is of a few m.s^{-1}, the flight duration is of about 0.3 s.

3. Metallography of as-quenched and annealed droplets.

3.1. AS-QUENCHED DROPLETS.

AlCuFe droplets ranging in diameter from 0.2 to 1.1 mm are elaborated in the same run. They size between the small AlCuFe droplets (< 25 μm) produced by high pressure gas atomisation apparatus (7) and bulk samples processed in a levitation-melting device (10). Both the external aspect and the bulk microstructure of our droplets depend strongly on the cooling rate through the size effect. Indeed, they are sieved before examinations (fig.1ab).

Fig 1 : Typical surface aspect of small (a) and large (b) droplets.

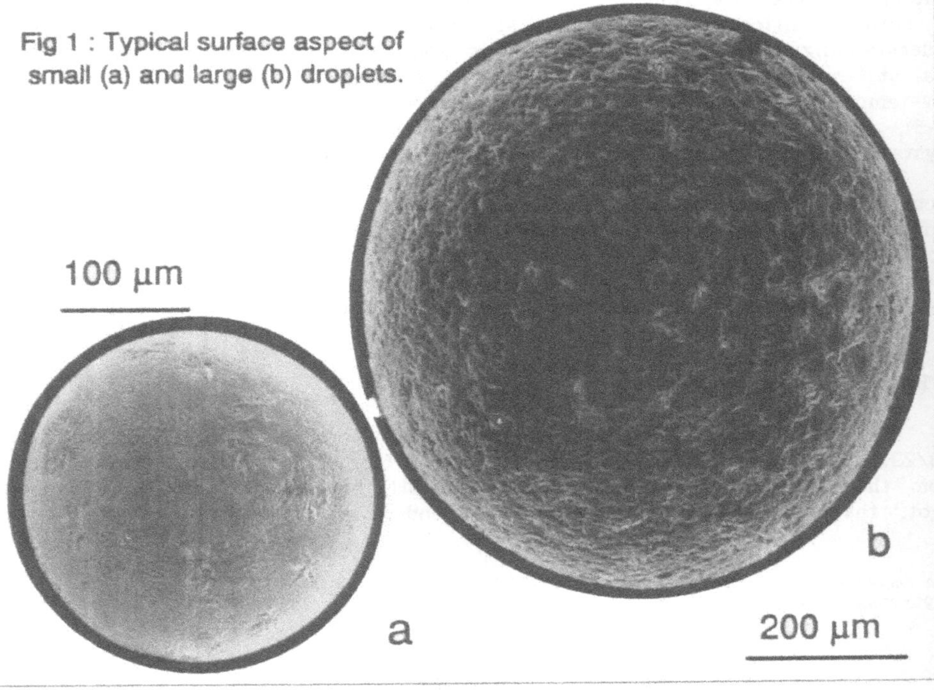

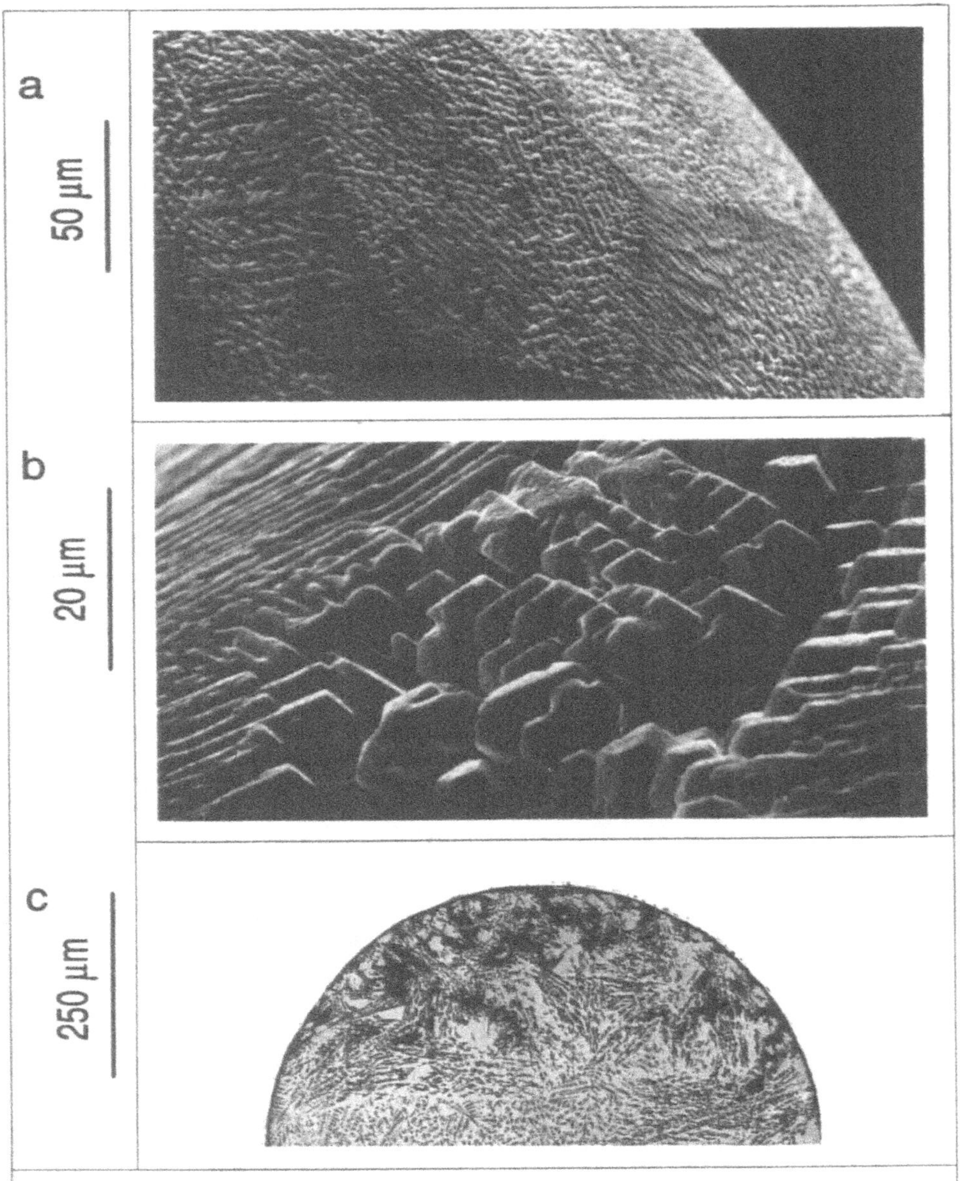

Fig.2 : small droplets : pentagonal facet (a) and dendrite tips (b).
large droplets : cross-section with black platelets (c).

The surface of the shiny and smooth small droplets (< 0.5 mm) exhibits pentagonal facets and dendrite tips (fig.2ab). Cross sections reveal the presence of two phases. In the dull and rough larger droplets (> 1 mm) a third phase, in form of platelets (c), is found. Sparse pentagonal motives can be still observed.

X-ray diffraction patterns indicate that the small droplets only contain a so-called icosahedral phase (i', primary dendritic phase) and a cubic β phase (Al_3Fe like, interdendritic phase) with lattice parameter a = 0.2924 nm. This result is independent of the alloy composition, except for the relative amount of each phase. The third phase, in the larger droplets, is the $Al_{13}Fe_4$ like monoclinic phase (λ).

The microstructural evolutions certainly reflect different amounts of undercooling (ΔT) prior to solidification. ΔT can be evaluated by combining cooling rate and free-fall time. From thermal balance calculations (8), the cooling rate is shown to vary with the droplet diameter Φ (in mm) as,

$$\dot{T} \simeq 770\ \Phi^{-1.8} \qquad (\Delta t_{max} \simeq 0.3\ s)$$

that is, from $\simeq$ 14000 $K.s^{-1}$ for the smaller droplets down to 650 $K.s^{-1}$ for the larger. The solidification of the millimetric droplets may thus occur from a ΔT of $\simeq$ 180 K, *i.e* slightly above the upper temperature limit of the (i'+β) domain ($\simeq$ 1153 K), according to the phase diagram given by Quivy et al (11). This straightforward calculation explains the appearance of the λ phase in the larger droplets, which seems of primary solidification. As higher cooling rates allow an increase of undercooling, the smaller droplets can be solidified within the (i'+β) domain.

3.2. ANNEALED DROPLETS.

In order to test the thermal stability of the detected phases, the droplets are annealed at 1073 K for 96 h in a vacuum sealed silica tube, *i.e* near the limit, usually assumed, for the *i*-phase stability. Independently of the sample size, the treatment produces a massive transformation into **monophase droplets (fig.3), eliminating β and λ phases. Moreover, inter and**

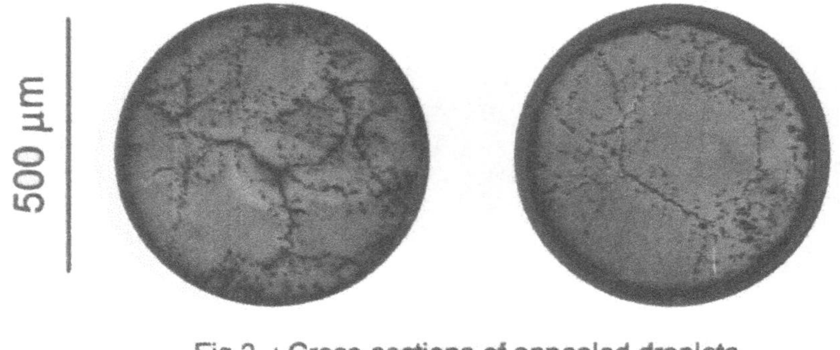

Fig.3 : Cross-sections of annealed droplets.

intragranular porosities appear in the bulk, making us believe to a densification process during annealing. Such a formation of holes has been already observed by Audier, Guyot and Brécher (6) during an electron beam annealing within a microscope.

4. Metallurgical analysis.

other phases				as–quenched droplets				annealed samples			
β	Al	Cu	Fe	i'	Al	Cu	Fe	i	Al	Cu	Fe
this work	51.3	44.5	4.2	this work	67.1	15.8	17.1	this work	63.9	23.2	12.9
Liu	49.5	44.3	6.2	Ebalard	71.1	12.9	16.0	Ebalard	63.7	23.6	12.7
λ	Al	Cu	Fe	Liu	69.3	14.0	16.7	Dong	64	23	13
this work	71.5	6.6	21.9					Calvayrac	65.5	22.5	12.0
Dong	73	5	22					Tsai	65	20	15
Liu	73.6	5.4	21.0					Bradley	65.0	22.5	12.5

This table summarizes the microprobe analysis of both as-quenched and annealed samples. A good agreement is found with the reported atomic composition for β (7) and λ (4,7) phases. The composition of the icosahedral phase (i) that appears in the annealed droplets squares well with that published for the stable (low-temperature) icosahedral phase (1-5). (i) and (i') phases have clearly different composition, making us believe that the as-quenched phase is certainly an approximant of the true i-phase obtained after annealing treatment. The difference in composition could explain the appearance of porosities, the average atomic mass of the as-quenched phase (37.7 g) being lower than that of the annealed one (39.2 g).

Figure 4 shows the evolution of X-ray diffraction patterns (CoKα, λ = 0.1789 nm) from as-quenched (a) to annealed small droplets (b).

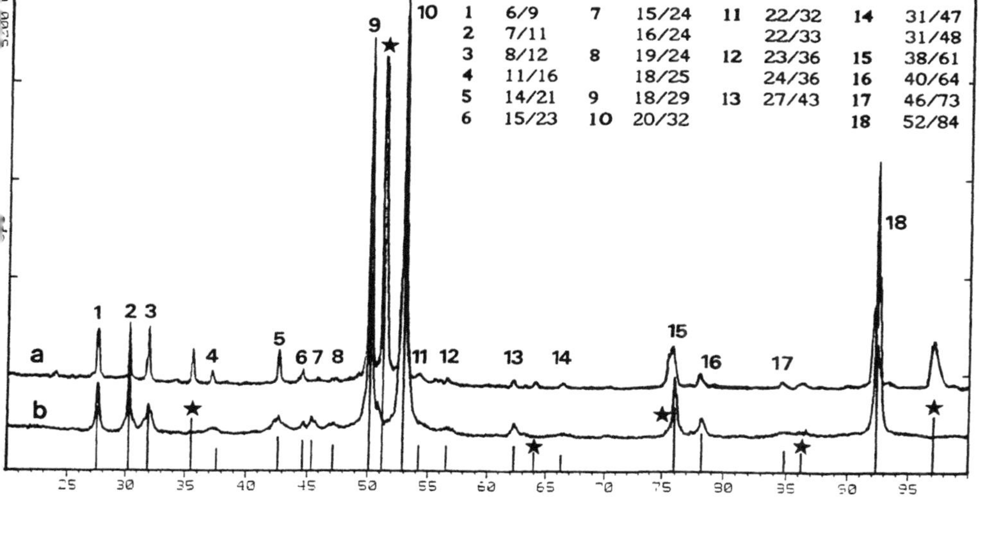

The disappearance of the cubic β phase (*) is easily observed. Conversely, the transformation of the approximant form into the icosahedral phase can hardly be detected by a slight widening of X-ray lines. Complementary MET examinations should be a valuable asset to distinguish properly between these two phases. These patterns can be very well interpreted using the N/M indexing system which corresponds to a 6D fcc lattice with parameter a = 1.264 nm (5).

5. Conclusions.

These experiments have demonstrated that short-drop-tube processing may contribute to the understanding of the complex transformations that are involved during the elaboration of AlCuFe quasicrystal-forming alloys. The wide spread of cooling rates depending on the droplet diameter allow solidifications on both sides of the upper stability limit of the i'+ β domain, leading to a palette of microstructures. This system has also been investigated using a 6.5 m high drop tube (12), leading to a mixing of our i and i' phases. Thus a modification of secondary cooling by increasing, at least, free-fall time can be of consequence for the final structure of the droplets. This underlines the complexity of solidification paths.

This study shows that the true quasicrystalline phase is obtained after annealing treatment of the as-quenched samples in agreement with several authors (2-5). The as-quenched form is very likely a crystalline approximant with different composition and lower density.

ACKNOWLEDGEMENTS: This work has been conducted within the framework of the GRAMME agreement between CEA and CNES. The authors would like to thank William Capello and Isabelle Touet for their assistance.

References.

(1) A.J.Bradley, H.J.Goldschmidt; J.of Inst.Met., 65 (1939) 389.
(2) A.P.Tsaï, A.Inoue, T.Matsumoto; Jap.J.of Appl.Phys., 21 (1987) 1505.
(3) S.Ebalard, F.Spaepen; J.Mater.Sci., 5 (1990) 62.
(4) C.Dong, M.De Boissieu, J.M.Dubois, J.Pannetier, C.Janot; J.Mater.Sci. Let., 8 (1989) 827.
(5) Y.Calvayrac, A.Quivy, M.Bessière, S.Lefebvre, M.Cornier Quiquandon, D. Gratias; J. de Physique, 51 (1990) 417.(6) M.Audier, P.Guyot, Y.Brechet; Phil.Mag.Let., 61 (1990) 55.
(7) P.Liu, A.P.Tsaï, A.Inoue, L.Arnberg; J.Mater.Sci., 26 (1991) 963.
(8) B.Vinet, L.Cortella, J.Coméra, Ann.Chim.Fr., 15 (1990) 443.
(9) C.Voltz; Thèse CNAM, (1992) Université de Reims (F).
(10) K.Urban, D.Holland-Moritz, D.M.Herlach, B.Grushko; NATO Advanced Materials Workshop on Undercooled Metallic Melts, (Il Ciocco, Italy, 1993) Mat.Sci.Eng.A, in press.
(11) A. Quivy, F.Faudot, J.Devaud-Rzepski, D.Gratias, Y.Calvayrac; IIIième Colloque Français sur les Quasicristaux, (1991) Grenoble (F).
(12) S.Ebalard, F.Spaepen, R.F.Cochrane, A.L.Greer; Mater.Sci.Eng.A, 133 (1991) 569.

SOLIDIFICATION OF REFRACTORY MATERIALS PROCESSED IN THE ULTRAHIGH VACUUM DROP TUBE AT THE CEREM - GRENOBLE

S.Tournier, B.Vinet and E.Cini.
Commissariat à l'Energie Atomique
DTA/CEREM/DEM/ Section d'Etudes de la Solidification
CEN, 17 rue des Martyrs, 38054 Grenoble Cedex 9, France

ABSTRACT. Most of the undercooling experiments performed in the 48 metre high Grenoble drop-tube have essentially dealt with pure refractory metals. Here, material processing for studies on alloys is examined. Emphasis is put on general aspects of surface and bulk microstructures of solidified droplets.

1. Introduction.

Solidification of deeply undercooled melts can lead to the same variety of solid state metastabilities as do methods of rapid quenching of the liquid onto cold substrates. The difference in Gibbs free-energy between the metastable liquid and the stable solid state provides the driving force for rapid growth after nucleation far from equilibrium. Since undercooling of bulk samples is realized at slow cooling rates, containerless processing is often needed to avoid external nucleation sources. If any of several levitation techniques is able to balance the gravitational forces (1), an extension towards high undercooling regimes is expected through a transfer into the space environment. Indeed, weightlessness conditions allow further refinements, as for instance an additional purification step of the melt under ultrahigh vacuum. On earth, such a true containerless state is realized for few seconds by letting single droplets fall through an evacuated high drop-tube (2). Since they cool down exclusively by radiation, such a facility is devoted to refractory materials (T_m > 2000 K).

2. Material processing.

At the top of the facility, droplets a few millimetres in diameter are prepared from a wire by using the pendant drop technique with electron bombardment heating. They are received at the bottom onto a damping plate. As turbomolecular and ionic pumps maintain pressures below 10^{-7} mbar in the facility (3), efficient outgassing of the sintered wires may occur, yielding unusually good reproducibilities of both the masses (0.6 %) and the initial temperatures (a few K) of the released droplets (4). Since very few refractory alloys are obtainable commercially, an in-situ alloying method is used which

M. A. Otooni (ed.), Science and Technology of Rapid Solidification and Processing, 179–184.

consists in shifting the composition of commercial wires by adding locally small quantities of pure elements or alloys (fig.1a). With alloy droplets, the melting temperature of the droplet (T_{md}) may be significantly (< 400 K) less than that of the unalloyed wire (T_{ms}). If the droplet temperature at release is close to T_{md}, the persistance of an unmelted piece of the support wire inside the droplets precludes any undercooling. Experience with the W-Re system has shown that this problem is avoided when the droplet temperature at release is the same as that for unalloyed samples, a condition favoured by slow melting of thin wires (≃ 0.5 mm). It must be noted that this overheating of the alloy droplet (by T_{ms}-T_{md}) reduces the time available for undercooling. Measured compositions are very close to nominal, after determination of the critical mass at which the alloyed droplet separates. Unfortunately, this method suffers from experimental restrictions when the overheating is too large, as shown on fig 1b for a Ta-Zr alloy.

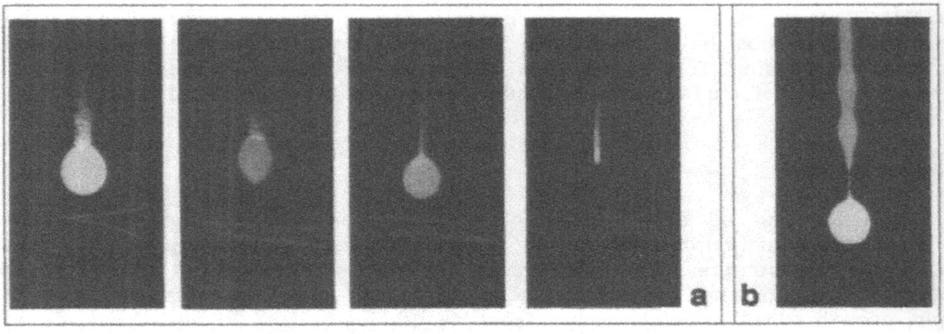

Fig 1: Different stages of the alloying process (a) W- Re (b) Ta- Zr.

After nucleation there is recalescence and the monitoring of this event by silicon detectors offers two independent ways to determine the nucleation temperature T_n. Measuring the duration of radiant cooling until nucleation, T_n may be calculated via a numerical integration of the cooling rate which takes into account linear evolutions of liquid density, heat capacity and hemispherical emissivity in the undercooled state. T_n can also be inferred from the height of the recalescence peak in the single-colour pyrometer signal, assuming that the temperature T_f at the end of the recalescence reaches a known value (5).

A useful agreement between these two methods is found for pure metals (T_f = T_m), the differences in undercooling determinations being smaller than the uncertainty of either method. However, a hypercooling regime (T_f < T_m) was reached for Re, in which the second method is inapplicable, showing that it is desirable to use several methods. For refractory alloys, parameters are often not well known. The density and heat capacity can be approximated, but the liquidus (T_l) and solidus (T_s) lines are often imprecise and the emissivity is hard to estimate. Up to now, we have used an average value for emissivity based on atomic percentage. Surface tension measurements via the pendant drop method could allow better estimations by indicating the proportion of each element at the droplet surface. However, for a Mo-50wt%Re commercial alloy (T_{md} ≃ T_l ≃ T_s), the two methods give the same value for the undercooling

($\Delta T \simeq 400$ K) in spite of Mo and Re having very different thermophysical properties. Thus, the procedure developed for temperature measurement on pure metals appears reasonably good for studies on alloys.

3. Solidification of droplets.

3.1. Surface aspects.

Seven refractory metals (W, Re, Ta, Mo, Nb, Ir and Zr), have been already studied, statistical analysis of undercoolings giving confidence in a homogeneous nucleation process for all of these, except Zr (6). The droplet solidification is thus realized from the highest achievable amount of undercooling that ranges from $\simeq$ 300 K for Mo up to 860 K for Re.

Nucleation probabilities indicate that a single event is likely to initiate solidification. A surface nucleation centre, with a 6-fold symmetry, is observed only for Re (7) and Zr (3). In the case of Re, some droplets show a double recalescence due to a transitory appearance of a metastable fcc phase (8) ; in these droplets, no nucleation centre is found on the surface even though ΔT ($\simeq$ 830 K) is somewhat smaller. Thus, heterogeneous nucleation probably occurs in the liquid bulk. In the case of Zr, there is a large spread of undercoolings and thus different appearances of the droplet surfaces. However, a nucleation centre is observed at the surface of the deepest undercooled droplets. In a wide composition range of the Mo-Re and W-Re systems, nucleation centres are also distinguished with very different appearances (fig 2ab). It is notable that a nucleation centre (with a different symmetry) appears on the Cu droplets prepared by processing in a short drop-tube (9). This emphasizes the complementarity of long and short drop-tube facilities in fundamental studies of solidification processes (2).

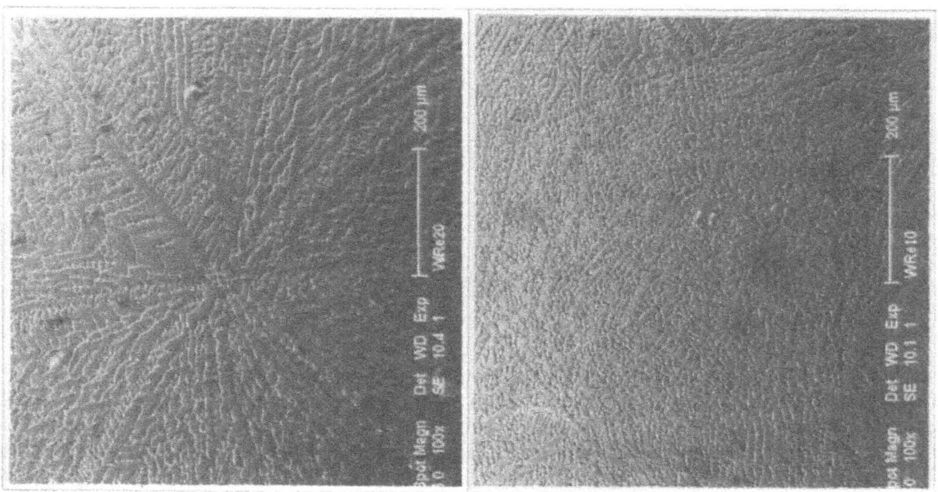

Fig 2: Nucleation 'area' at the surface of W-Re droplets (a) 26wt%Re, (b) 35wt%Re.

W, Ta, Mo and Nb droplets are shiny while Re, Ir and Zr ones are dull. In the case of alloys, for the same composition, the appearance may depend on the achievable ΔT. For instance, deeply undercooled (≃ 400 K) Mo-50wt%Re droplets are dull and exhibit a nucleation centre. Conversely, the slightly undercooled samples (≃ 300 K) are shiny, like Mo droplets. The appearance is closely related to the surface relief. Both dull and shiny domains may appear on the same droplet, as shown for a Mo-69wt%Re alloy (ΔT ≃ 430 K) (fig 3a).

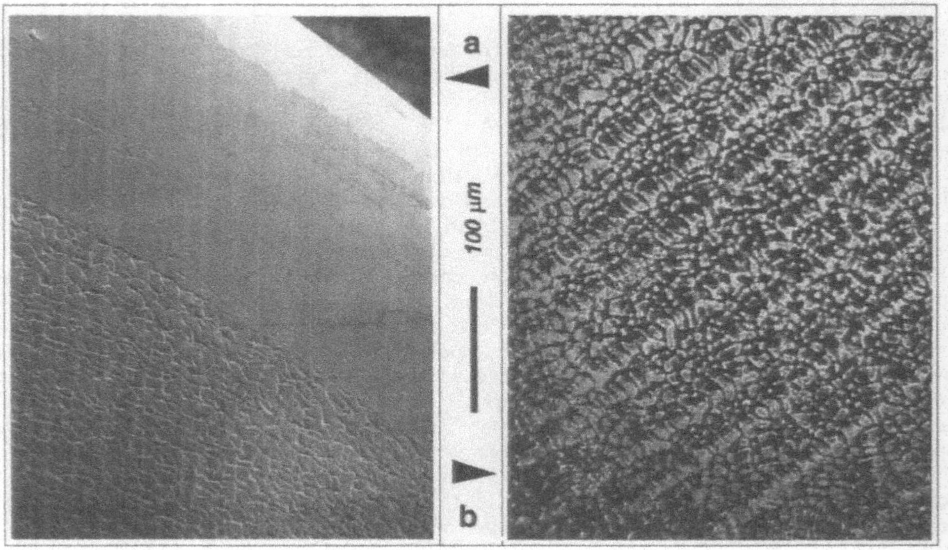

Fig 3: Surface aspects of (a) Mo-69wt%Re and (b) W-26wt%Re droplets.

If solidification is initiated from a single nucleus, a monocrystalline droplet is expected. As discussed below, this result is found only for W. In the case of W-Re alloys having compositions lying within the (W) bcc solution range, the monocrystallinity of the as-solidified droplets is well established from cross-sectional examinations. The solidification front radiates from a surface nucleation centre, forming regular arrays of dendrites which sweep across the droplet (3). At the surface, series of circles are observed which correspond to the emergence of these dendrites (fig 3b).

3.2. Bulk microstructure.

The droplets are usually polycrystalline. For Re, when there is a single nucleation centre on the surface, the bulk microstructure exhibits equiaxed polycrystals which are unconnected with the dendritic surface (3). Successive etchings of the dendritic directional microstructure of a W-26wt%Re droplet gradually reveal grain boundaries (fig 4). We have to conclude that there is recrystallization (with subsequent grain growth) which would correspond to the relaxation of the internal stresses stored in the as-solidified samples (fig 5a). A total correspondence between surface and underlying microstructure is found only for poorly undercooled samples. Recrystallization is favoured by receiving (quenching) the droplets in solid tin or by reducing the time between recalescence event and final impact. As W solidifies after 0.85 s of free-fall time, the thermal stress is certainly smaller, explaining why the original monocrystalline structure is preserved (with noticeable distortions of the crystallographic lattice, determined from X-ray Laue photographs).

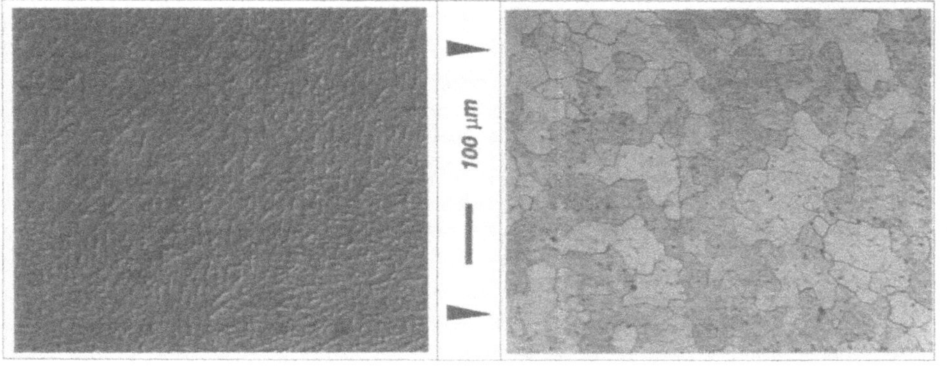

Fig 4: Chemical etching of the dendritic microstructure of a W-26wt%Re droplet.

Numerical simulations performed with our 3D enthalpic non-Newtonian code SORGO (11) show that a W droplet can solidify in two typically different ways depending on the relative amount of undercooling ($\theta = \Delta T/T_m$) prior to nucleation : one with a "shell" solidification front for small θ (< 25 %) and the other with a "quasi-planar" front for larger θ (> 25 %). In the case of the Re droplets (θ = 28 %), the microporosity does not exceed 0.8 % as determined by density measurements. When a shell solidification is expected,

Fig 5: (a) Solidification stress (Ta-Zr) ; (b,c) Free-volume of W droplets.

both the shape and size of internal porosities vary accordingly : the poorly undercooled droplets have a central, quasi spherical cavity whereas the highly undercooled ones show flat cavities and interdendritic microvoids. For the W-26wt%Re alloy, porosity decreases from 7 % down to 2 % with increasing undercooling. Moreover the question remains to explain the presence of lengthened flat cavities (covered with dendrites) within a spherical sample, even though numerical simulations can suggest such a shape (fig 5bc).

4. Conclusion.

Studies on nucleation and solid state metastabilities can be undertaken on refractory materials (T_m > 2000 K) by processing in a high drop-tube. The application of a simple in-situ alloying method to the pendant drop technique permits to study of a wide variety of phase diagrams. The procedure developed for temperature measurements on pure metals appears reasonably good for alloys. This is a valuable asset to study the formation of non-equilibrium microstructures which are often similar to those observed in rapid quenching.

The surface appearance of the droplets is closely related to the achievable amount of undercooling. Due to the relaxation of the internal stresses stored in the as-solidified droplets, recrystallization causes polycrystallinity. Information on the effect of secondary cooling on microstructure can be gained by quenching the droplets in solid tin at the end of free-fall.

Combining moderate cooling rates and large undercoolings, solidification velocities lie between crucible and quenching methods. Potential for metastable phase formation in our facility has been already demonstrated when giving the first evidence of an undercooling-induced metastable phase transformation in pure transitory metals (8). Furthermore, supersaturated solid solutions have been reported for Ta-Zr (10) and Mo-Re (3) alloys.

This work has been concluded within the framework of the GRAMME agreement between CEA and CNES. Thanks are due to J.C. Idelon and P. Sibellas.

References.

(1) D.M.Herlach, R.F.Cochrane, I.Egry, H.J.Fecht, A.L.Greer; 1993, Int.Mater.Rev., **38**, 273.
(2) A.L.Greer, 1994, Mater.Sci.Eng., Nato-Advanced Research Workshop on "Undercooled Metallic Melts", Il Ciocco (I), (1993).
(3) E.Cini, B.Vinet; 1993, 13th Int.Plansee Seminar, Reutte (A), 1, 203.
(4) B.Vinet, J.P.Garandet, L.Cortella; 1993, J.Appl.Phys., **73**, 3830.
(5) W.H.Hofmeister, R.J.Bayuzick, M.B.Robinson; 1989, Int.J.Therm., **10**, 279.
(6) L.Cortella, B.Vinet; accepted in Phil.Mag.A.
(7) B.Vinet, L.Cortella, J.J.Favier, P.Desré; 1991, Appl.Phys.Lett., **58**, 97.
(8) L.Cortella, B.Vinet, P.Desré, A.Pasturel, T.Paxton, M.Van Schilfgaarde; 1993, Phys.Rev.Lett., **70**, 1469.
(9) Ch.Voltz; 1992, Thèse CNAM, Reims (F).
(10) E.Cini, S.Tournier, B.Vinet; 1993, Mater.Sci.Eng., **A173**, 389.
(11) J.D.Semeria, F.Ternay; 1988, Rapport CISI.

THIXOFORMING OF SPRAY CAST MATERIALS

A. ZAVALIANGOS and A. LAWLEY
Department of Materials Engineering
Drexel University, Philadelphia, PA19104

ABSTRACT: Metallic systems which possess a non-dendritic microstructure behave thixotropically and can be formed to net shape in the semisolid state. Semisolid processing provides several advantages over conventional casting, forging, and P/M techniques, but has not been accepted widely. Problems related to the methods for producing non-dendritic thixoworkable microstructures, and to process design and control must be addressed. In order to evaluate and characterize the ability of semisolid spray cast materials to be thixoformed a combined theoretical-numerical-experimental program is under way at Drexel University. In this paper we review the current understanding of the flow/deformation behavior of semisolid materials with emphasis on spray cast alloys. Outstanding issues of material characterization, modeling, process design and control, and on-going experimental work are discussed.
KEYWORDS: Spray Casting / Thixoforming / Semisolid Processing

1. Introduction

Thixotropy[1] in semisolid alloy slurries was first observed by Flemings and his co-workers in the early seventies [1]. In particular, it was found that if the alloy is agitated vigorously during slow cooling from the melt, the dendritic microstructure is destroyed and replaced by a microstructure consisting of equiaxed grains. These alloys behave thixotropically with apparent viscosities in the range 10^{-2} - 10^{4} Pa·s. In general, metallic systems which possess a non-dendritic microstructure can be formed to net shape in the semisolid state. In addition, this microstructure is maintained during solidification and partial remelting of the alloy. This property allows the decoupling of the forming step from the process that produces the equiaxed microstructure[2]. Solidified billets can be transported, stored, and sectioned to smaller sizes. Reheated billets can then be formed in the semisolid range. In this case the forming process is termed *thixoforming*.

---

[1]Thixotropy is defined as material behavior characterized by a reversible decrease of apparent viscosity due to the application of shear.

[2]In this case there is no need to match the rate of production of the two processes. The process that produces the equiaxed microstructure can still benefit from the economy of scale, while at the same time small batch forming can be viable economically.

M. A. Otooni (ed.), Science and Technology of Rapid Solidification and Processing, 185–193.

The processing of alloys and composites in the solid+liquid range provides several advantages over conventional casting, forging, and P/M techniques, as identified in Table I. Recently, interest in semisolid processing has increased, since it is now realized that it can provide an attractive route for fabricating particle or fiber reinforced composite materials [2]. Although the feasibility of semisolid processing has been demonstrated, it has not been accepted widely. To realize the advantages of semisolid processing, issues related to the methods for producing non-dendritic thixoworkable microstructures, and to process design and control must be resolved.

In order to evaluate and characterize the ability of semisolid spray cast materials to be thixoformed a combined theoretical-numerical-experimental program is under way at Drexel University. Spray cast materials are ideally suited to semi-solid processing, since they exhibit a fine equiaxed grain structure devoid of dendrites, a prerequisite for processability in the semi-solid range. The combination of spray casting and thixoforming offers a unique opportunity to tailor and couple the preferred intrinsic microstructural characteristics of spray cast materials with the net-shape capability of thixoforming. The objectives of our study are to: (i) develop an experimental database for the flow/deformation behavior of spray cast materials in the semisolid range, (ii) formulate mathematically rigorous three-dimensional constitutive models that will portray accurately the flow of spray cast materials under thixoforming conditions, and (iii) evaluate the predictive capabilities of these models via prototype experiments.

In this paper the current understanding of the flow/deformation behavior of semisolid materials is reviewed. In particular, emphasis is placed on the ability of spray cast materials to deform thixotropically. Outstanding issues of material characterization, modeling, process design and control, and on-going experimental work are discussed. Our preliminary modeling results appear in [3].

2. The Production Of Thixoworkable Microstructures

The production of non-dendritic slurries has been the subject of research since the early work of Flemings and coworkers [1]. Their work involved vigorous mechanical stirring during solidification from the melt and was abandoned due to wear problems. Subsequent efforts involving mechanical stirring were inefficient and susceptible to gas entrapment, and contamination.

In an effort to eliminate problems associated with mechanical stirring, magnetohydrodynamic (MHD) stirring was introduced [4]. Careful control of the process is crucial. If stirring is not vigorous enough, the resulting final grain size is large (>200 μm). On the other hand, excessive stirring leads to macro-turbulence which destabilizes the process, and has a negative effect on product quality [5]. Limits on the size of the billets and the processing parameters are dictated by the available power.

Another approach has been developed [6], in which the semisolid slurry is forced by pressure to pass through the tortuous path of a static mixer ensuring high shear while it solidifies. Spherical grains approximately 100 μm can be produced by this method.

The intense shearing action required to produce the necessary equiaxed microstructure can be provided by an injection molding type of system [7]. Cold metallic pellets are injected through a volumetric feeder into a reciprocating screw. The material is processed thermomechanically by the rotating screw. Material of 30-50% volume fraction of primary solid is injected into a mold.

The fragmentation of previously deformed elongated solid grains by the formation of liquid phase during partial melting, is the central idea of another class of techniques (e.g., strain induced melting activation [4] and recrystallization and remelting techniques [8]). These methods, although they offer simplicity and efficiency, have size limitations (e.g., maximum billet diameter ~50 mm). Moreover, the optimum processing conditions for different alloys are not well understood.

3. Thixoworkable Microstructures Via Spray Casting

Spray deposition is recognized as an attractive technology for the production of near net shape components of a variety of alloys [9]. The process involves sequential atomization and droplet consolidation at deposition rates up to 2.5 kg/s. Depending upon substrate configuration and motion, it is possible to fabricate disks, billets, tubes and strip/sheet. Process fundamentals, advantages and commercial applications of the spray casting process have been documented [9, 10].

Under optimum conditions, 50-70w/o of the alloy solidifies during flight with attendant cooling rates ~10^4 °C/s [11]. A mushy/liquid spray impacting on a mushy deposit gives the optimum condition for preform formation. Except for a small transient zone, the temperature of the surface of the preform must be held approximately constant, in order to maintain a high fraction of solid which provides sufficient resistance to deformation of the preform, while the liquid fills the interstices between particles. Formation of the microstructure during spray deposition begins with the solidification of individual droplets during flight and is completed after the consolidated deposit is fully solidified. Solidification of the droplets in flight results in a fine-scale microstructure due to the high cooling rate. Subsequent solidification of the remaining liquid in the deposit results in equiaxed grains in the size range 20-40 μm. Recent studies of spray casting have shown that, unlike conventional casting, more than 90% of the solidification time in spray forming is spent in the presence of a high volume fraction of solid [12]. In consequence, the coarsening process is retarded in the absence of bulk liquid diffusion.

Attractive features of spray cast microstructures include:

- *Fine Grain Size:* Spray casting results in small equiaxed grains ~20-40μm. In order to produce a similar grain size via ingot casting, a prohibitively large number of working/annealing cycles must be performed. It is possible to achieve a small grain size by powder processing but there are complications associated with powder handling and attendant contamination.

- *Equiaxed Grains*: In the optimum situation the small droplets solidify in flight. The deposited material cools at a relatively low rate such that much of the local solidification time is spent in the company of a high volume fraction of solid (f_S>75%). The resulting microstructure is non-dendritic. Isothermal coarsening, which is dominant at low volume fractions of solid, is now suppressed and the operative coarsening mechanisms are the migration of liquid films which wet the grain boundaries, and Ostwald ripening of liquid droplets [12]. Thus, microstructures produced by spray casting are inherently suitable for semisolid processing.
- *"Clean" Solid Grains*: In spray cast materials entrapment of the liquid phase is minimal. Any liquid entrapped within solid grains does not contribute to flow during thixoforming. Thus spray cast microstructures are superior to those produced by other methods in which a significant quantity of the liquid is trapped in the solid grains.
- *Microsegragation*: The solidified liquid forms a solute rich layer around the already solid grains [11]. Such microstructures, when reheated into the semisolid range, give the required amount of liquid at relatively low temperatures because the solute rich layer melts very quickly.

Spray casting is also becoming attractive as an alternative production route for the fabrication of particulate reinforced composites [10]. Current methods of fabricating metal matrix composites include casting, pressure infiltration and powder metallurgy. There are several intrinsic limitations to each of these processes which result in material deficiencies; in particular, the composites are characterized by segregation, porosity, and interfacial reactions, all of which impair load-bearing integrity.

4. The Thixoforming Process

The first critical stage in thixoforming consists of heating the billet to the desired processing temperature. Slow heating results in significant grain growth, which impairs processability and final properties. In contrast, rapid heating, can cause significant temperature gradients due to the transient response of the heating system and/or the low thermal conductivity of the semisolid material. As a result of the high sensitivity of the flow of semisolid materials to temperature, temperature gradients can cause non-uniform flow behavior, uneven flow, or premature flow. Induction heating is the preferred method of choice due to its fast and controllable action. Careful design and optimization are required, however, in order to minimize the temperature gradients [13]. If conventional heating systems are to be used (e.g., resistance heating) a two step heating schedule is necessary with the first step being a long soak just below the eutectic temperature to minimize the effect of transients.

The forming operation must be performed at the lowest possible temperature in order to minimize the porosity that usually appears during solidification. In addition, low processing temperatures result in energy savings, longer die life, and a reduction in the number of defective parts. Such conditions permit easy handling of the billet, which retains its shape for a period of time. Elimination of the need to handle liquids simplifies dramatically the equipment requirements. At the same time, the selected temperature must be sufficiently high for the flow resistance to be low enough to permit filling of complex shaped dies without excessive force requirements.

Forming can be performed either by injection molding or by open or closed die forging. Issues pertinent to the optimization of processing parameters (e.g., ram velocity, die temperature, maximum load and dwell time) are not clearly understood. The sensitivity of the flow of semisolid materials to temperature, and their strong history-dependent flow behavior necessitate careful selection of the process variables. Moreover, semisolid processing is often unsuccessful due to flow instabilities and problems associated with the combined flow/solidification process, e.g. turbulent flow, solid-liquid segregation, premature freezing of the advancing slurry. It is important to develop the necessary processing science base consisting of constitutive models describing the rheology of the semisolid state in terms of material and process parameters. In turn, the design and optimization of semisolid processes can benefit directly from modeling.

5. Current Understanding of the Flow of Semisolid Materials

The ability of semisolid materials to deform easily, even at a high volume fraction of solid, is central to thixoforming. At a medium volume fraction of solid (~0.2-$0.3<g_S<0.6$), increased resistance to flow, with respect to the liquid and non-linear viscous history-dependent behavior, is exhibited due to strong interactions between the solid particles. Hydrodynamic interactions exist because of the relative motion of solid particles. Agglomeration of the solid particles is attributed to a welding-type mechanism. Recent work on the behavior of semisolid materials is based on the assumption that the resistance to deformation of the sheared material reflects the level of agglomeration of the solid particles [14]. Thixotropy has been explained on the basis of a disagglomeration process, namely the disruption of the solid particle bonds. At *a high volume fraction of solid* ($g_S>0.6$), a skeleton of solid particles is formed and the system can be considered as a porous solid saturated with liquid.

Research has focused on the behavior of solid slurries with medium volume fractions of solid. However, many thixoforming operations are performed at much higher volume fractions of solid (up to 80-90%). High volume fractions of solid are preferable, since they facilitate the handling of the billet and minimize shrinkage porosity, as long as they provide relatively low resistance to flow. Currently there is a limited understanding of the physics of the flow at high volume fractions of solid. Preliminary models, based of the assumption of the existence of a solid skeleton (e.g., [3]), may be useful for first order assessment of several phenomena observed during the deformation of semisolid materials. For example, the relative motion of the liquid phase can lead to extensive liquid-solid segregation. The permeability of the porous skeleton depends on the square of a characteristic length scale [15], typically of the order of the center-to-center distance of the solid grains. Therefore fine scale microstructures will reduce solid-liquid segregation due to the associated low permeability. These phenomena have been examined theoretically and the effects of macroscopic strain rate and microstructure have been explained qualitatively [3].

The solid skeleton model, however, does not portray accurately grain-to-grain interactions, which are governed by the presence or absence of a thin wetting layer of the liquid phase, and its contribution to macroscopic flow resistance. Energy considerations can provide the criterion for wettability of the solid grains [16] but neither the kinetics and dynamics of the problem nor the evolution of the structure during flow are currently understood. One of the striking features of thixoformed components is that the final microstructure remains equiaxed. Such microstructure may be expected at medium volume fractions of solid but not at high volume fractions of solid when geometric interactions necessitate the deformation of the solid grains in order to accommodate macroscopic flow.

6. Experimental Work

The limited data that exist on the behavior of semisolid spray cast materials have been derived from feasibility studies on thixoforming [17,18]. It is necessary to establish an experimental data base for the response of spray cast material in the semisolid range. Experiments that produce results unbiased by the measuring instrument are essential in the development of constitutive models. Of particular importance are the transient effects and very high strain rates (10 to 1000 s^{-1}). Thixoforming operations are performed at high die velocities and high strain rates, over a cycle time of a few seconds. *Therefore the effect of the transient response will be dominant in the processing of semisolid materials and must be carefully characterized.*

In order to characterize the behavior of spray cast materials in the semisolid range the following types of experiments are currently performed:

(i) Structural effects and the determination of the temperature range of interest for each alloy can be probed by using a penetrometer device. It consists of a small diameter tube with a hemispherical end which penetrates into a block of semisolid material when subjected to constant stress. The velocity response as a function of temperature can be converted into viscosity at different temperatures. Interpretation of these results is not simple because the strain rates at the tip of the penetrometer vary substantially and are typically lower than those expected in thixoforming operations. However, structural effects due to grain size, and shape can be explored. This test will provide guidance in narrowing the range of temperatures which are of interest for further examination.

(ii) At high volume fractions of solid, simple compression tests are possible. Careful lubrication is required in order to impose homogeneous deformation on the sample. Careful control of the temperature field in and around the specimen is necessary in order to interpret correctly the experimental data. Results from these tests are expected to provide information on the resistance to flow of semisolid materials under conditions similar to those appearing in thixoforming. A variation of these experiments for low to medium volume fractions of solid has been used in [19] in a manner similar to squeeze flow viscometry tests performed on polymers.

(iii) Because of the nature of the semisolid materials and their similarity with fully saturated soils, it is expected that the macroscopic and interstitial pressure may have a significant effect on the macroscopic flow behavior of semisolid materials. The only method that can shed light on this dependency is a triaxial stress experiment similar to that routinely perfomed on soils [20]. Performing these experiments on semisolids metals is not trivial, because of the inherent difficulty of dealing with pressurized systems at high temperatures, and the need to enclose the specimen in a deformable sealing envelope.

7. Acknowledgments

Support for this work was provided by the Mechanical and Structural Systems Division of the National Science Foundation (Grants No MSS-9007392, MSS-9313254).

8. References

[1] D. B. Spencer, R. Mehrabian, and M. C. Flemings, "Rheological Behavior of Sn-15%Pb in the Crystallization Range", *Met. Trans.*, **3**, pp. 1925-1932, 1972.

[2] F. A. Girot, L. Albingre, J. M. Quenisset and R. Naslain, "Rheocasting Al Matrix Composites", *J. Metals*, **33**, 11, pp. 18-21, 1987.

[3] A. Zavaliangos and A. Lawley, "Numerical Simulation of Thixoforming", TMS Fall Meeting, Pittsburgh, 1993, to appear in *J. Materials Engineering and Performance*.

[4] M. P. Kenney, J. A. Courtois, R. D. Evans, G. M. Farrier, C. P. Kyonka, A. A. Koch, and K. P. Young, *ASM Metal Handbook*, 9th edition, ASM International, Metals Park, Ohio, **15**, pp. 327-338, 1988

[5] J.-P. Gabathuler, D. Barras, Y. Krahennbuhl, J.-C. and Weber, *Proceedings of the 2nd International Conf.. on Processing of Semi-Solid Alloys and Composites,* M. Flemings, and S. Brown edts. M.I.T., Cambridge, pp. 33-46, 1992.

[6] P.L. Antona and R. Moschini, "A New Foundry Process for Production of Light-metals in a Semi-liquid Doughy State", *Foundry Trade Journal International*, pp. 173-178, 1987.

[7] L. Pasternak, R. Carnahan, R. Decker, R. Kilbert, "Semi-Solid Production Processing of Magnesium Alloys by Thixomolding", ibid [5], pp. 159-169, 1992.

[8] P. Kapranos, and D. H. Kirkwood, "Semisolid Processing of Alloys", *Metals and Materials*, **5**, 1, pp. 16-19, January 1989.

[9] A.G. Leatham, L. G. Elias, T. Itami, Y. Kawashima, P. Brooks, K. Humert, D. E. Tyler, H. P. Cheskis, R. P. Dalal, and P. D. Prichard, in *Spray Forming: Science, Technology and Applications*, Powder Metallurgy World Congress, pp. 59-112, San Francisco, Metal Powder Industries Federation, 1992.

TABLE I : Comparison of Forging, Die Casting and Semisolid Processing

	Advantages	Disadvantages
Forging	Favorable microstructure Structural integrity	High loads Die wear Low shape complexity
Die Casting	Complex shape capability Low force requirements	Turbulent flow Undesirable microstructure Residual porosity Entrapped gas Inferior mechanical properties
Semisolid Processing	Complex shape capability Low force requirements Low porosity Acceptable microstructure Ease of handling Laminar flow	Requires careful control Liquid segregation possible

[10] P.C. Mathur, D. Apelian, and A. Lawley, "Analysis of the Spray Deposition Process, *Acta Metallurgica*, **37**, 2, pp. 429-443, 1989.

[11] A. J. W. Ogilvy, "The Metallurgy of the OSPREY Process", ibid[9], pp. 33-46, 1992.

[12] S. Annavarapu, R.D. Doherty, D. Apelian, A. Lawley, and P. Mathur, "Fundamental Aspects Of Consolidation And Microstructure Development During Spray Casting", paper No 4 in *Proceedings of 1st Int. Conf. on Spray Forming* (ICSF), OSPREY Metals Ltd., Wales, U.K., 1990.

[13] J. A. Dantzig and S. P. Midson, "Billet Heating For Semi-solid Forming", ibid [5], pp. 105-118., 1992.

[14] P. Kumar, C. Martin, and S. B. Brown, "Shear Rate Thickening Behavior of Semisolid Slurries", *Met. Trans.,* **24A**, pp. 1107-1116, 1993.

[15] T. S. Piwonka, and M. C., Flemings,"Pore Formation in Solidification", *Trans. of AIME*, **236**, p. 1157, 1966.

[16] P. Kapranos, D. H. Kirkwood, C. M. Sellars, "The Microstructure of Thixotropic Alloy Slurries", *Proc. EUROMAT 90 Conf.*, pp. 165-168, Aachen, 1990.

[17] Osprey Metals Ltd., European Patent No 0200349, 1985. U.S. Patent No. 4804034, 1987.

[18] Henderson N. G., McLelland A. R. A., Atkinson H. V., and Kirkwood D. H., "High Temperature Rheology of Novel Matrix Composites Based on Aluminum/High Silicon Alloys", ibid [5], pp. 349-358, 1992.

[19] M. Mada, and F. Ajersch, "Viscometry Measurements of A356-15% Sic Semi-solid Alloys Using a Squeezing Flow Viscometer", ibid [5], pp. 276-289, 1992.

[20] A. W. Bishop, and D. J. Henkel, "The Measurement of Soil Properties in the Triaxial Test", Edward Arnold (Publishers) Ltd., London, 1957.

HIGH TEMPERATURE DEFORMATION OF RAPIDLY SOLIDIFIED 7091 P/M ALUMINUM ALLOY

H.N. AZARI, G.S. MURTY AND G.S. UPADHYAYA
Department of Materials and Metallurgical Engineering
Indian Institute of Technology
Kanpur -208016, India

ABSTRACT. The elevated temperature ductility of rapidly solidified 7091 P/M Al alloy was explored in the temperature range of 523 - 723 K making use of strain rate change and constant crosshead speed tensile tests. The alloy exhibited a maximum strain rate sensitivity of 0.3 and 190 % elongation despite its initial fine grain size of 2 μm. This relatively low ductility is attributed to partially recrystallized microstructure and preferential distribution of fine oxide particles. The alloy was subjected to a thermomechanical treatment involving solutionizing treatment (763 K / 3 h) followed by water quenching, overaging (603 K / 36 h) and warm rolling at 423 K to ~ 80 % reduction in area. With this treatment, a maximum of 450 % elongation at 573 K and 623 K and a strain rate of 8 X 10^{-5} S^{-1} was observed. The optimal superplastic behavior is associated with continuous recrystallization and the consequent fine grain size.

1. Introduction

Rapid solidification processing of 7xxx Al alloys has led to improved mechanical properties and corrosion resistance [1] of conventional I/M alloys through microstructural refinement, chemical homogeneity and extended solubility [2,3]. The 7091 P/M Al alloy is a rapidly solidified Al-Zn- Mg-Cu alloy with 0.4 wt% Co, which forms finely dispersed Co_2Al_9 intermetallic particles [4]. In view of its fine grain size, superplastic forming of the alloy may be possible. Previous studies on the superplastic response of this alloy however indicate a relatively low ductility [5,6]. The objective of this study has been to evaluate the superplastic response of the alloy in the as-received condition and after a suitable thermomechanical treatment aimed at improving its elevated temperature ductility.

M. A. Otooni (ed.), Science and Technology of Rapid Solidification and Processing, 195–203.

2. Experimental

The nominal composition in wt % of the 7091 P/M Al alloy processed by Alcoa through rapid solidification followed by hot extrusion is : Zn(5.7-7.1), Mg(2.0-3.0), Cu(1.1-1.8), Co(0.2-0.6),Si((0.12),Fe(0.15),O(0.2-0.5),others(0.15) and Al(balance). Flat specimens of 3 X 3 X 6 mm gage section were machined for high temperature tensile testing in the range of 523 - 773 K. Strain rate change and constant crosshead speed tests were performed to assess strain rate sensitivity index 'm' ($= d \log\sigma / d \log\dot{\epsilon}$) and tensile ductility. Details of thermomechanical schedules, optical metallography and TEM observations are given elsewhere [7].

3. Results and Discussion

3.1. AS-RECEIVED ALLOY (T7E69 TEMPER)

The microstructure of the alloy is shown in Fig.1, wherein optical micrographs on three orthogonal planes are presented. The alloy exhibits a grain morphology with mean intercept lengths of 2.0, 1.8 and 1.7 μm in the longitudinal (L), long-transverse (LT) and short-transverse (ST) directions respectively of the as-received billet. TEM observations indicate a partially recrystallized substructure with a high dislocation density [8]. Fine Co_2Al_9 particles of 0.05-0.1 μm size are uniformly distributed, whereas oxide particles of size < 0.05 μm are preferentially distributed along grain boundaries.

The flow stress and 'm' as a function of strain rate are shown in Figs.2 and 3 for the test temperature range of 523- 773 K. The observed maximum value of 'm' is 0.3 at 773 K. From the data presented in Table 1, the maximum elongation is seen to be 190 % at 723 K and it drops to a lower value at the highest temperature of 773 K.

The relatively low 'm' and ductility can be attributed to the partially recrystallized initial microstructure and the fine oxide particles at the grain boundaries. Such a microstructure is not conducive for grain boundary sliding during high temperature deformation. Observations on deformed specimens indicate fine dynamically recrystallized grains following deformation at 573 K and a strain rate of 2 X 10 $^{-4}$ S^{-1}, whereas grain growth occurs at test temperatures of 723 K and 773 K. These observations suggest the possibility of improving the elevated temperature ductility of this alloy through a thermomechanical treatment to obtain fully recrystallized and fine grains. Such a possibility has been explored as per the details given below.

3.2. THERMOMECHANICALLY TREATED ALLOY

Three different schedules of thermomechanical treatments [7] were attempted for the above purpose (Fig.4). These involve (i) solution treatment for 3 h at 763 K and water quenching (ii) overaging at 603 K or 653 K from 2 h - 36 h and air cooling to room temperature, and then (iii) warm rolling at 423 K to 80-82 % reduction in area (RA) in

Table 1. Tensile test data of the as-received alloy

Temperature (K)	Strain rate (S^{-1})	Elongation (%)	% RA
573	5×10^{-5}	115	86
	2×10^{-4}	95	71
	5×10^{-3}	71	91
673	5×10^{-4}	116	93
	5×10^{-3}	100	83
	1×10^{-2}	168	91
773	5×10^{-3}	165	79
	1×10^{-2}	160	84

Table 2. Tensile test data of the alloy with TMT-C schedule

Temperature(K)	Strain rate (S^{-1})	Elongation(%)	% RA
573	8×10^{-5}	445	97
	9×10^{-4}	319	97
	5×10^{-3}	165	99
673	9×10^{-4}	105	78
	5×10^{-3}	126	77
773	9×10^{-4}	90	61
	5×10^{-3}	58	59

20 passes. While TMT-A schedule resulted in an unrecrystallized microstructure, TMT-B yielded a fully recrystallized structure with coarse grains of 100 μm in size. The TMT-C schedule with the overaging treatment of 603 K / 36 h led to the formation of large $MgZn_2$ precipitate particles in the size range of 0.7-2.4 μm (Fig.5). Warm working of the overaged alloy resulted in a structure with fine subgrains and a high dislocation density. Subsequent discontinuous recrystallization of the alloy with TMT-C schedule led to a fully recrystallized grains in the size range of 0.5-10.0 μm. As the grain size of this material is observed to be unstable with indications of grain growth and non-uniform grain size, its superplastic response was explored in the warm rolled condition prior to the discontinuous recrystallization step.

The rate sensitive flow characterization of the warm rolled alloy (TMT-C) is presented in Figs. 6 and 7. A maximum of 450 % elongation was noticed at 573 K and a strain rate of 8×10^{-5} S^{-1} (Table 2). While static annealing of the warm rolled alloy at 573 K did not cause recrystallization, TEM observations following 450 % elongation indicate equiaxed grains of 1-3 μm in size. Thus annealing and concurrent straining led to high angle grain boundaries through continuous recrystallization. Extensive grain boundary sliding and high tensile ductility are favored when continuous recrystallization occurs. Further details on the superplastic response of the TMT-C alloy are presented elsewhere [9].

4. Conclusions

1. The 7091 P/M Al alloy in the as-received condition exhibits a maximum 'm' value of 0.3 and 190 % elongation at 723 K and a strain rate of 10^{-2} S^{-1}. The partially recrystallized starting microstructure and preferentially distributed oxide particles seem to limit the rate sensitivity of flow stress and tensile ductility.

2. A TMT schedule consisting of solutionizing (763K / 3 h), overaging (603 K / 36 h) and warm rolling (80-82 % RA at 423 K) results in a structure with fine subgrains and a high dislocation density. The alloy with this treatment exhibits a maximum of 450 % elongation when tested at 573 K and a strain rate of 8×10^{-5} S^{-1}. Continuous recrystallization and the resulting fine grains are associated with the optimal superplastic response of the alloy.

REFERENCES

1. Y.W. Kim, W.M. Griffith and F.H. Froes: J. Metals, 1985, 32, (8), 27.

2. F.H. Froes and J.R. Pickens: J. Metals, 1984, 36, (1), 14.

3. H. Jones: 'Rapid solidification of metals and alloys', Monograph No.8, The Institute of Metals, London, 1982.

4. P.K. Domalavage, M.G. Grant and Y. Gefen: Metall. Trans., 1983, 14A, 1599.

5. Y.C. Zhang and N.J. Grant: Mater. Sci. Eng., 1984, 68, 119.

6. A.K. Ghosh, C.H. Hamilton, M. Kumar and J. Waldman: in 'Superplastic forming of structural alloys', (ed. N.E. Paton and C.H. Hamilton), 359, 1982, TMS, Warrendale.

7. H.N. Azari, G.S. Murty and G.S. Upadhyaya: Berg - und Huttenmannische Monatshefte, 1993, 138, (9), 347.

8. H.N. Azari, G.S. Murty and G.S. Upadhyaya: Prac. Metallography, 1993, 30, (4), 186.

9. H.N. Azari, G.S. Murty and G.S. Upadhyaya: Metall. Trans., (Accepted for publication).

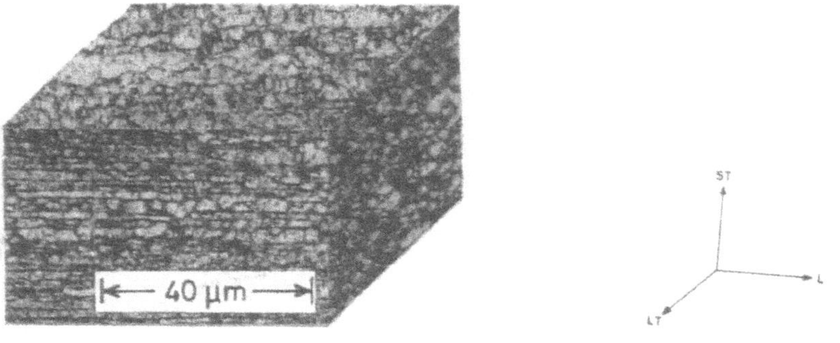

Fig.1. Optical micrograph of the as-received alloy on three orthogonal planes

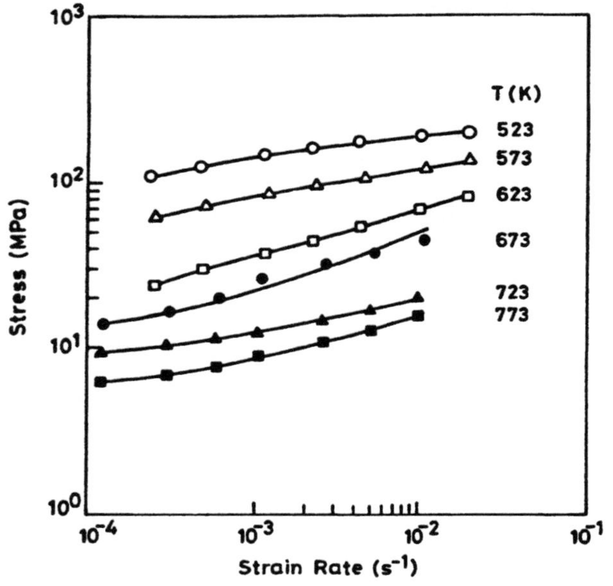

Fig.2. Flow stress vs. strain rate plots of the as-received alloy

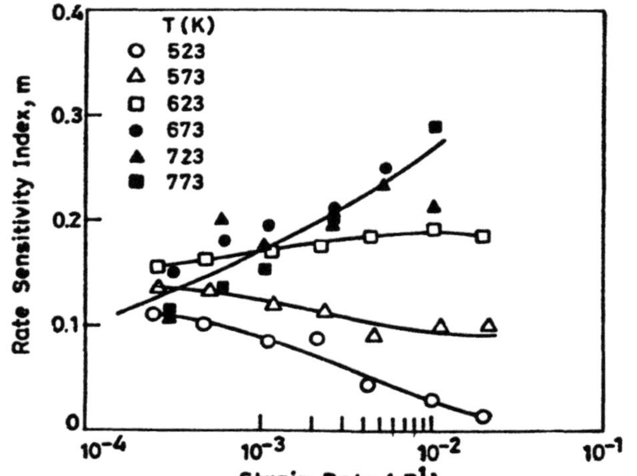

Fig.3. 'm' vs. strain rate plots of the as-received alloy

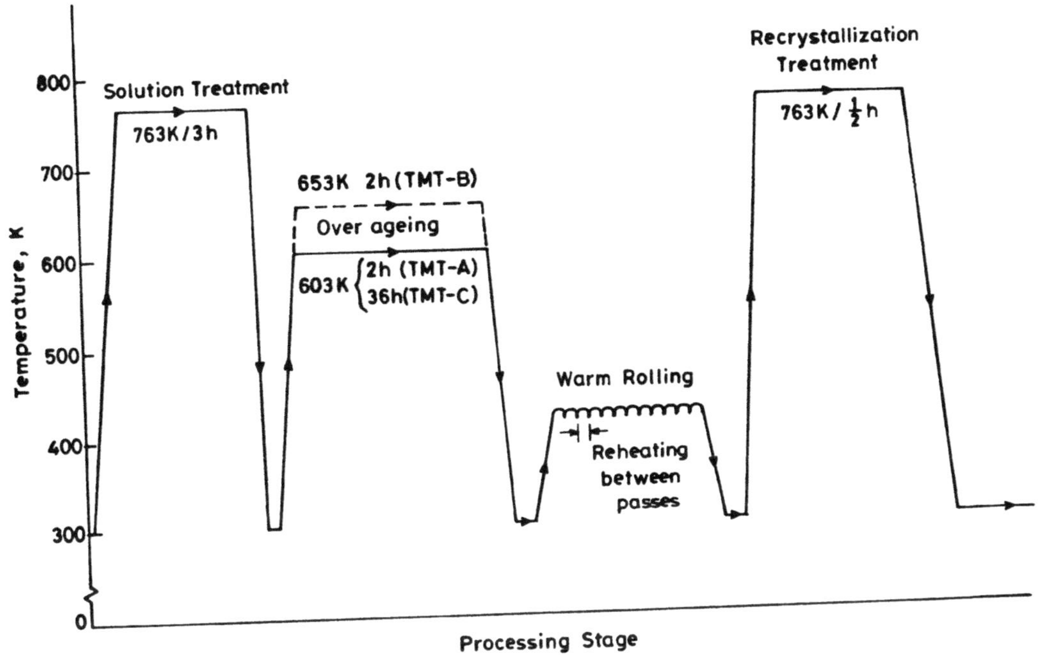

Fig.4. Schematic diagram showing various thermomechanical treatment schedules

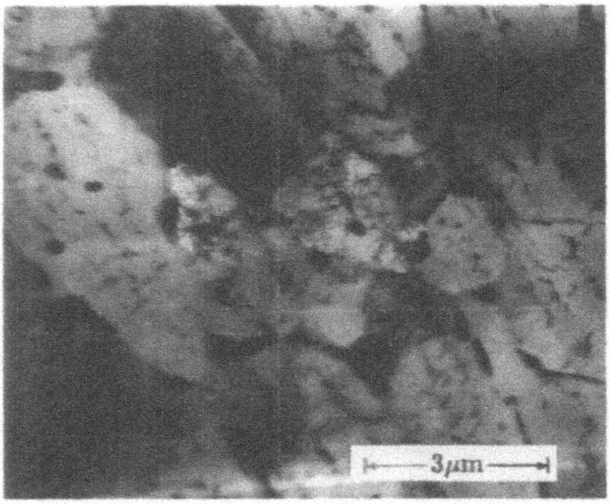

Fig.5. TEM micrographs of TMT-C treated alloy in the overaged condition

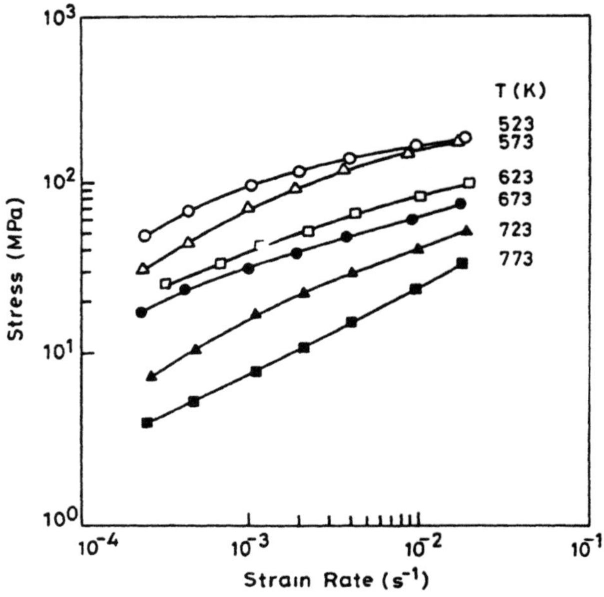

Fig.6. Flow stress vs. strain rate plots of TMT-C treated alloy tested after warm rolling step.

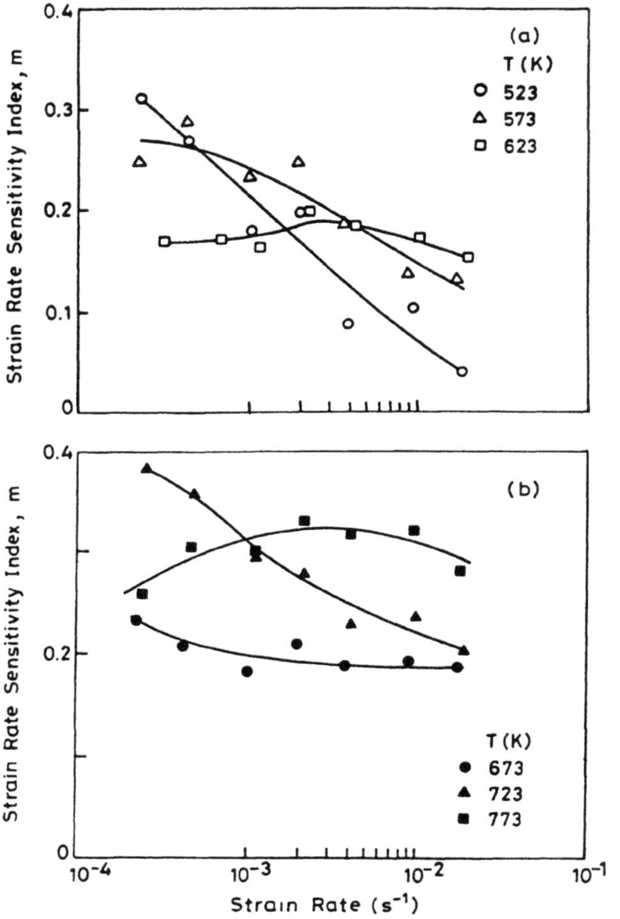

Fig.7. 'm' vs. strain rate plots of TMT-C treated alloy tested after warm rolling step

MICROSTRUCTURES, TENSILE PROPERTIES AND FATIGUE CRACK GROWTH BEHAVIOR OF THE ZIRCONIUM MODIFIED 2024 ALLOYS PROCESSED BY LIQUID DYNAMIC COMPACTION

XIAOMAN DUAN, ALI FARAH, IBRAHIM UCOK AND NICHOLAS J. GRANT
Department of Materials Science and Engineering
Massachusetts Institute of Technology
Cambridge, MA 02139

ABSTRACT. In this study, Liquid Dynamic Compaction (LDC) was used to develop rapidly solidified 2024 aluminum alloys modified with small additions of zirconium. Three concentrations of zirconium (0.12, 0.28 and 0.53 wt%) were selected to examine the effect of zirconium on the microstructures, tensile properties and fatigue crack growth behavior of the modified 2024 alloys. The results show that all as-spray deposited materials have acceptable densities (91-95 %) and fine equiaxed microstructures with uniform distribution of fine secondary phases, mainly Al_2Cu. Additions of zirconium (0.12 to 0.53 wt%) to the LDC 2024 alloy significantly refined the grain structures. The LDC-produced fine microstructures were retained during the subsequent hot rolling and T4 treatment when the Zr content was greater than about 0.28%. The increase in Zr concentration from 0.12 to 0.28 and 0.53 wt% significantly increased the tensile strength of the modified 2024 alloys. A small amount of Zr (< 0.3 wt%) also improved the ductility. The zirconium modified LDC 2024 alloys exhibited fatigue crack growth resistance superior to those for standard 2024-T3 and experimental LDC 7150-FS, a 7150 alloy with a relatively high concentration of Fe and Si. These beneficial effects are believed to come mainly from the additional precipitation of fine, stoichiometric Al_3Zr particles and refinement of microstructures.

1. Introduction

High strength aluminum alloy 2024 is widely used commercially, including aircraft structures. Because of its commercial importance, alloy 2024 has been studied extensively. Among such studies rapid solidification processing has demonstrated a potential for structural enhancement leading to improved mechanical properties [1-5]. In previous studies [5-7], alloy 2024 was modified with additions of 2% Fe and 2% Ni (wt%) to demonstrate the ability of the Liquid Dynamic Compaction Process to supersaturate 2024 with large additions of insoluble refractory alloying elements, in a highly beneficial way.

It has also been reported[8-12] that small additions of zirconium (up to 1%) to high strength aluminum alloys retards recrystallization and refines the grain structure. Zirconium has also been reported to improve toughness, stress-corrosion resistance and the quench sensitivity of aluminum alloys [10, 13, 14]. It has further been established that the improved mechanical properties of the Zr-bearing aluminum alloys are due to a high density of very fine precipitates of the cubic Al_3Zr phase [11,15,16].

In this study, three concentrations of zirconium (0.12, 0.28 and 0.53 wt%) were used to investigate the effects of zirconium on the R.S. as-spray deposited microstructures of modified LDC 2024 alloys, as well as the effects on the hot and cold workability and

M. A. Otooni (ed.), Science and Technology of Rapid Solidification and Processing, 205–216.

mechanical properties of the alloys in subsequent thermomechanical processing and heat treating.

2. Alloy Preparation and Experimental Procedures

Master alloys of 2024 with or without additions of Zr were supplied by Alcoa. The compositions of the master alloys are given in Table 1.

Table 1. Chemical compositions of 2024 and 2024+Zr (wt%) alloys

	Al	Cu	Mg	Mn	Zr
2024*	Bal.	4.4	1.5	0.6	---
2024+0.12Zr	Bal.	4.4	1.5	0.6	0.12
2024+0.28Zr	Bal.	4.4	1.5	0.6	0.28
2024+0.53Zr	Bal.	4.4	1.5	0.6	0.53

* Basic composition (Cu, Mg, Mn) of master alloy

The alloys were melted in a graphite crucible and heated to 1073 K (800°C) or 1103K (830°C, for alloy 2024+0.53Zr) under an argon atmosphere by induction heating at a slight positive pressure. Atomization of the liquid alloys was achieved through a circular ultrasonic gas atomizer under an argon pressure of 0.69 MPa with a metal pouring tube 5 mm in internal diameter and a spray distance of 356 mm. The high speed, atomized liquid droplets were impacted against a water-cooled mild steel substrate to achieved the desired rapid solidification rates. The details regarding the experimental equipment are given elsewhere [17,18].

Thermomechanical processing and heat treatments were carried out on the LDC deposits. Samples were cut from appropriate sections of the deposits to examine the structures and properties of the processed material. The samples were hot rolled at 673K (400°C) in 3 or 4 passes, each with a reduction approaching 20%, with intermediate, very short term reheating. After hot rolling to a thickness of 2 mm, a solution treatment and a subsequent natural aging treatment (T4) were carried out on these three modified 2024 alloys, as well as on the reference 2024 alloy with the standard composition. The heat treatment and aging used for these alloys was: solution heat treatment at 490°C/0.5 hour, water quench, plus 20°C/15 days aging.

Optical and electron microscopy (SEM and TEM) were employed for microstructural characterization at each stage of processing. X-ray diffraction and electron diffraction were used to determine the phases present in these alloys.

Tensile testing at room temperature was done with an Instron testing machine at an initial strain rate of 1×10^{-4} per sec. Fatigue crack growth rates were measured for alloys 2024+0.12Zr and 2024+0.53Zr at room temperature. Center pre-cracked panel specimens were used and the tests were carried out on an Instron 1331 electro-hydraulic testing machine. In all the tests, a frequency of 10 Hz, in air, at a stress range of 35.5 MPa, with a R ratio of 0.05, was used.

3. Experimental Results and Discussion

3 1 MICROSTRUCTURES

Systematic microstructural characterization studies were conducted on the as-deposited 2024 alloys with and without additions of zirconium. The as-deposited materials showed high density (94-95% of theoretical) in the center sections of the deposits and acceptable density (91-92% of theoretical) at the edges of the deposits. Figure 1 shows microstructures of the 2024, 2024+0.12Zr, 2024+0.28 Zr and 2024+0.53Zr (wt%) alloys in the as spray-deposited state, taken near the edge of the each deposit. The material from the edges of the LDC deposits indicated fine, equiaxed, fully recrystallized microstructures, with very fine, uniform distribution of secondary phases. The porosity, argon at very low

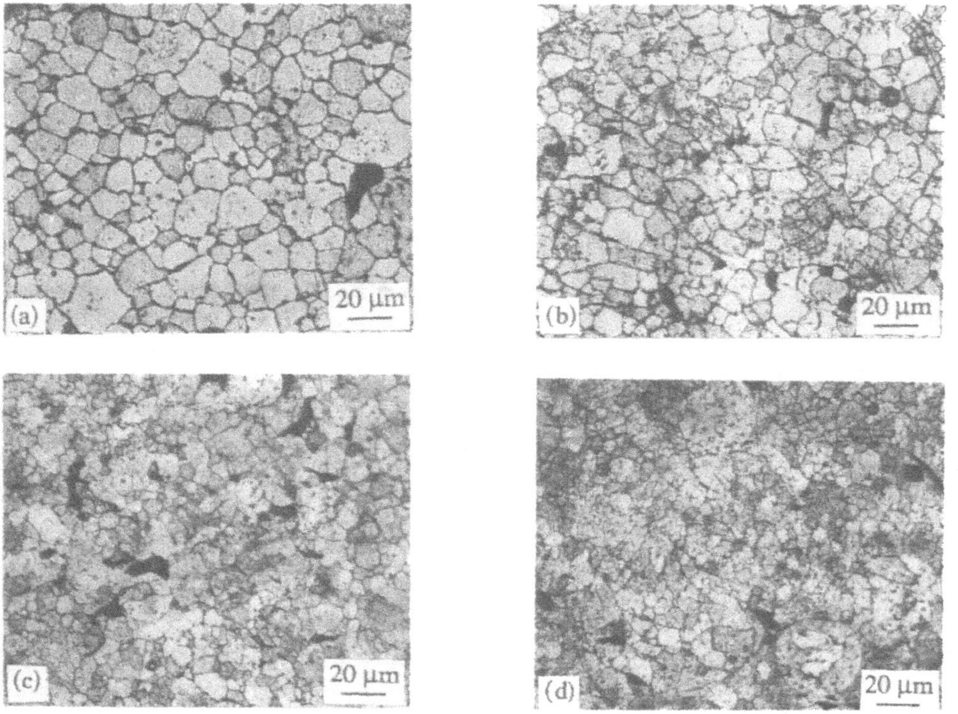

Figure 1. As-spray deposited microstructures.
(a) 2024; (b) 2024+0.12Zr
(c) 2024+0.28Zr; (d) 2024+0.53Zr

pressure, formed during spray deposition, is fine and well scattered. Prior-droplet boundaries are relatively infrequently seen in the microstructures, and no large splat structures were observed. For the LDC 2024 alloy, the as-deposited microstructures consist of fine, equiaxed grains with an average size of 20-30 μm (Figure 1a). A small addition of Zr (0.12 wt%) to the 2024 alloy resulted in a finer grain structure (10-15 μm) (Figure 1b). With an further increase in zirconium concentration from 0.12 to 0.28 and 0.53 wt%, the microstructures of the Zr containing alloys become progressively finer as shown in Figure 1c and 1d. The average as-deposited grain size for the 2024+0.53Zr alloy is only about 5 μm.

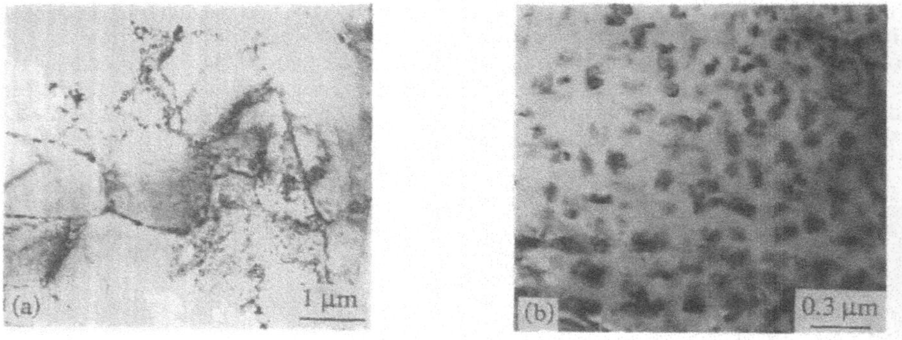

Figure 2. TEM micrographs of the as-deposited 2024+0.53Zr alloy.
(a) showing precipitates on subgrain boundaries.
(b) precipitates in the matrix

High magnification via TEM of the as-deposited 2024+0.53Zr alloy revealed fine subgrains (0.5-1.5 μm, as shown in Figure 2a) containing a high density of dislocations on the boundaries, and limited amounts of fine second phase precipitates both on the grain boundaries (Figure 2a) and in the grain interiors (Figure 2b). Most of the second phase particles are of the order of 0.05 μm in size, and are distributed uniformly throughout the matrix. Selected area electron diffraction indicated the great majority of the second phases to be Al_2Cu.

The XRD patterns for LDC 2024+Zr alloys show only a few weak second phase peaks (Figure 3), indicating a high degree of supersaturation in the as-deposited materials. Indexing of the minor peaks for all LDC alloys indicated the presence of Al_2Cu in the as-deposited material, with no obvious peaks for the phase $S(Al_2CuMg)$ normally observed in ingot processed 2024. Minor peaks for Al_3Zr are poorly identifiable in the as-deposited 2024-Zr alloys even when the concentration of Zr is as high as 0.53 wt% (Figure 3b).

Figure 4 shows microstructures of the LDC 2024 and 2024+Zr alloys after hot rolling at 673°K (400°C) for 80-90% reduction in several rolling passes with a 10-20% reduction in each pass. All alloys show a uniformly worked, ultra fine microstructure. The minor porosity observed in the as-deposited state has been completely eliminated. A

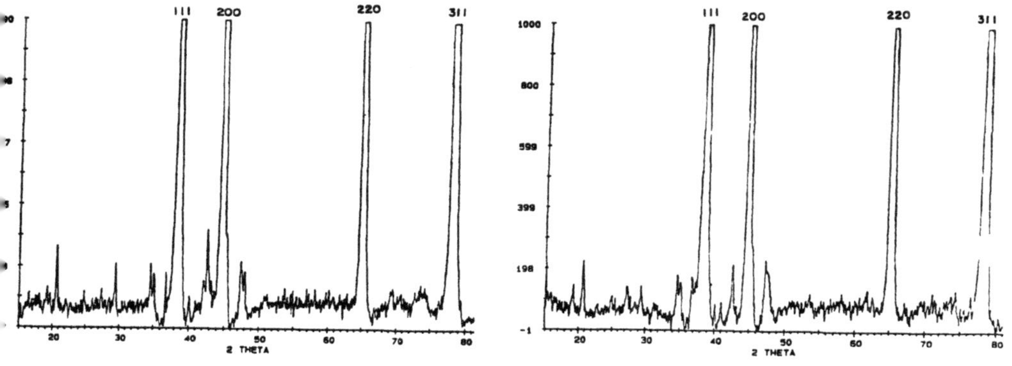

(a) (b)

Figure 3. XRD patterns for the LDC alloys in as-deposited state.

more uniform microstructure resulted from the LDC 2024 alloy after an additional rolling pass at 673°K (400°C) with a reduction of 10% (Figure 4a).

Figure 5 shows microstructures of the LDC 2024 and 2024+Zr alloys after solutionizing at 763°K (490°C) for 0.5h and aging at room temperature for 15 days. For the LDC 2024 alloy of standard composition, the as-spray deposited microstructure was not remained during the solution treatment. A significantly coarser recrystallized grain structure with high grain aspect ratios (Figure 5a) was produced. Though a small amount of the zirconium (0.12 wt%) addition to 2024 alloy did produce a somewhat finer T4 microstructure (Figure 5b) compared to the LDC 2024 alloy, the significant grain refinement effect of zirconium can be seen only when the concentration of Zr is increased to about 0.28% (Figure 5c). The higher the concentration of zirconium, the finer is the T4 microstructure; See Figure 5d.

Figure 6 shows TEM micrographs of the LDC 2024 and 2024+0.53Zr alloys in the T4 condition. For the LDC 2024 alloy, Al_2Cu (θ') and Al_2CuMg (S') precipitates were revealed (Figures 6a). For the LDC 2024+0.53Zr T4 alloy, not only are much finer Al_2Cu and Al_2CuMg precipitates observed, but also some needle-shaped precipitates (Figure 6b) were present which are believed to be another type of Al_3Zr (β') according to the SAD pattern analysis and other references [9,19]. XRD diffraction results further confirmed the presence of the Al_3Zr phase in the LDC 2024-Zr alloys in the T4 condition. The fraction of Al_3Zr increased with increasing concentration of zirconium. Thus the grain growth retardation effect (refer to Figure 5) is believed to come mainly from the very fine Al_3Zr precipitates which inhibit recovery and recrystallization during thermomechanical processing and heat treatment.

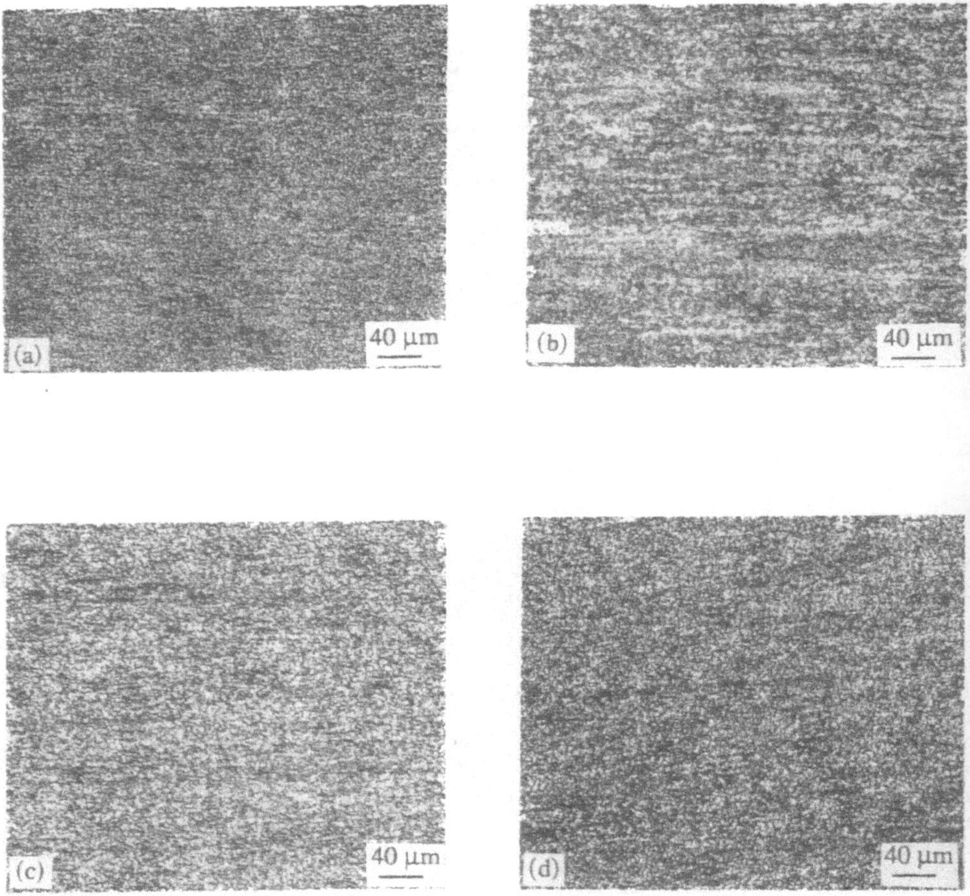

Figure 4. As-rolled microstructure of LDC alloys.
(a) 2024, 400°C/91%; (b) 2024+0.12Zr, 400°C/81%
(c) 2024+0.28Zr, 81%; (d) 2024+0.53Zr, 400°C/81%

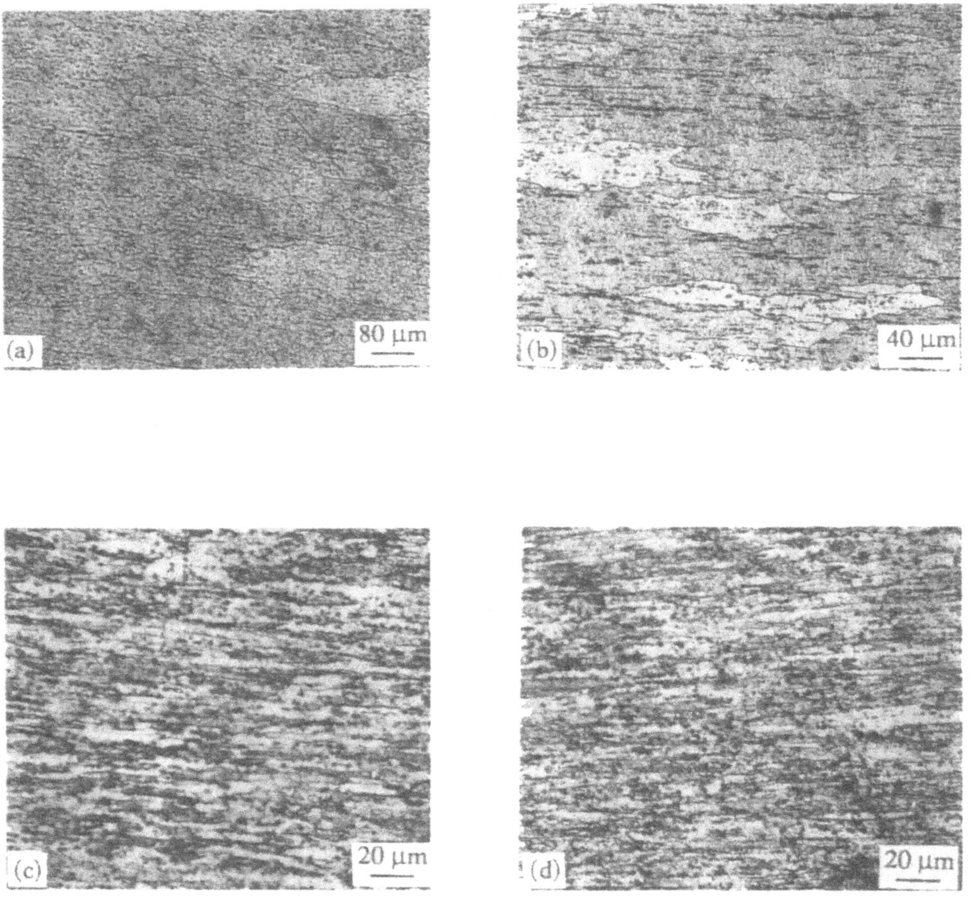

Figure 5. T4 microstructures of LDC alloys.
(a) 2024, (b) 2024+0 12Zr,
(c) 2024+0.28Zr; (d) 2024+0.53Zr

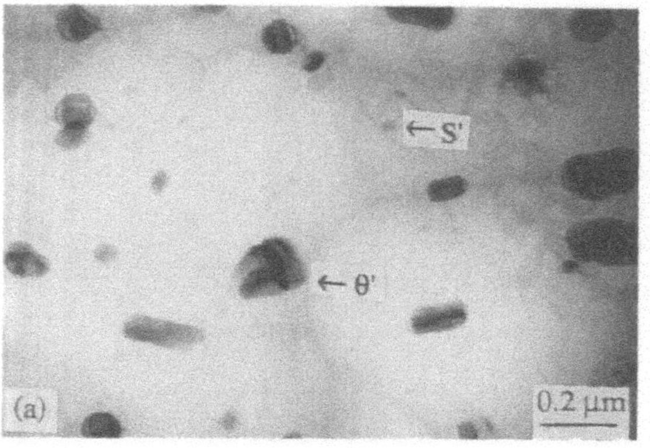

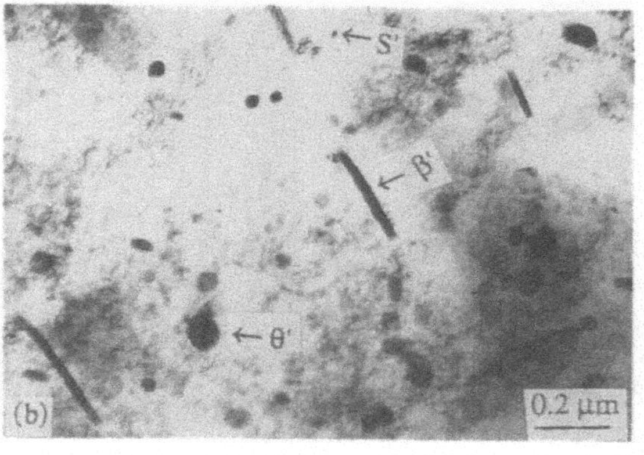

Figure 6 TEM micrographs of LDC processed, hot rolled and T4 treated alloys. (a) 2024, showing presence of second phases Al_2Cu (θ') and Al_2CuMg (S'), (b) 2024+0.53Zr, showing second phases Al_2Cu (θ'), Al_2CuMg (S') and Al_3Zr (β').

3.2 TENSILE PROPERTIES

Tensile tests were performed at room temperature; the effect of zirconium on the tensile properties of LDC 2024 alloy is shown in Figure 7. The increase in Zr concentration from 0.12 to 0.28 and 0.53 wt% significantly increased the tensile properties, both ultimate tensile strength (UTS) and yield strength (YS), of the LDC 2024 alloy. Compared to LDC 2024 without additions of zirconium (YS = 42 ksi, UTS = 63 ksi), the 2024+0.12Zr has a higher yield stress of 47 ksi and UTS of 66 ksi; the 2024+0.28Zr has a further increased yield stress of 49 ksi and UTS of 68 ksi; and the 2024+0.53Zr has the highest yield stress of 55 ksi and UTS of 73 ksi. Furthermore, the addition of Zr to 2024 does not result any loss of ductility. As a matter of fact, it can be seen that a small amount of Zr (<0.3 wt%) can improve ductility.

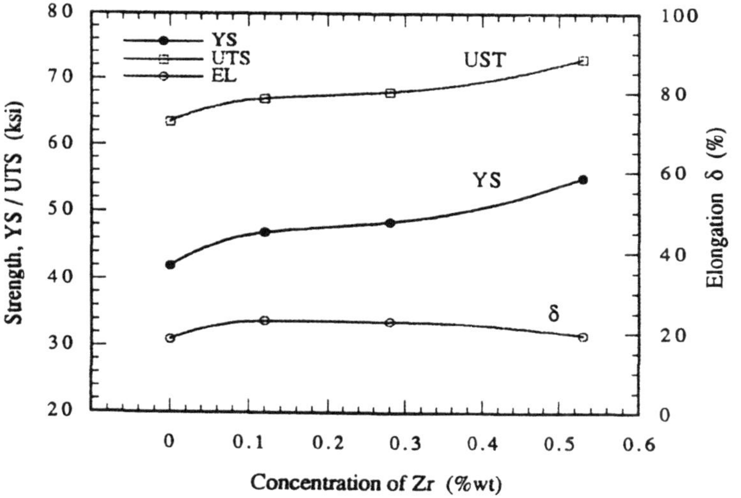

Figure 7. Effects of additions of Zr on UTS, YS and elongation of LDC 2024.

The improved properties of LDC 2024-Zr alloys may be explained in terms of the following mechanisms: (1) the fine precipitates which formed during LDC deposition or during subsequent thermomechanical processing result in a higher density and more uniform distribution of dislocations during deformation, which can be seen in Figure 6b; (2) the fine particles also have a pinning effect on grain and subgrain boundaries during annealing and solution treatment; (3) the high density of dislocations and fine subgrains lead to more refined microstructures.

3 3 FATIGUE CRACK GROWTH BEHAVIOR

Fatigue crack growth rates were measured for 2024-0.12Zr and 2024-0 53Zr. The da/dn data were derived from the 2a-N data by a seven-point-polynomial approximation ΔK is given as follows:

$$\Delta K = \Delta\sigma\sqrt{\pi a}\sqrt{\sec\left(\frac{2\pi a}{W}\right)}$$

where $\Delta\sigma$ is the nominal stress range, a is half the crack length and W is the width of the sample.

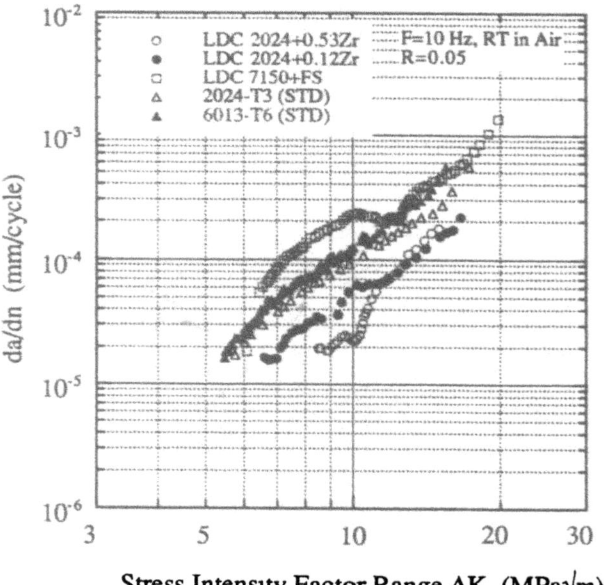

Figure 8. Fatigue crack growth rate da/dn vs ΔK for LDC 2024+0 12Zr and 2024+0.53Zr, plus comparisons of fatigue crack growth rate da/dn values vs ΔK for LDC 2024+0.12Zr and 2024+0 53Zr with standard 2024-T3 and 6013-T6, and the experimental LDC 7150+FS.

Figure 8 shows the da/dn vs. ΔK curves of two RS-LDC 2024-Zr alloys at R = 0.05. It can be seen that the two 2024-Zr LDC alloys have significantly higher fatigue crack growth resistance, particularly in the low stress intensity factor regime. The figure also provides comparison of the fatigue crack growth resistance of these two Zr-modified 2024 alloys and that for a standard 2024-T3 alloy and the newly-developed 6013-T6 alloy, as well as for the LDC 7150-FS alloy. It is clear that the new LDC 2024 alloys with 0.12Zr to 0.53Zr alloys show fatigue crack growth resistance considerably superior to that for the standard 2024-T3 and 6013-T6 alloys as well as LDC 7150-FS alloy. The small increase in crack growth resistance for the 2024-0.53Zr alloy compared with that for the 2024-0.12Zr alloy in the low stress intensity factor range is believed to come mainly from large quantity of precipitation of fine Al_3Zr particles.

4. Conclusions

Small additions of zirconium (0.12 to 0.53 wt%) to the LDC 2024 alloy can significantly refine the microstructures; the higher the concentration of the Zr, the finer the microstructure. The LDC-produced fine microstructures were retained during the subsequent hot rolling and T4 treatments when the Zr concentration is greater than about 0.28%. Additions of Zr to the LDC 2024 alloy increase not only the conventional short time tensile properties, without any loss in ductility, but also significantly increase the fatigue crack growth resistance compared to other high strength standard aluminum alloys. The gains in mechanical properties mainly come from a higher volume of stable precipitates of very fine Al_3Zr which stabilized the fine grain structures produced by rapid solidification, minimizing grain coarsening during solution heat treatments.

Acknowledgements

The authors deeply appreciate the preparation and gift of these special, modified alloys from the Aluminum Company of America, with special thanks to Dr. John Benjamin and Frank Baker.

References

1. M. Lebo and N. J. Grant; *Met. Trans.* 5 (1974) 1547.

2. D. Webster; *Met. Trans.* 10A (1979) 1913.

3. B. Van den Brandt, P. J. Van den Brink, H. F. de Jong and L. Katgerman; *J. Mat. Sci. Lett.* 2 (1983) 67.

4. L. Katgerman and B. Van den Brandt; *High Strength Powder Metallurgy Aluminum Alloys II*, (G. J. Hildeman and M. J. Koczak, eds.), TMS-AIME, Warrendale, PA (1985) 65.

5. X. Duan, T. Ando and N. J. Grant; *Advances in Powder Metallurgy Vol 6 Aerospace, Refractory and Advanced Materials*, Proc. of the 1991 Powder Metall. Conf. and Exhibition, Chicago, IL (1991) 75.

6. X. Duan, I. Ucok, T. Ando and N. J. Grant; *Melt-spinning and Strip Casting: Research and Implementation*, (Eric F Matthys, ed.) TMS-AIME, Warrendale, PA (1992) 33.

7. X. Duan, T. Ando and N. J. Grant; *The First Pacific Rim International Conference on Advanced Materials and Processing (PRICM-1)*, (Changxu Shi, Hengde Li and Alexander Scott, eds.), TMS-AIME, Warrendale, PA (1992) 315.

8. E. Ness and N. Ryum; *Scripta. Metall.* 5 (1971) 987.

9. E. Ness; *Acta Metall.* 20 (1972) 499.

10. E. Di Russo; *Alluminio-Nuova Metall.* 37 (1967) 349.

11. N. Ryum; *Acta Met.*, 17 (1969) 269.

12. M. Sundberg, R. Sundberg and B. Jacobson; *Jernkont. Annlr,* 155 (1971) 1.

13. E. Di Russo; *Alluminio-Nuova Metall.* 33 (1964) 505.

14. M. Conserva, E. Di Russo and O. Caloni; *Metall. Trans.* 2 (1971) 1227.

15. B. Thundal and R. Sundberg; *J. Inst. Metals* 97 (1969) 160.

16. O. Izumi and D. Oelschlagel; *Z. Metallk.* 60 (1969) 845.

17. E. J. Lavernia and N. J. Grant; *Met. Powder Report* 41 (4) (1986) 255.

18. G. Rai, E. J. Lavernia and N. J. Grant; *J. of Metals* 37 (8) (1985) 22.

19. J. M. Galbraith, M. H. Tosten and P. R. Howell; *J. Mater. Sci.* 22 (1987) 27.

PROCESSING AND STRUCTURE OF A RAPIDLY SOLIDIFIED Mg-8.5 wt.%Li-1wt.%Ca-1wt.%Y ALLOY VIA LIQUID DYNAMIC COMPACTION

İbrahim Üçok, Kohei Kubota†,
Xiaoman Duan and Nicholas J. Grant
Department of Materials Science and Engineering
Massachusetts Institute of Technology, Cambridge, MA 02139
†*R&D Center, Mitsui Kinzoku Corp., Ageo, Saitama 362, Japan*

ABSTRACT. A magnesium- 8.5 wt.% lithium-1 wt.% calcium-1 wt.% yttrium alloy was rapidly solidified using the Liquid Dynamic Compaction technique. Microstructures and properties of the alloy were studied in the as-deposited state and after hot and cold rolling and annealing treatments. Different combinations of hot and cold work and annealing treatments were done to refine the microstructure. Room temperature hardness and tensile tests were performed on rolled and annealed specimens.

Introduction

Magnesium-lithium alloys are being considered for use in aerospace, and more recently for automotive applications because of their low density, high ductility and high specific elastic modulus (E/ρ) [1-7]. Low strength of Mg-Li binary alloys have been a drawback for their use in structural applications. Mg-Li alloys exhibit two phase structures between 5 and 11 wt.% Li contents consisting of the α (hcp) magnesium rich and β (bcc) lithium rich phases at room temperature. The α single phase structure exists for less than 5 wt.% Li and β single phase for greater than 11 wt.% Li contents. The α single phase alloys were shown to overage at temperatures between 339 and 394 K [4]. The β single phase alloys do not have attractive strength values, i.e. a yield strength of 96 MPa (14 ksi) for Alloy LA141A [4]. Two phase ($\alpha+\beta$) Mg-Li alloys were found to have better mechanical properties at room temperature and after intermediate temperature exposures. The better strength values of the two phase Mg-Li alloys were attributed to selective hardening by the α and β phases [4].

Previous reports for the two phase Mg-Li alloys cover ingot metallurgy, foil metallurgy, rapid solidification and thermomechanical processing for improving strength values. Mg-Li alloys produced by the ingot metallurgy route were either hot rolled plus cold rolled and annealed [4] or warm rolled [3], with heavy reductions in thickness to refine the grain size and to improve the mechanical properties. Foil metallurgy [1-3] processing involves cold rolling of the as-cast alloy to foils with low temperature recovery anneals and low temperature press bonding of foils. It was shown that the final grain size was inversely proportional with the thickness reduction during cold rolling via the foil metallurgy route. Single roll and twin roller quenching [5,6] were commonly used Rapid Solidification Processing (RSP) methods to improve the properties of Mg-Li alloys.

Meschter et al [5,8] showed that Mg_2Si dispersoids in Mg-8.4 wt.%Li-1 wt.%Si alloy flakes, produced by twin roller quenching, were effective in grain boundary pinning for annealing temperatures close to 573 K. In a review paper, Hehmann and Jones [6]

M. A. Otooni (ed.), Science and Technology of Rapid Solidification and Processing, 217–225.

reported that RSP, combined with alloying additions of Mn, Si, Ca and rare earths, improved the mechanical properties of Mg alloys in general. Higashi et al [3] demonstrated that Y additions to two phase Mg-Li alloys produced by ingot (I/M) and foil metallurgy (F/M) routes were effective in grain refinement and improved the yield strength. Despite the advantages of RSP in Mg-based alloys there are various drawbacks such as oxidation of rapidly solidified particles during handling and consolidation. The difficulties in handling very fine reactive magnesium powders is another major problem. In addition, RSPM and/or single or twin roll quenched products require lengthy and costly processing routes with relatively low efficiency.

Liquid dynamic compaction (LDC), on the other hand, eliminates several processing steps. LDC is known to provide preforms with essentially unsegregated, non-dendritic refined grain sizes and fine dispersoids. LDC products do not require a protective atmosphere during hot rolling and thermomechanical treatments.

The purpose of this study was to investigate the effects of rapid solidification and thermomechanical treatments on the microstructure, and the room temperature mechanical properties of a Mg-8.5wt.% Li-1wt.% Ca-1wt.%Y alloy processed by LDC. The additions of calcium and yttrium were made to improve the strength values of the Mg-8.5 Li alloy both at room and elevated temperatures. The Mg-8.5 Li-1 Ca-1 Y alloy is a two phase alloy with approximately 40% α and 60% β phases.

Experimental Procedure

The Mg-8.5Li-1Ca-1Y alloy, prepared as a direct-chill cast ingot by Mitsui Kinzoku Corp., Saitama, Japan, was processed by LDC and subsequent thermomechanical processing. The alloy was induction melted in a pre-evacuated and Ar-filled chamber. Atomization was done using argon gas and a circular ultrasonic gas atomizer. The spray was collected on a 1 cm thick copper, rectangular moving substrate. For a detailed description of the experimental parameters, see Table 1. Sections 7 mm by 7 mm and 50 mm long were machined out of LDC preforms.

Table 1. LDC parameters used in the processing of Mg-8.5 Li-1 Ca-1 Y alloy deposits.

Experimental parameters	
Pour tube diameter (mm)	5
Pouring temp., K	1013
Atomizing gas	Argon
Aspiration pressure, psi	0.49
Atomizing gas pressure (psig)	70
Nozzle-substrate distance (cm)	35
Substrate material (grit blasted, no cooling)	copper
Substrate speed (cm/sec)	0.679
Gas mass flow rate (kg/min)	0.864
Avg. melt flow rate (kg/min)	1.35
Gas/Melt mass flow ratio	0.64

Because of the small sizes of the alloys, the samples were sandwiched between two steel plates before hot rolling. Upon reaching the temperature, Mg-Li alloy sandwiches were subjected to hot rolling at 473, 498 and 523 K in two or three passes with total thickness reductions between 50 and 80%. By employing this method, the temperature of the alloy was kept constant during rolling. Cold rolling was done either in the as-deposited state or after an initial hot rolling (to reach full density) with thickness reductions between 50 and

80%. Intermediate anneals were made at 403 K to allow recovery to occur after each cold rolling pass, about 40% reduction in thickness. Annealing of cold rolled samples were done at temperatures between 403 and 493 K for 10 to 60 minutes. The purpose of hot and cold rolling and annealing was to close up the porosity and to refine the grain size. Table 2 summarizes the processing conditions. Density measurements were made on the as-

Table 2. Hot and cold rolling and annealing treatments given to LDC deposits.

Type	Thermomechanical Treatments
I	Hot rolling 70 and 78% at 473, 498 and 523 K.
II	Hot rolling 52 and 74% at 498 K plus cold rolling† 50 to 80% at 298 K plus annealing at 423 to 463 K.
III	Cold rolling† 90% at 298 K plus annealing between 423 and 493 K.

†Recovery anneals were made at 403 K for 30 minutes between each cold rolling pass (about 40% reduction in thickness).

deposited and rolled (densified) samples using Archimedes' principle. Microstructures were observed, using an optical microscope, in the as-deposited, hot and cold rolled and annealed conditions. Hardness measurements were made using the Rockwell-15T scale on cold rolled plus annealed samples. Tensile specimens of 14 mm gage length were machined from hot rolled, hot rolled+cold rolled, hot rolled+cold rolled+annealed, and cold rolled +annealed LDC deposits parallel to the rolling direction. Room temperature tensile tests were performed using an Instron universal testing machine with a cross-head speed of 2.1×10^{-2} mm/s.

Results and Discussion

In these early trials, optimum shapes were not spray deposited. The deposits showed a near Gaussian type thickness profile across the width. A cross sectional view of a piece cut from the Mg-8.5 Li-1 Ca-1 Y alloy deposit is shown in figure 1. Spraying direction was downwards and parallel to the paper plane. The peak thickness, which was along the center line of the deposit, was about 60 mm. The densities of the pieces cut from the different sections of the deposit were found to be in the range of 92 to 98% of the full density of 1.51 for deposit #1 and 87 to 95% for deposit #2, with the highest densities being at the center line of the deposit. As-deposited microstructures exhibit fine equiaxed $\alpha+\beta$ grains in the size range of 20 to 40 microns at the midthickness region of the deposit (about 30 mm away from the substrate), see figure 2. The matrix is the β phase, with a volume fraction about 60%, and the α phase (with a volume fraction of 40%) is dispersed in the β matrix in a randomly oriented fashion.

In each hot rolling pass about 50% reduction in thickness was achieved without any edge cracking. Samples were rolled at 498 and 523 K and subjected to 78% reduction, whereas at 473 K, the sample was subjected to 70% total reduction in thickness in two rolling passes. Density measurements and optical microscopy on hot rolled samples indicated that the deposits were fully densified after about 50% hot reduction in thickness. Figure 3 shows the microstructures of samples rolled at 473, 498 and 523 K with thickness reductions between 70 and 78%. Microstructures of the hot rolled samples were relatively similar, see Figure 3 a, b, c. Deformation of the α phase shows a tendency towards

Figure 1. Cross sectional view of a Mg-8.5Li-1Ca-1Y LDC spray-deposit: 9 cm wide by 15 cm long.

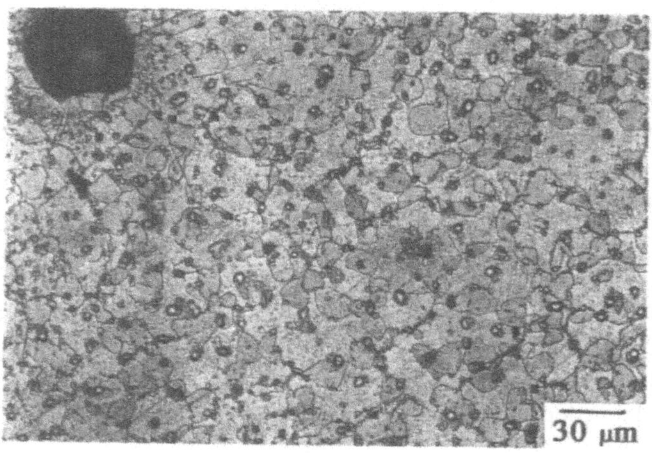

Figure 2. As-deposited microstructure of Mg-8.5Li-1Ca-1Y alloy. Dark region at the upper left corner is a pore.

banding within the β matrix upon hot rolling. The grain aspect ratio for the α phase was between 2:1 and 5:1. Hot rolled samples showed a linear intercept band spacing of 9 μm.

Some of the LDC samples were subjected to hot rolling reductions of 52 to 74% at 498 K prior to cold rolling (see Table 2, Type II) with reductions of 50 to 80%. Samples subjected to the Type II processing sequence exhibited finer structures than those of hot rolled deposits. Banding was also observed after both cold rolling and annealing steps in Type II processed LDC samples, see Figure 4. Average band spacing was 2.5 μm for 52% hot rolled plus 80% cold rolled samples.

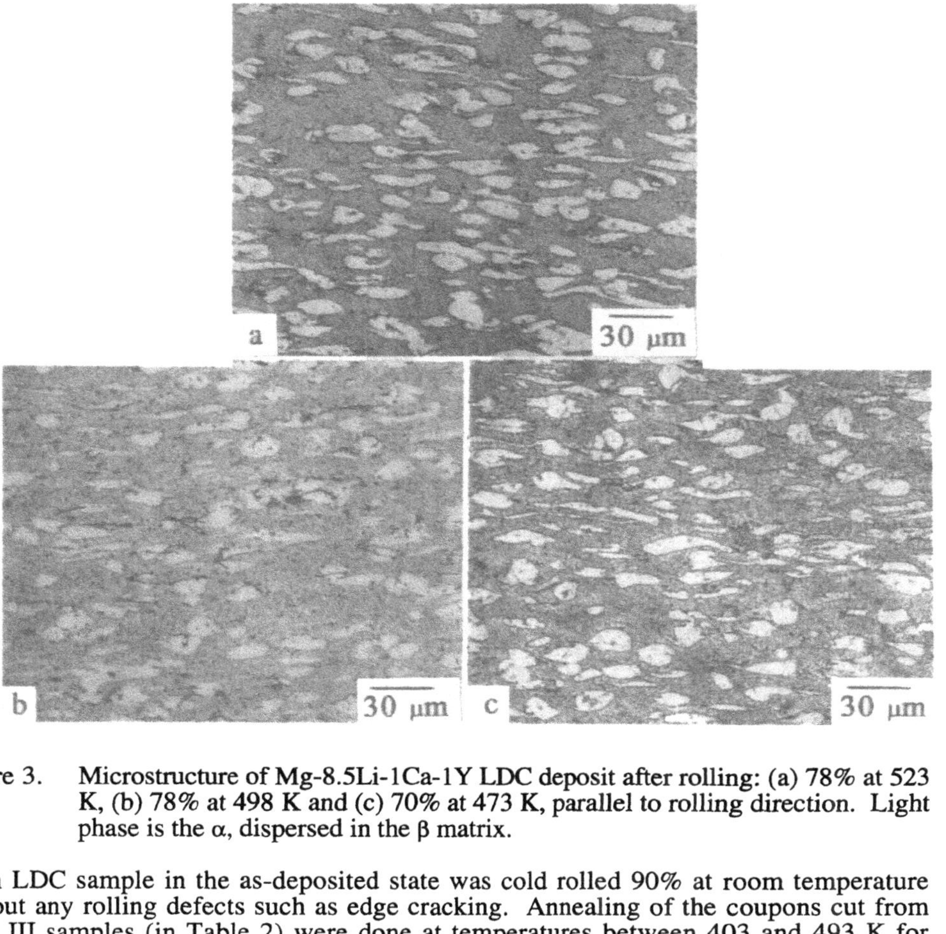

Figure 3. Microstructure of Mg-8.5Li-1Ca-1Y LDC deposit after rolling: (a) 78% at 523 K, (b) 78% at 498 K and (c) 70% at 473 K, parallel to rolling direction. Light phase is the α, dispersed in the β matrix.

An LDC sample in the as-deposited state was cold rolled 90% at room temperature without any rolling defects such as edge cracking. Annealing of the coupons cut from Type III samples (in Table 2) were done at temperatures between 403 and 493 K for different times up to 60 minutes. Hardness measurement data given in Figure 5 show the hardness change in the Type III samples during isothermal anneals. It should be noted that the hardness value of the LDC deposit was 72 on the Rockwell-15T scale after 90% cold rolling at 298 K. There was no significant hardness change observed for annealing temperatures of 403, 423 and 453 K indicating that recovery was mainly the cause for the hardness drop. Even at 473 K anneals, no microstructural evidence of recrystallization was seen. At 493 K anneals, samples showed a sharp decrease in hardness about 13 Rockwell-15T in the initial 15 minute period, indicating that some recrystallization occurred. Continued isothermal holding at 493 K resulted in a gradual hardness decrease of about 6 Rockwell-15T. Figure 6 shows the microstructures of 90% cold rolled and 493 K annealed samples. Measured band spacing was about 2 μm for 90% cold rolled samples. After annealing at 493 K for 15 minutes, the average band spacing was about 3 μm, Figure 6 (b). Further annealing at the same temperature for 60 minutes resulted in a band spacing of 4 μm, see Figure 6 (c). This indicates relatively high level of structural stability of the cold rolled plus annealed Mg-8.5Li-1Ca-1Y spray deposited samples.

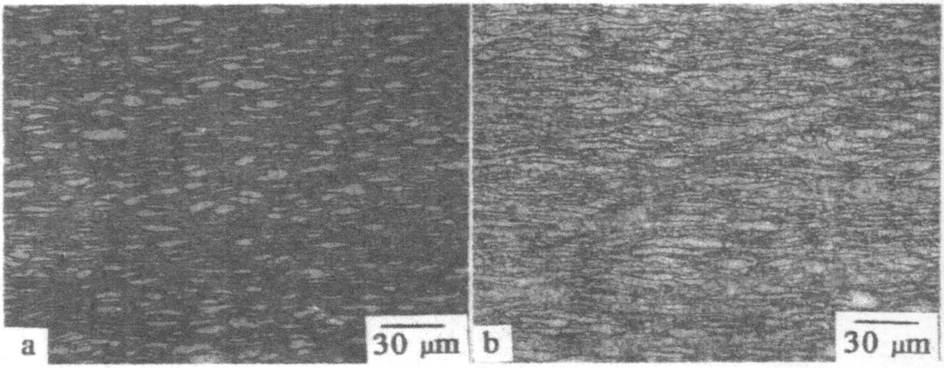

Figure 4. Microstructures of Mg-8.5Li-1Ca-1Y LDC sprayed deposits after (a) 52% rolling at 498 K plus 80% rolling at 298 K, transverse to rolling direction, (b) same processing as in (a) plus annealing at 423 K for 10 minutes, parallel to rolling direction.

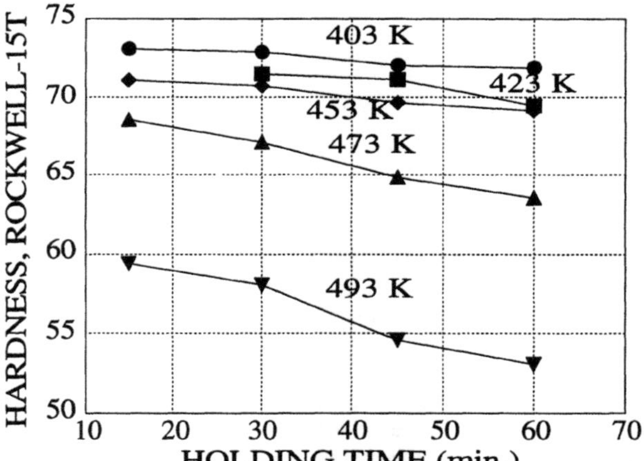

Figure 5. Hardness of Mg-8.5Li-1Ca-1Y LDC alloy after 90% cold rolling at 298 K and subsequent isothermal annealing. Type III samples, see Table 2.

Room temperature tensile tests were performed using specimens of the LDC Mg-8.5Li-1Ca-1Y alloy after thermomechanical treatments summarized in Table 2. Results of the tensile tests along with tensile data from the literature are given in Table 3. Strength values of the LDC Mg-8.5Li-1Ca-1Y alloy were superior to those of similar alloys processed by ingot and foil metallurgy techniques. The yield strength value for hot rolled LDC Mg-8.5Li-1Ca-1Y alloy of 152 MPa and elongation at a ductility level of 39% are higher than those for the annealed condition(YS=90 MPa, elongation= 64%),as well as the hot rolled ingot condition (YS=143 MPa, elongation= 35%). The highest strength values measured belong to a Type II sample which was hot rolled 52% at 473 K plus 80% cold rolled at 298

Table 3. Mechanical properties of LDC processed Mg-8.5Li-1Ca-1Y alloy along with properties of I/M and F/M alloys.

Processing	YS, 0.2% MPa	UTS MPa	e %
78% rolling at 523 K (1)	152	172	39
78% rolling at 498 K (1)	158	179	28
70% rolling at 473 K (1)	158	179	25
74% rolling at 473 K+50% rolling at 298 K+423 K/15 min. (1, 2)	172	193	30
74% rolling at 473 K+50% rolling at 298 K+463 K/10 min. (1, 2)	165	186	31
52% rolling at 473 K+80% rolling at 298 K (1, 2)	234	262	17
52% rolling at 473 K+80% rolling at 298 K+423 K/15 min. (1, 2)	207	234	18
52% rolling at 473 K+80% rolling at 298 K+463 K/10 min. (1, 2)	165	186	23
52% rolling at 473 K+73% rolling at 298 K+423 K/15 min. (1, 2)	186	214	23
52% rolling at 473 K+73% rolling at 298 K+463 K/10 min. (1, 2)	165	193	28
90% rolling at 298 K+423 K/30 min. (1, 2)	193	234	23
90% rolling at 298 K+453 K/15 min. (1, 2)	193	220	24
90% rolling at 298 K+473 K/15 min. (1, 2)	193	220	20
Mg-8.5Li ingot+623 K anneal+86%rolling at 473 K+623 K/30 min. (3)	86	121	65
Mg-8.5Li-1Y ingot+623 K anneal+86%rolling at 473 K+623 K/30 min. (3)	90	121	64
Mg-5.5Li ingot+623 K anneal+86%rolling at 473 K+623 K/30 min. (3)	70	132	52
Mg-5.5Li-1Y ingot+623 K anneal+86%rolling at 473 K+623 K/30 min. (3)	111	145	36
Mg-9Li ingot+170:1 rolling at 298 K+(3-10:1) press bonding at 453-503 K/30 min. (4)	133	-	22
Mg-10Li-1.8Al-1Ce ingot+hot rolled (5)	143	184	35
Mg-10Li-1.8Al-1Ce ingot+hot rolled+84% rolled at 298 K (5)	148	207	27

(1) Mg-8.5Li-1Ca-1Y alloy processed by LDC.
(2) A recovery anneal at 403 K for 30 minutes was made between each cold rolling pass (after about 40% reduction in thickness).
(3) see reference [3].
(4) Processed by foil metallurgy, see reference [1].
(5) see reference [4].

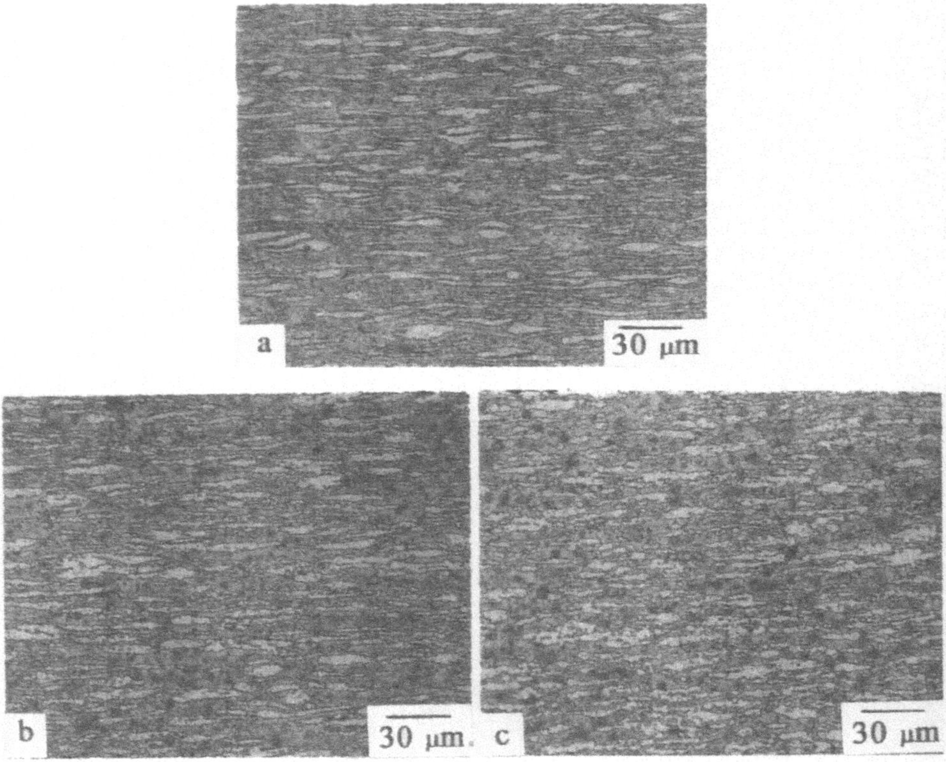

Figure 6. (a) Microstructure of Mg-8.5Li-1Ca-1Y LDC deposit after 90% cold rolling at room temperature, (b) plus annealing at 493 K for 15 minutes, (c) same processing as in (a) plus annealing at 493 K for 60 minutes. Note recrystallized structure with fine banding in rolling direction.

K (YS=234, UTS=262 MPa, e= 17%). It should be noted that the strength and elongation values for the last LDC sample were higher than the values for an I/M processed Mg-10Li-1.8Al-1Ce alloy (YS of 148, UTS of 207 MPa and E of 27%) in a similar condition. The strength values of the samples subjected to the Type II treatment decreased with a corresponding increase in ductility when the final annealing time or temperature was increased, indicating that the recovery and recrystallization processes were taking place. LDC samples subjected to cold rolling (in the as-deposited state) plus annealing treatments (Type III) also showed high strength values with very good ductility. The YS of 193 MPa and elongation of 20-24% were higher than those of a foil metallurgy processed ultrafine grained Mg-9 Li alloy (YS of 133 MPa, elongation of 22%). Higher strength values for the LDC processed alloy may be attributed to the refined structures and finer dispersoids of Y and Ca as a result of rapid solidification.

Conclusions

1. A Mg-8.5Li-1Ca-1Y alloy, rapidly solidified by LDC, showed improved strength values 152 (YS) MPa, 172 (UTS) MPa with 39% elongation after hot rolling at 523 K.
2. Cold rolling of the LDC Mg-8.5Li-1Ca-1Y alloy with/without an initial hot rolling step produced the highest strength values: 234 (YS) MPa, 262 (UTS) MPa with 17% elongation.
3. Annealing of cold rolled LDC Mg-8.5Li-1Ca-1Y alloy resulted in fine band spacings of 3 to 4 μm with a grain aspect ratio of 10:1. Strength values as high as 193 (YS) MPa, 220 (UTS) MPa with 20% elongation were achieved.

Acknowledgements

We are pleased to acknowledge the generous assistance of Mitsui Kinzoku Corp of Japan, and its staff for supplying the master alloys and support for this highly interesting study.

References

[1] G. Gonzalez-Doncel, J. Wolfenstine, P. Metenier, O.A. Ruano and O.D. Sherby, J. Mater. Sci., 25 (1990) 4535-4540.
[2] P. Metenier, G. Gonzalez-Doncel, O.A. Ruano, J. Wolfenstine and O.D. Sherby, Mater. Sci. and Engng., A125 (1990) 195-202.
[3] K. Higashi, K. Kubota and G. Neite, Paper presented at Symp. Int. Magnesium Assoc.,Germany, 1992.
[4] R.K. Wyss, in Advances in Magnesium Alloys and Composites, Proc. of Conf., 1988 TMS Annual Meeting, Phoenix, AR, TMS, Warrendale,PA, 1988, 25-40.
[5] P.J. Meschter and J.E. O'Neal, Met. Trans., 15A (1984) 237-40.
[6] F. Hehmann and H. Jones,"The Status and Potential of Rapid Solidification of Magnesium Alloys", Proc. MRS Symposium, Boston, MA, Dec. 1985.
[7] A. Joshi and R.E. Lewis, in Advances in Magnesium Alloys and Composites, Proc. of Conf., 1988 TMS Annual Meeting, Phoenix, AR, TMS, Warrendale,PA, 1988, 89-103.
[8] P. Meschter, Met. Trans., 18A (1987) 347-50.

Rapid Solidification Processing of Discontinuously-Reinforced Metal Matrix Composites

T. S. Srivatsan
Department of Mechanical Engineering
The University of Akron
Akron, Ohio 44325-3903, USA

T. S. Sudarshan
Materials Modification Inc.,
2929-P1, Eskridge Center
Fairfax, Virginia 22031, USA

E. J. Lavernia
Department of Mechanical and Aerospace Engineering
University of California
Irvine, California 92717, USA

Abstract

High strength, high stiffness, high performance and lightweight structural materials in the form of discontinuously reinforced metal- matrix composites provide attractive combinations of microstructure and mechanical properties through the use of rapid solidification technology. Hybrid materials based on metallic and intermetallic matrices have combined the standard alloys of metals with a wide variety of discontinuous reinforcements such as particulates, whiskers and chopped fibers of ceramic materials. In this paper, several of the rapid solidification techniques for the processing of discontinuously-reinforced metal matrix composites are highlighted and the salient features of each technique are presented and discussed. The variables involved with each processing technique are elucidated. Several spray processing techniques have engendered considerable technological interest, and a few key spray-based methods are examined and discussed. These include spray atomization and deposition processing, low pressure plasma deposition, modified gas welding and the high velocity oxyfuel thermal spraying technique.

M. A. Otooni (ed.), Science and Technology of Rapid Solidification and Processing, 227–248.

1. INTRODUCTION

The continuing need for lightweight and high-performance materials to satisfy the demands set by the aerospace and automotive industries in both structural and nonstructural applications have provided the necessary impetus for the development of many innovative combinations of materials. In recent years, continuous attempts have been made particularly in the areas of alloy design and the use of novel processing techniques to synthesize high performance hybrid materials or composites as attractive alternatives to the traditional and newer generation of unreinforced metallic alloys. A large majority of these composite materials are metallic matrices reinforced with high strength, high modulus and often brittle second-phases which can be either continuous in the form of fibers or discontinuous in the form of whisker, platelet, particulate and short-fiber reinforcements. Ceramic reinforcements in a metal matrix have been shown to offer improvements in elastic modulus, wear resistance, strength, structural efficiency, reliability and control of physical properties such as density and coefficient of thermal expansion, thereby, providing improved mechanical performance in comparison to the unreinforced matrix [1-4].

Use of discontinuous reinforcements is advantageous as the resultant composites can be synthesized by powder metallurgy (P/M), ingot metallurgy (I/M) and mechanical alloying (MA) methods. The properties derived from these methods are a function of the processing technique. Rapidly solidified powders offer the advantage of a reduced probability for deleterious reactions between the metal matrix and the reinforcement phase while simultaneously extending the range of potential matrix-reinforcement combinations. Several powder metallurgy-based processes have been used for the production of Discontinuously-Reinforced Metal Matrix Composites (DRMMCs) and traditional fabrication methods such as rolling, forging and extrusion can be used to deform the ingots or billets of these composites to produce a wide-range of product forms. While DRMMCs exhibit near isotropic behavior unless the reinforcement is aligned, continuously-reinforced matrices are difficult to fabricate and exhibit anisotropic behavior.

In automotive and aerospace applications DRMMCs based on particulate, whisker, nodule, platelet and chopped-fiber reinforcements have emerged as viable commercial materials [5-8]. These composites are preferred because they offer a number of advantages such as a 15-40% increase in strength and a 30-100% increase in stiffness compared to the unreinforced alloys [9-10], while generally maintaining receptiveness to processing and characterization techniques used for the conventional unreinforced counterparts. For example, an increase in modulus of greater than 100% has been reported for aluminum alloys reinforced with a volume fraction of 40% silicon carbide (SiC). From a design perspective, the attractiveness stems from an improvement in specific modulus, i.e., density compensated increase in elastic modulus. The moduli obtained are greater than those of typical titanium alloys and only marginally less than those of steels. Furthermore, the DRMMCs based on particulate reinforcements are attractive and valuable material systems because they can be made with properties that are isotropic in three orthogonal directions, or in a plane. The DRMMCs based on whisker-reinforcements offer the potential for enhanced properties but suffer from whisker breakage and damage during secondary fabrication [11]. DRMMCs are also not homogeneous and material properties are sensitive to properties of the continuum matrix phase, interfacial properties, secondary processing and geometrical constitution of the reinforcement.

In this paper, we consider the family of discontinuously reinforced metal-matrix composites [DRMMCs]. Depending on the nature of the reinforcing phase, DRMMCs can be categorized as: (a) dispersion strengthened, (b) particulate strengthened, or (c) short fiber-reinforced. Typical matrices are aluminum, titanium and magnesium, but copper, nickel-base superalloys and lead have also been considered for specific applications. Typical reinforcements can be in the form of continuous fibers like boron, silicon carbide, alumina and graphite or wires like tungsten or discontinuous fibers like alumina or graphite or even whiskers/particles. The reinforced metal matrices offer advantages in applications where lightweight, low density, high strength and high stiffness are of primary concern. The incorporation of high strength and high stiffness filaments or reinforcements into a ductile metal matrix can result in a composite material of relatively low density that exhibits good to excellent mechanical properties at ambient and elevated temperatures (Figure 1).

2. DISCONTINUOUSLY-REINFORCED RAPIDLY SOLIDIFIED COMPOSITES

In principle, the rapid solidification processing methods utilized to manufacture discontinuous particulate-reinforced MMCs can be grouped according to the temperature of the metal matrix during processing. Accordingly, the processes can be classified into categories of:
(1) Solid-phase processes.
(2) Liquid-phase processes.
(3) Two-phase (solid/liquid) processes.
The salient features of the solid-phase processes and the two-phase processes are presented and discussed.

2.1 SOLID PHASE PROCESSES

A: Powder Metallurgy

Solid phase processes involve the blending of rapidly solidified alloy matrix powders with particulates, platelets or whiskers through a sequence of steps summarized in Figure 2. The steps include:
(a) Sieving of the rapidly solidified particles.
(b) Precision blending of the whiskers or particulates with the powdered alloy matrix.
(c) Compressing the reinforcements and metal powder mixture to approximately 75 percent density.

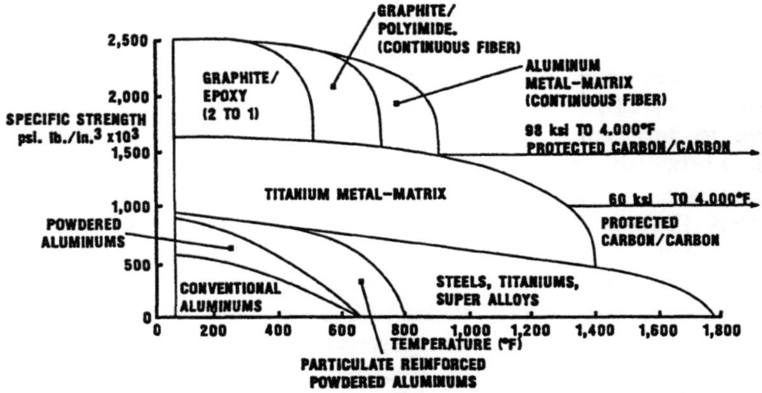

Figure 1. Elevated temperature performance of advanced materials.

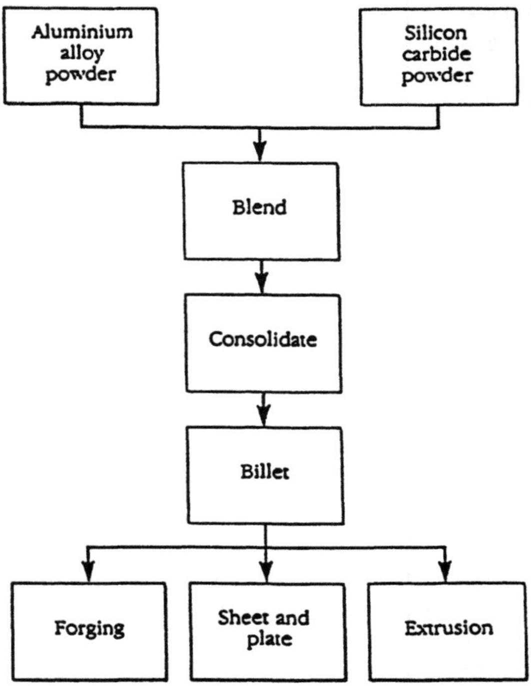

Figure 2. Flow chart showing the key fabrication steps for a powder metallurgy processed aluminum alloy metal-matrix composite.

(d) Metal powder-reinforcement mixture is loaded into a die, heated, subsequently degassed to remove absorbed and adsorbed gases and then consolidated to get a billet.
(e) Billet is cooled, homogenized and scalped prior to extrusion, forging, rolling or drawing.

Secondary fabrication of the billets can involve all normally applied metal working practices [12-14] and have been developed to various degrees of success by several commercial manufacturers, to include: Advanced Composite Materials Corporation (Greer, SC); Aluminum Company of America (ALCOA: Pittsburgh, PA); Ceracon Inc. (Sacramento, CA); and DWA Composite Specialties Inc. (Chatsworth, CA).

The most important step in the manufacturing sequence of DRMMCs involves a proper selection of the discontinuous reinforcement phase and the matrix alloy. The key criteria for the selection of a reinforcement phase include [15]:

- elastic modulus,
- tensile strength,
- density,
- melting temperature,
- thermal stability,
- compatibility with the matrix,
- thermal coefficient of expansion,
- size and shape, and
- cost.

The various possible discontinuous ceramic reinforcements and their properties are summarized in Table 1. For example, a spectrum of silicon carbide particulate sizes and shapes are available as reinforcements. Typically alpha and beta silicon carbide crystal structures have been examined as possible reinforcements [15]. A typical particle size distribution for the silicon carbide particulate is shown in Figure 3. Selection of an appropriate silicon carbide particulate (SiC_p) size is determined by the ratio of appropriate size of silicon carbide particulate to metal powder size [12,13].

In the classical powder metallurgy (PM) approach, the reinforcement material is blended with rapidly solidified metal alloy powders. The blend is then consolidated into billets by cold compaction, outgassed and hot isostatic or vacuum hot pressed to form starting products (billets) for subsequent fabrication. The cold compaction density is controlled in order to maintain an open, interconnecting porosity that facilitates degassing and/or outgassing during the pressing operation. The process of outgassing involves removal of the adsorbed gases, chemically combined water, and any other volatile species, through the synergistic action of heat, vacuum and inert gas flushing. For SiC_p-reinforced aluminum alloy-based MMCs, the outgassing helps in the removal of adsorbed water from both the carbide particle reinforcement phase and the aluminum alloy matrix, as well as chemically combined water from the matrix. PM techniques have been successfully applied to a large number of metal-ceramic combinations [14-18] and to age-hardenable alloys belonging to the 2XXX (2124) and 6XXX (6061, 6092) series. These alloys were prepared as elemental or pre-alloyed helium inert gas atomized powders. Other matrix compositions that have also been investigated include the: (i) 7090, 7091 alloys, (ii) high temperature Al-Fe-

Ce alloys, and (iii) lightweight Al-Li-Cu-Mg alloys. All these matrix alloys can utilize the full benefits of rapid solidification but require some modifications in both consolidation and processing procedures [15].

Table 1 Properties of selected ceramic reinforcements.

Ceramic	Coefficient of expansion (10-6/ °F)	Strength (MPa)	Elastic Modulus (GPa)
BeO	4.10	24 (2000°F)	190 (2000°F)
MgO	6.45	41 (2000°F)	150 (2000°F)
ThO_2	5.30	190 (2000°F)	200 (2000°F)
UO_2	5.30	-	175 (2000°F)
ZrO_2	6.67	85 (2000°F)	130 (2000°F)
CeO_2	6.90	600 (75°F)	200 (75°F)
Al_2O_3	4.40	220 (2000°F)	380 (2000°F)
$TaSi_2$	6.00	-	340 (2300°F)
$MoSi_2$	4.85	280 (2000°F)	280 (2300°F)
WSi_2	5.00	-	250 (2000°F)
TiB_2	4.60	-	410 (2000°F)
ZrB_2	4.50	-	503 (75°F)
TiC	4.22	55 (2000°F)	270 (75°F)
ZrC	3.70	90 (2000°F)	360 (75°F)
HfC	3.70	-	320 (75°F)
VC	3.98	-	430 (75°F)
NbC	3.80	-	340 (75°F)
Mo_2C	3.23	-	230 (75°F)
B_4C	2.83	2800 (75°F)	670 (75°F)
SiC	3.38	-	450 (75°F)
AlN	3.00	2100 (75°F)	325 (75°F)
TaC	3.59	-	360 (75°F)

In terms of final microstructural refinement, the PM approach is superior and preferred as the rapid solidification experienced by the powders permits the development of novel matrix materials that are outside the compositional limits dictated by equilibrium thermodynamics in conventional solidification processes [19].

B: High Energy Rate Processes

High energy high rate (HEHR) processing has been successfully utilized to consolidate rapidly quenched powders containing a fine distribution of ceramic particulates [20-21]. In this technique, consolidation of a metal-ceramic mixture is achieved using high energy and high rates from a homopolar generator (HPG). For example, at consolidation energy requirements of 800 kJ/kg, an existing 60 MJ energy source can easily consolidate a 20 kg billet. The homopolar generator converts the stored rotational kinetic energy into electrical energy using the Faraday effect. The HPG is a low voltage, high current device that is best operated in the pulse mode [22]. In this particular process, a degassed powder is subjected to a pulsed high-current discharge under low pressure. The current pulse is very high; reaching a maximum of 1 MA after 200 milliseconds and lasting for about 1-2 seconds. At the onset of the pulse, pressure is stepped-up to ensure electrical conductivity of the powder and to assist in powder consolidation by applying a mild forging. The bonding which occurs during consolidation stems from the pulse resistive heating which is produced at interparticle surfaces. The hot and rapidly applied thermal cycle facilitates consolidation without the occurrence of coarsening, when compared to the conventional consolidation techniques. The high energy-high rate consolidation process can produce DRMMCs with densities greater than 95 percent of the theoretical density. However, some local interparticle melting contributes to densification at the highest energy levels [21].

C: Dynamic Powder Compaction

Refined and sometimes metastable microstructures obtained by the use of rapid solidification methods can be lost during consolidation as prolonged thermal exposure results in coarsening of the finer microstructural features. Dynamic compaction avoids thermal excursions [23] and static compaction in the speed with which the compaction occurs, but also in the physical mechanisms taking place [24,25].

The process involves an unconventional, extremely rapid fabrication process with many direct and indirect sub-routes [26]. An intense shock wave is passed from one side to the other of the material to be compacted. Very high pressures are produced and the resultant high densities achieved facilitates little or no requirement for long range diffusion. In addition, the occurrence of a shearing effect between the particles results in an extensive heating of the surface, thereby, promoting the breaking of the interparticle oxide film on the powder surface and enhancing the effective bonding between the particles. This is quite analogous to explosive bonding or friction welding. As there is little need for diffusion, the dynamic compaction technique is useful for the compaction of thermally unstable materials [25].

In practice, the confined, precompacted powder encounters a high pressure shock wave which is generated by one of the following methods [27]:

- Explosives,
- Gas driven gun, and
- Electromagnetic force.

The pressure wave exceeds the yield strength of the powder mass, thereby densifying the powder as it proceeds through the compact [27]. Once consolidated to full density, conventional press equipment is employed for mill shape production. Direct compaction methods tend to differ from one another with respect to:

- the compact size and recovery method used, and
- the degree to which the amplitude and duration of the pressure pulse can be controlled.

Despite their promise, dynamic powder consolidation methods have two serious limitations [28]:

(i) Cracking may occur either during compaction or during the release of high pressure, due to the generation of high local stresses which are predominantly tensile in nature.

(ii) The nature of intrinsic microstructural features changes significantly as the shock compaction progresses through the powder.

3. Spray Processing

A variety of processing techniques have been developed that are viable alternatives to optimize both the microstructure and mechanical properties of DRMMCs. The technique of spray processing offers an opportunity to synergistically utilize the benefits associated with fine particulate technology, that is, refinements in intrinsic microstructural features, modifications in alloy chemistry and improved microstructural homogeneity coupled with in-situ processing and, in some cases, near-net shape manufacturing. A variety of spray-based methods include:

(a) Spray Atomization and Deposition processing [29-30].
(b) Low Pressure Plasma Deposition [31,32].
(c) Modified Gas Welding technique [33].
(d) High Velocity Oxyfuel thermal spraying [34].

All these techniques involve the mixing of discontinuous reinforcements with the matrix material under nonequilibrium conditions. Consequently, these processes offer:

(A) an opportunity of modifying and enhancing the properties of existing alloy systems, and
(B) developing novel alloy compositions.

3.1 Spray Atomization and Deposition Processing

The general principle of the spray atomization and deposition technique is to atomize a stream of molten metal using high velocity inert gases such as: argon or nitrogen, and to direct the resulting spray of molten metal onto a shaped collector which can be either a mold or substrate. On impact with the substrate, the particles flatten and weld together to form a hot, high density preform which can be readily forged to form a fully dense product. The surface of the substrate must be prepared in such a way that the first layer of atomized droplets will adhere to form a smooth quenched layer. Early studies by Singer were performed using conventional atomization equipment on aluminum alloys [35] and the atomized droplets were approximately 100-150 micron in diameter. The process parameters used in these studies were sufficient to atomize a melt stream and to cool the particles to temperatures near the solidus during a flight path of about 0.45m (18in.). A primary problem associated with this process of strip making was that the thickness of the as-deposited strip was not uniform across the width [36]. Several years later, Osprey Metals Ltd. (Neath, South Wales) successfully implemented the early ideas of Singer for the production of forging preforms.

3.1.1 THE OSPREY™ DEPOSITION PROCESS

The widespread interest and resultant development of the Osprey™ process can be attributed to several interactive factors, the most important being "flexibility". The Osprey™ process technique can be best described as incorporating the advantages of powder metallurgy without the disadvantages associated with consolidation. The primary advantages of this technique are [37,38]:

- the rapid formation of a range of semi-finished products such as tubes, billets or strips directly from the melt.
- use of an inert gas environment avoids oxidation of the droplets.
- creation of composite materials which can contain up to and over 20 volume percent of the reinforcement by injection of the particulates into the atomized stream of molten alloy.
- the processing of "difficult to work" conventional alloys and novel alloy compositions as a consequence of rapid solidification.
- large melts can be handled without nozzle erosion problems.
- the gas/metal ratio and gas pressure can be adjusted to suit new alloy compositions.
- potential for the economic manufacture of a spectrum of products as a direct consequence of the high rates of metal deposition that are available coupled with the concurrent elimination of several of the processing steps required in the ingot metallurgy (IM) and powder metallurgy (P/M) techniques.
- facilitates secondary processing routes without prolonged soaking times and/or the large deformation ratios required for fine grain-size, segregation-free microstructures.

The four key stages in the Osprey™ technique are:

- melting and dispensing,
- gas atomization,
- deposition, and
- collector manipulation.

The melting is undertaken by induction heating in a furnace which is directly linked with a tundish. The molten metal flows from the tundish into the gas atomizer. The melting and dispensing operation is normally carried out in a vacuum chamber (Figure 4).

Production of metal-matrix composites is accomplished by injecting the reinforcing ceramic particulates into the atomized stream of molten metal leading to co-deposition with the atomized metal onto the substrate. Careful control of both the atomizing conditions and the particulate feeding conditions is required and essential in order to ensure a homogeneous distribution of the reinforcing ceramic particulates in the microstructure. The primary attraction of this technique is the rapid production of semi-finished composite products in a combined atomizing, particulate injection

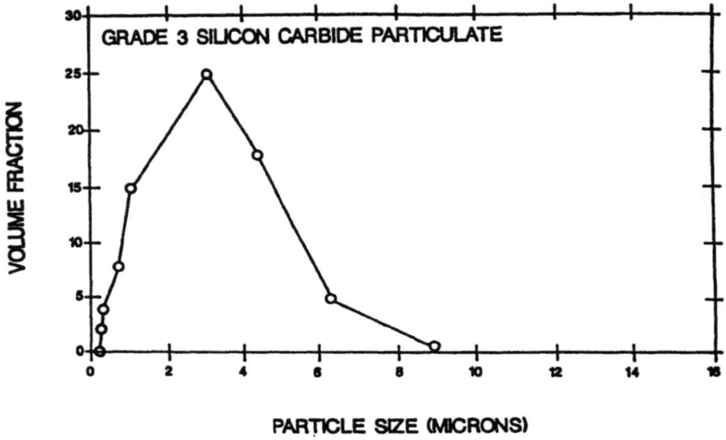

Figure 3. Distribution of silicon carbide particulate size (Ref. 12).

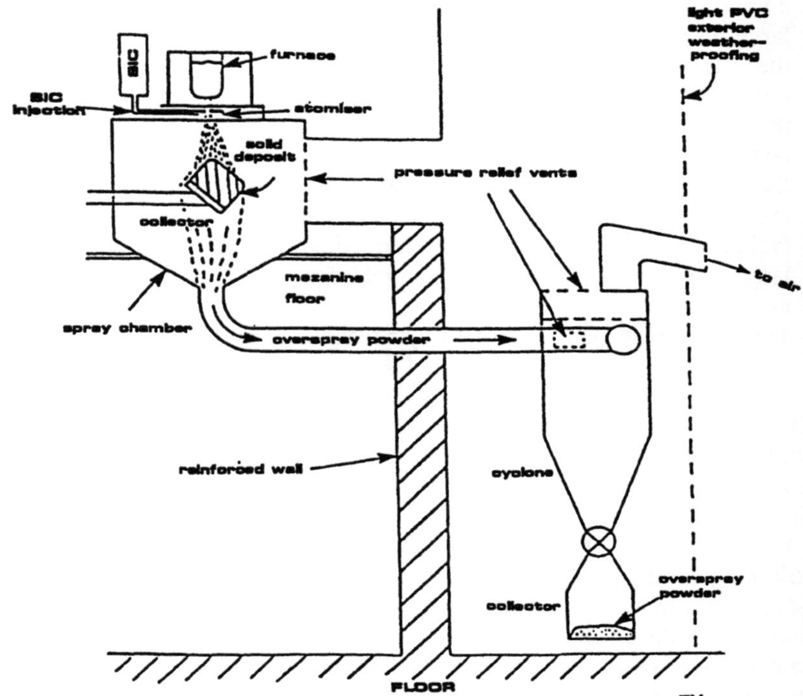

Figure 4. Schematic diagram of the modified Osprey^{TM} technique (Ref. 12)

and deposition operation. This method offers cost savings compared to other RSP techniques for the manufacture of MMC materials.

The spray atomization and deposition process inherently avoids the extreme thermal excursions, and the concomitant degradation of mechanical and interfacial properties and extensive macrosegregation, normally associated with conventional casting processes. Furthermore, this approach also eliminates the need to handle fine reactive particulates as is necessary with PM processes. A series of studies based on the early spray atomization and collection processes but using high pressure gas atomization for the spraying of fine, rapidly quenched droplets, in a controlled atmosphere, resulted in the development and emergence of several other spray deposition process that has been termed variable co-deposition of multiphase materials (VCM) [29,39]

3.1.2 Variable Co-deposition of Multiphase Materials (VCM).

In the Variable Co-deposition of Materials [VCM] process, the matrix material is disintegrated into a fine dispersion of droplets using high velocity inert gas jets (Figure 5 a). Simultaneously, one or more jets of the strengthening phase(s) are injected into the atomized spray at a prescribed spatial location (Figure 5 b). Interfacial control is achieved by injecting the reinforcing ceramic particulates at a spatial location where the atomized matrix spray contains a limited amount of the volume fraction of liquid. Hence, contact time and thermal exposure of the ceramic particulates with the partially solidified matrix is minimized and the interfacial reactions are closely controlled. In addition, tight control of the environment during processing minimizes oxidation and other environmental effects [29,39]. For situations where reactivity between the matrix material and the reinforcement is negligible, the reinforcing phases are introduced into the liquid alloy prior to spray deposition.

A modified spray atomization and deposition set-up incorporates a series of pressure transducers and thermocouples coupled with pneumatic control of the deposition surface. The pressure transducers are used to monitor and control:

- the gas atomization pressure,
- the chamber pressure, and
- the instantaneous position of the deposition surface.

The thermocouples aid in monitoring:

- the temperature of the melt, and
- the temperature of the nozzle and the deposition surface during the experiments.

During spray atomization and deposition, the matrix material is disintegrated into a dispersion of droplets, the average size denoted by d_{50}, using high-velocity inert gas jets. Concurrently, fine ceramic particulates are co-injected into the atomized spray at a previously determined spatial location (Z_i) and angle (Figure 6). The injection distance (Z_i) is determined on the basis of numerical analysis of the fraction of solid contained in the atomized matrix as a function of flight distance. This synthesis

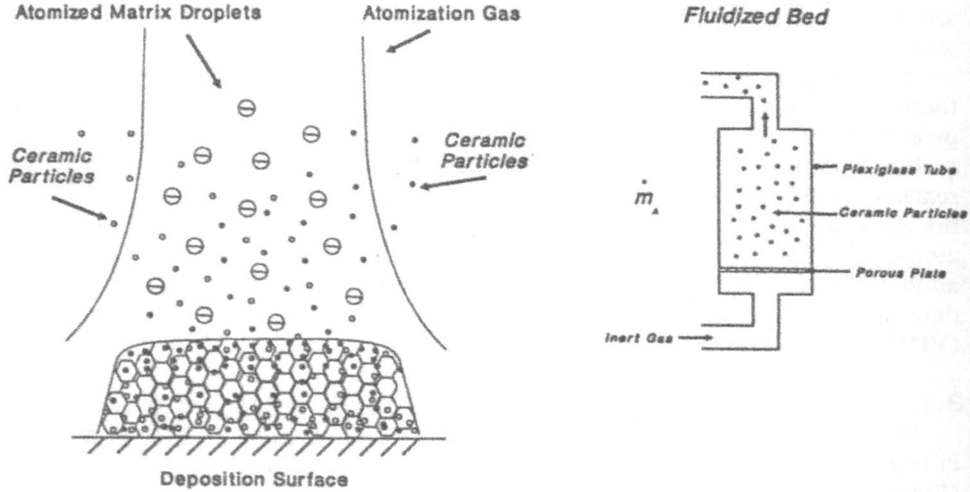

Figure 5 (a) Schematic diagram of the variable co-deposition technique (Ref. 29).

(b) Schematic of the fluidized bed used in the variable co-deposition technique (Ref. 29).

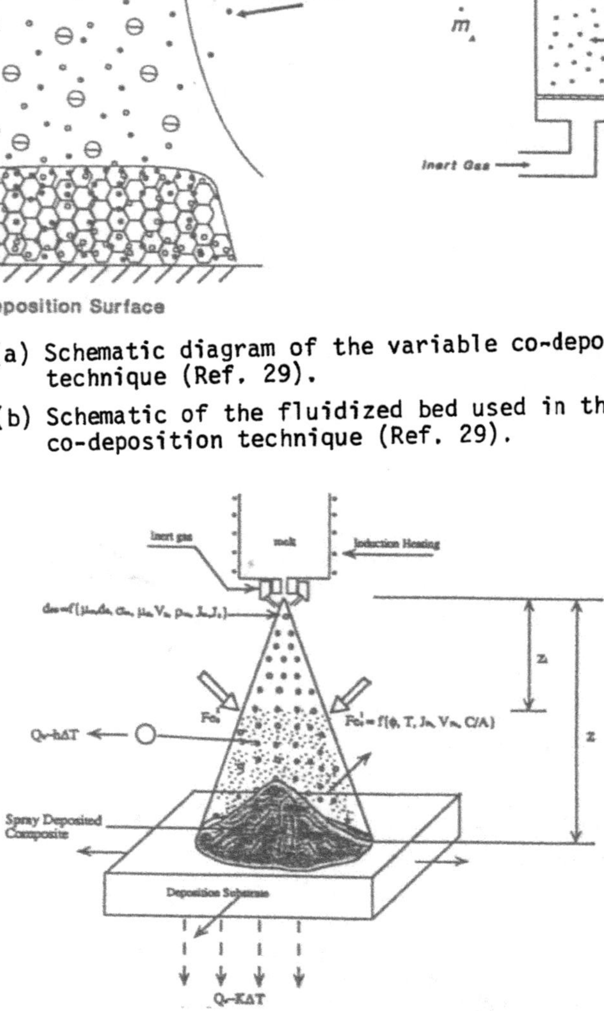

Figure 6. Schematic diagram showing the experimental arrangement for spray deposition processing of metal-matrix composites (Ref. 29, 39).

approach attempts to achieve interfacial control by injecting the reinforcing particulates at a spatial location where the atomized spray contains a limited amount of volume fraction of the liquid. In this manner, the contact time, thermal exposure of the reinforcing particulates to the partially solidified matrix, and interfacial reactions are minimized. In order to prevent extensive oxidation of the matrix during processing, the experiments are conducted in an environmental chamber.

Microstructural characteristics of the spray-deposited material depends to a great extent on the conditions of the droplets prior to their impact, that is, on:

(a) the relative proportions of liquid and solid phases present,
(b) temperature,
(c) velocity, and,
(d) size of the intrinsic microstructural feature in both the partially-solidified and fully-solidified droplets.

It is also important to know the size and distribution of the droplets, since the amount of heat dissipated by the droplets during flight is strongly dependent on their size. Furthermore, during the deposition stage, extensive fragmentation of the dendrite arms formed during solidification occurs as a result of the repeated impact of the partially-solidified droplets, first with the deposition surface, and subsequently, with each other. The presence of dendrite fragments, the development of strong convective currents in the liquid during impact coupled with the formation of a large number of solid nuclei have all been proposed as factors contributing to the development of an equiaxed grain morphology during spray atomization and deposition [40,41].

In summary, the spray atomization and deposition processes can be appropriately classified as rapid quenching processes since many of the microstructural features present in spray deposited materials are characteristic of rapid solidification technology. These include [29,41]:

- fine grained microstructures,
- metastable phase formation,
- increased solid solubilities,
- non-equilibrium phases, and
- an absence of macrosegregation.

However, the extent of microstructural control achievable with spray atomization and deposition processes is not as extensive as that available with other rapid quenching techniques. The elimination of particulate handling concurrently with particulate oxidation can result in significant microstructural benefits [41].

In view of the complex thermal, fluid and solidification phenomena involved during spray atomization and co-deposition it is highly improbable that the observed reduction in grain sizes can be attributed to a single mechanism. Most likely, the observed changes in microstructure can be attributed to the mutually interactive influences of several mechanisms. An approach that is utilized to provide an insight into the effects of the ceramic particulates in microstructural evolution during spray deposition, is to decouple the thermal, momentum and solidification phenomena into three distinct mechanisms: (i) solidification effects, (ii) thermal effects, and (iii) solid state cooling effects.

3.2. Low Pressure Plasma Deposition

Synergising the two independent operations of powder synthesis and its consolidation into a single step or operation led to the emergence of the technique of plasma processing. Around the same time, a new process was developed by the General Electric Company for net shape processing by plasma spraying in a reduced pressure environment. This technique was called Low Pressure Plasma Deposition [LPPD].

The LPPD combines particle melting, quenching and consolidation in a single operation. The process involves an injection of a variety of powder particles ranging from metallic to ceramic, and including mixtures, into a plasma jet stream created by heating an inert gas in an electric-arc confined within a water-cooled nozzle. The particles injected into the plasma jet, at temperatures of 10,000 K and higher, undergo rapid melting and are accelerated towards the workpiece. Rapid quenching of the molten particles occurs when the droplets impact the substrate. Cooling rates are very high, of the order of 10^5-10^6 K/sec, and the resultant microstructure is fine grained and homogeneous. Compared to other rapid solidification processing approaches, plasma spraying offers the advantages of: (i) large throughput (Kg/hr.), (ii) high density, (iii) an ability to deposit objects of complex shape, and (iv) to even produce near-net shape bulk forms.

The technique of conventional plasma spraying (CPS) is normally carried out at atmospheric pressure. The resultant deposit contains oxidation products together with some porosity due to the conjoint influence of incomplete melting and wetting or fusing together of the deposited particles. The problem of oxidation is overcome or minimized by either shielding the plasma arc in an inert gas atmosphere, or by enclosing the entire plasma spraying unit in an evacuated chamber, which is maintained at about 30-60 torr inert gas pressure by high speed pumping. Under the latter operating conditions of LPPD, gas velocities are high (typically in the Mach 2-3 range) due to the higher permissible pressure ratios. When the inert gas or environment is vacuum, the technique is referred to as vacuum plasma spraying [VPS] [31,32]. Other noteworthy advantages of low pressure inert gas deposition or Vacuum Plasma Spraying (VPS), compared with Conventional Plasma Spraying (CPS), include:

- Higher particle velocities, which give rise to denser deposits (often > 98% of theoretical density).
- Broad spray patterns, which produce large areas of relatively uniform deposition characteristics.
- Automatic regulation of the technique to produce controlled deposits of complex geometries at reasonably high deposition rates (up to 50 kg/hr.).

The four basic components of a typical plasma gun, namely: (a) cathode, (b) arc chamber, (c) throat, and (d) exit nozzle, are shown in Figure 7 [30]. The cathode and anode are located concentrically and are water cooled to prevent melting by the high gas temperatures. An arc gas (argon or a mixture of argon and helium) is fed from behind the cathode. The arc gas is fed tangentially, creating a vortex that acts to stabilize the electric arc. Powder injection ports are located either within the nozzle (anode) or downstream of the nozzle, depending primarily on [30]:

(a) the characteristics of the plasma gun, and
(b) the type of powder being sprayed.

Protection against environmental interactions after particle melting has occurred is achieved by injecting a shroud of inert gas around the periphery of the jet and/or around the substrate. The reduced presence of the inert gas environment ensures that the particles and droplets are protected from oxidation. By minimizing environmental interactions, high quality metallurgically sound deposits, which are well bonded to the substrate, are achieved by the low-pressure inert gas spraying technique. The resultant deposits typically have fine microstructures and contain varying amounts of retained porosity and inclusions, depending on deposition parameters, such as:

- power level,
- arc gas composition,
- jet transport properties, and
- level of reduced pressure.

The protected atmosphere minimizes both the number of oxide-containing inclusions and the number density of pores. The resultant deposit contains fewer inclusions and the density achieved under conditions of low-pressure plasma spraying is about 98% of the theoretical density.

3.3 Modified Gas Welding Techniques

In this technique the gas metal arc (GMA) welding torch is modified, wherein, aluminum or aluminum alloy wire feed stock is melted and combined with either silicon carbide particulates (SiC_p) or silicon carbide whiskers (SiC_w) entrained in an inert gas. Upon striking a substrate or mold, the mixture of aluminum alloy and carbide particle reinforcing phase solidifies into a composite structure [33]. A schematic of the deposition process is shown in Figure 8.

An arc between the end of the aluminum wire and a water cooled copper cathode produces a stream of droplets about 1 mm in diameter. At about 230 Amps the melting rates are on the order of 2 kg/hour. The thermal history of these droplets are controlled to some extent by:

- the electrical parameters of the melting process,
- the shielding gas used, and
- the distance from the orifice to the mold or substrate.

Thermal control of the solidification is affected by both the thermal properties of the mold and the substrate. At any instant of time very little liquid is present, adequate enough to bond successive droplets and any reinforcement present. In this technique, the reinforcement material is not substantially affected by fluid flow and macrosegregation during solidification. Using appropriate molds and substrates, complex parts and near-net shapes of a uniform composite structure can be produced. The process could be developed into a welding technique for MMCs that would provide the weld metal of approximately the same composition as the base material, thereby, reducing the need for mechanical fasteners and adhesives in the fabrication of large structures. Several advantages of this technique include:

- An ability to melt, by the arc, any metal, including the very high temperature alloys.
- An ability to entrain any reinforcement in the shielding gas, thus providing a wide variety of reinforced metallic matrices or composite materials.
- Melting is more convenient than conventional melting methods, since it can be turned on and off with relative ease.
- An ability to produce large droplets, thereby, minimizing explosion hazards associated with finely divided aluminum powder, and,
- The modified gas metal arc torch is compact, inexpensive and controllable.

For the process to be successful, proper shielding of the droplet stream and solidifying metal are both important and essential in order to produce a good surface finish while keeping internal porosity at a low level. With a high-quality matrix coupled with an ability to add variable amounts and kinds of reinforcement phases, this processing technique facilitates microstructural control of MMCs.

3.4 High Velocity Oxyfuel Spraying

The HVOF thermal spraying process uses an internal combustion (rocket) jet to generate hypersonic gas velocities of the order of 1,830 m/sec [6000 ft/sec] - more than five times the speed of sound. Some of the combustion fuels used included propylene, acetylene, propane and hydrogen gases for spraying carbide and noncarbide coating materials. When burned in an atmosphere or in conjunction with pure oxygen, these fuels produce gas temperatures greater than 2760°C [5000°F]. A schematic of the high velocity oxy-fuel gun is shown in Figure 9.

The combustion ignition, gas control and powder feed are fundamentally simple in the HVOF spraying system. A pilot flame, typically operating on hydrogen and oxygen, is ignited manually, and flow rates of two main jet gases are controlled by a flow meter. Electrically operated solenoids activate the main combustion jet. The flow of powder is electrically controlled and feed rates are monitored automatically. The powders deposited using HVOF thermal spraying include pure metals, alloys, carbides, certain ceramics and even plastics. In fact, early use of the HVOF thermal spraying technology was for carbide coatings on aircraft gas turbine engine components, which experienced wear from abrasion, adhesion, erosion, fretting and corrosion. The key parameters influencing wear resistance of carbide coatings are flame temperature and particle velocity. An additional advantage of this spraying technique is that the tungsten-carbide coatings produced are better than those obtained using plasma synthesis in terms of:

- higher hardness,
- higher coating bond strength,
- lower oxide content and porosity, and
- improved wear resistance combined with low residual stress in the coating.

Furthermore, the microstructures resulting from HVOF spraying are equal to, or better than, those of the highest quality plasma sprayed coatings.

The most significant and noteworthy advantages of the HVOF process over conventional plasma spraying (CPS) are:

- higher bond strength of the coating,
- lower oxide content, and
- improved wear resistance (predicted based on the higher hardness of the HVOF-sprayed coating).

HVOF spraying also has a deposit efficiency of 75 percent compared with 45 percent for plasma spraying. Besides, HVOF spraying has only one-half as many spraying parameters to control compared with CPS. Thus, the process is simpler while concurrently enhancing coating reproducibility [34].

4. Summary

There exists a number of techniques using the principle of rapid solidification for the processing of discontinuously reinforced metal-matrix composites. The results of studies, documented by researchers, from both academia and industry, have enriched our understanding of the importance of the interrelationships between processing-microstructural characteristics and macroscopic behavior. The primary limitations that forces constraints on the rapid advancement of the field of processing lies in the development of techniques that promote wettability between the metal matrix and ceramic reinforcement coupled with high speed compaction techniques that would result in minimum exposure of the composite to elevated temperatures.

A description of the various processing techniques used for discontinuously-reinforced metal matrix composites demonstrates the feasibility and applicability of spray atomization and deposition as a viable route for the production of large quantities of metal-matrix composites, since it circumvents many of the difficulties associated with particulate handling. Using the spray deposition technique a spectrum of MMCs can be produced. The technique offers advantages over the competitive techniques in terms of microstructural refinement, property enhancement and overall efficiency of the process.

A rapid utilization of discontinuously-reinforced metal-matrix composite materials produced using rapid solidification techniques is dictated by the competing influences of:

(a) economics of production of the final product, and
(b) a lack of reproducibility due to limited understanding of the wettability characteristics of the reinforcement by the matrix, and the metal-reinforcement (ceramic) interfaces.

As processing advances occur, it is expected that the reliability will be enhanced in certain families of composites. Continuous efforts at modelling several of the composite material systems coupled with innovations in the development of intelligent manufacturing systems and embedded sensors that can provide valuable information on the solidification process are also expected to be areas that will rapidly accelerate the economics of rapid solidification processing of composites. Research on the processing of these newer generation materials is still at an embryonic stage and concurrent improvements in both the processing and manufacturing of these materials is bound to have a significant impact on this technology in the years to come.

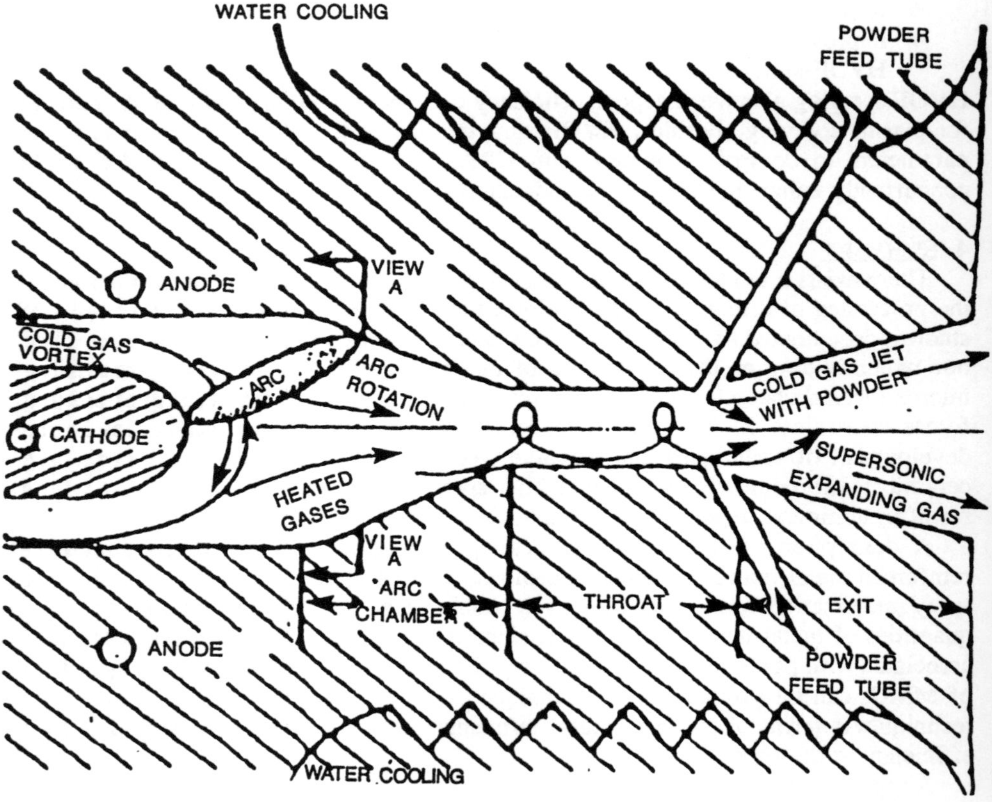

Figure 7. Cross-sectional schematic of the plasma gun showing the supersonic expanding gas jet (Ref. 30).

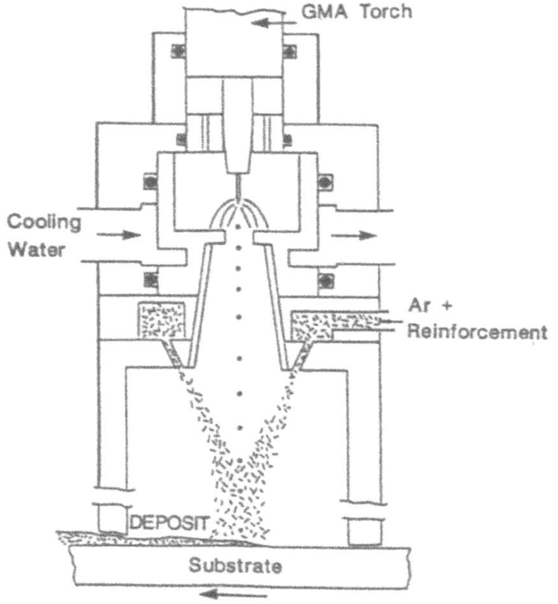

Figure 8. Schematic drawing of the modified gas metal arc welding deposition process (Ref. 33).

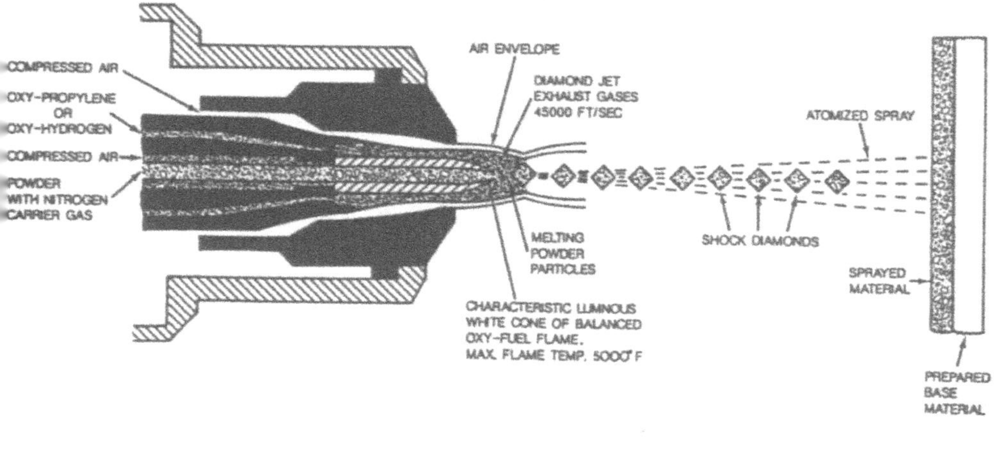

Figure 9. Schematic showing cross-section of the high velocity oxyfuel gun (Ref. 34).

5. REFERENCES

1. A.P. Divecha, S.G. Fishman and S.D. Karmarkar: **Journal of Metals**, Vol. 33, 1981, pp. 12-16.
2. S.V. Nair, J.K. Tien and R.C. Bates: **International Metals Reviews**, Vol. 30, No. 6, 1985, pp. 285-296.
3. M.Taya and R.J. Arsenault: in **Metal Matrix Composites: Thermomechanical Behavior**, Pergamon Press, Elmsford, New York, 1989.
4. T.S. Srivatsan and T.S. Sudarshan: in Rapid Solidification Technology: An Engineering Guide, Technomic Publishing Inc., PA, 1993, pp. 603-700.
5. S.G. Fishman: Role of Interfaces in Material Damping, ASM International, Materials Park, Ohio 1985, pp. 33-41
6. P.K. Liaw, H.G. Greggi and W.A. Logsdon: **Journal of Materials Science Letters**, Vol. 22, No. 5, 1987, pp. 937-940.
7. D.H. Grande, J.F. Mandell and K.C. Hong: **Journal of Materials Science**, Vol. 23, 1988, pp. 311-318.
8. D.F. Hasson and C.R. Crowe: in Strength of Metals and Alloys. ICSMA: 7, Oxford Press, United Kingdom,. Pergamon Press, 1985, pp. 1515-1520.
9. W.R. East: **Materials Engineering**, March 1988, pp. 33-36.
10. T.C. Willis: **Metals and Materials**, August 1988, pp. 485-489.
11. W.H. Hunt, Jr.: "Cost Effective High Performance Aluminum Matrix Composites for Aerospace Applications," presented at the International Conference on PM Aerospace Materials, November 1991, Lausanne, Switzerland.
12. H. J. Rack: in Advanced Materials and Manufacturing Processes, Vol. 3 (No. 3), 1988, pp. 327-340.
13. H.J. Rack and P. Ratnaparkhi: **Journal of Metals**, November 1988, p. 55-57.
14. H.J. Rack and P.W. Niskanen: **Light Metals Age**, February 1984, p. 9.
15. A.P. Divecha, S.G. Fishman and S.D. Karmarkar: **Journal of Metals**, 33(9), 1981, p. 12
16. W. H. Hunt, Jr., C. R. Cook, K. P. Armanie and T. B. Garganus: in Powder Metallurgy Composites (editors: P. Kumar, A. Ritter and K. Vedula), Metallurgical Society of AIME, Warrendale, PA, 1987.
17. H. J. Rack: in Processing and Properties of Powder Metallurgy Composites (editors: P. Kumar, K. Vedula and A.M. Ritter), TMS: The Minerals, Metals and Materials Society, Warrendale, PA, 1988, p. 155.
18. H. J. Rack: in Proceedings of the Sixth International Conference on Composite Materials (editors: F. L. Matthews, N.C.R. Buskell, J. M. Hodgkinson and J. Morton) Elsevier Applied Science Publishers, London, 1987, p. 2382.
19. H. J. Rack, P. Hood, P. Niskanen and J.L. Cook: in Proceedings of Conference on Discontinuously Reinforced Metal Matrix Composites Working Group, Santa Barbara, California, 1983.
20. H. L. Marcus, D. L. Bournell, Z. Eliezer, C. Prasad and W. F. Weldon:

Journal of Metals, Vol. 12, 1987.
21. G. Elkabir, L.K. Rakenberg, C.V. Prasad and H. L. Marcus: **Scripta Metallurgica**, Vol. 20, No. 10, 1986, p. 1411.
22. J. B. Walters and T. A. Aanstoos: **Metal Progress**, Vol. 127, 1985, p. 25.
23. W. S. Cebulak, E.W. Johnson and H.L. Marcus: **Materials Engineering Quarterly**, Vol. 16(4), 1976, p. 37.
24. D. G. Morris: Proceedings of Fourth Conference on Rapidly Quenched Metals, Vol. 1 (editors: T. Masumoto and K. Suzuki), Japan Institute of Metals, Sendai, Japan, 1982, p. 145.
25. D. G. Morris: Second International Conference on Rapid Solidification Processing (editors: R. Mehrabian, B.H. Kear and M. Cohen) Claitors Publishing Division, Baton Rouge, LA, USA, 1986, p. 1-23
26. C. R. A. Lennon A.K. Bhalla and J.D. Williams: **Powder Metallurgy**, Vol. 1, 1978, p. 19.
27. D. Raybould: **Journal of Materials Science**, Vol. 16, 1981, p. 589
28. M. L. Wilkins: "Dynamic Compaction", the Eight International Conference on High Energy Rate Fabrication (editors: I. Berman and J. W. Schroeder), San Antonio, Texas, 1984, p. 63.
29. M. Gupta, F. A. Mohamed and E. J. Lavernia: **Materials and Manufacturing Processes**, Vol. 5, No. 2, 1990, p. 165.
30. D. Apelian, D. Wei and B. Farouk: **Metallurgical Transactions** B, Vol. 20, 1989, pp. 251-262.
31. R. Tiwari, H. Herman, S. Sampath and B. Gudmundsson: Plasma Spray Consolidation of High Temperature Composites, Proceedings of the Symposium on Innovative Inorganic Composites (edited: S.G. Fishman), TMS, Warrendale, PA, 1991.
32. S. Sampath, R. Tiwari, B. Gudmundsson and H. Herman: "Microstructure and Properties of Plasma Sprayed Consolidated Two-Phase Nickel Aluminides, **Scripta Metallurgica**, Vol. 25, 1991, p.1425.
33. C. L. Buhrmaster, D.E. Clark and H.B. Smartt: **Journal of Metals**, Nov. 1988, p. 44 .
34. D.W. Parker and G.L. Kutner: **Advanced Materials and Processes**, April 1991, p. 68.
35. A.R.E. Singer: **Journal of the Institute of Metals**, Vol. 100, 1972, p. 185.
36. R.G. Brooks, A.G. Leatham, J.S. Coombs and C. Moore: **Metallurgy and Metal Forming**, Vol. 9, 1977, p. 1.
37. Leatham, R.G. Brooks: in Modern Developments in Powder Metallurgy, Vol. 15, 1984, pp. 157-173.
38. A.R.E. Singer and R.W. Evans: **Metals Technology**, Vol. 10, 1983, pp. 61-68.
39. X. Liang and E.J. Lavernia: **Materials Science and Engineering**, A153, 1992, pp. 646-653.

40. S. Annavarapu, D. Apelian and A. Lawley: **Metallurgical Transactions**, Vol. 19A, 1988, pp. 3077-3086.
41. M. Gupta, F.A. Mohamed and E.J. Lavernia: **International Journal of Rapid solidification**, Vol. 6, 1991, pp. 247-284.

PROCESSING AND PROPERTIES OF SPRAY FORMED 2XXX ALUMINUM ALLOYS

ALI I. KAHVECI
Alcoa Technical Center
Fabricating Technology Division
Alcoa Center, PA 15069

Introduction

Spray forming (S/F) process has been commercially used to fabricate large diameter tubes and roll preforms.[1] Monolithic and metal matrix composite extrusion and forging preforms have also been produced to develop an alternative preform fabrication process to P/M- or I/M-based processes.[2,3] Furthermore, reduction in the number of manufacturing steps and improvement in microstructure along with extended composition range; the spray forming process has emerged as a serious alternative primary process to I/M process as well as continuous casting processes.

Spray forming techniques such as the Osprey and Ultrasonic Gas Atomization/Liquid Dynamic Compaction (USGA/LDC) have been developed to produce homogeneous, fine grained and macrosegregation free microstructures directly from the melt.[3,4] The USGA/LDC technique differs from the Osprey technique only in the atomization area. Atomization of the liquid in the Osprey process is based on conventional gas atomization; a steady stream of high velocity gas disintegrates the liquid stream into droplets. In the USGA/LDC process, however, the liquid material is atomized by the impact of high velocity gas pulses generated by a train of shock waves with ultrasonic frequency. This atomization process is known as USGA and can be considered a modification of the conventional gas atomization process by using the USGA nozzle system. This technique has been claimed to provide higher solidification rates,[5-7] greater fraction of fines and more homogeneous size distribution than conventional gas atomization.[8] Others, however, reported essentially no difference in powder quality between conventionally and ultrasonically superimposed inert gas atomization.[9,10]

This work evaluates the microstructure and mechanical properties of 2XXX series aluminum alloys spray formed by the Osprey and LDC processes. The purpose of this study was to explore possible primary process, i.e., Osprey vs. USGA/LDC, related differences on the properties of spray formed materials. Some of the results were compared to those of the I/M controls.

M. A. Otooni (ed.), Science and Technology of Rapid Solidification and Processing, 249–269.

Experimental Details

ALLOYS AND PROCESSING

Nominal compositions of the alloys are given in Table 1. The 2618 and the N202 alloys were spray formed to cylindrical billets by Cospray Products Division of Alcan and Alusuisse-Lonza, respectively, using the Osprey process. Nitrogen gas was used for atomization. The atomized droplets were captured on an inclined rotating substrate or deposit surface.[11] As-received Osprey billets were 9-11 in. in diameter and 15-29.5 in. in length.

The other 2618 alloy was spray formed at Massachusetts Institute of Technology (MIT) using the USGA/LDC process with a circular nozzle. The alloy was first induction melted to 800°C in a positive argon atmosphere pressure in atomization chamber and then atomized using argon gas. Atomized droplets were deposited on a water cooled moving mild steel substrate. After atomization, edges of the as-sprayed plate were trimmed and the remaining material was used for microstructure characterization and further processing. The as-received USGA/LDC 2618 alloy plate was approximately 0.190 x 0.05 x 0.025 m^3 (7.48 x 1.96 x 0.98 in^3) in size. The I/M 2618 control billet was supplied by Alcoa.

Table 1. Chemical composition of 2618 and N202 alloys (in wt.%). The composition of the N202 alloy was taken from Reference 12.

Alloy	Cu	Mg	Fe	Ni	Si	ΣTi,Mn,Zn,Cr,Zr	Al
2618 (LDC)	2.28	1.55	1.09	1.09	0.17	0.12	bal.
2618 (Osprey)	2.32	1.45	1.17	1.02	0.19	0.18	bal.
I/M 2618	2.19	1.26	0.99	1.05	0.1	0.12	bal.

Alloy	Cu	Mn	Mg	Zr	V	Ti	Ag	Al
N202	6.3	1.8	0.4	0.5	0.25	0.25	0.4	bal.

The as-received 2618 billets (Osprey and I/M) were machined to sizes of 5-6 in. in diameter and 12 in. in length, and then extruded to 1 x 2 in^2 rectangular cross section bars at 725°F with an extrusion ratio of about 14/1 (Osprey 2618) and 93/1 (I/M 2618). The extrusion ram speed was about 2-3 ipm for the alloys.

The LDC 2618 and N202 alloys were machined to 0.59 (thickness) x 1.25 (width) x 7.5 (length) in^3 bar and 4 in. diameter billet, respectively, and then upset by 74%-87% at a ram speed of 1-3.5 ipm and at a temperature regime of 810°-850°F.

Γhe alloys were heat-treated to various T6 conditions: The 2618 alloys were homogenized ıt 980°F for 4 hr and then quenched to 140°-160°F water. Samples were artificial aged at 390°F for 20 hr for fatigue crack growth (FCG) tests and up to 40 hr for the room emperature tensile and fracture toughness tests. Heat treatment of the N202 alloy includes ıomogenization at 950°F for 1 hr followed by 130°-160°F water quenching, and then aging ıt 374°F for 5 hr.

Mechanical properties, tensile, fracture toughness, fatigue and fatigue crack growth rate, of he alloys were determined at room temperature. Two tests were conducted for each test ondition. The densities of the alloys were measured by Archimet's technique and the orosity of the deposits were determined metallographically. The microstructures of the amples were studied by Guinier x-ray diffraction, optical and electron microscopy echniques.

Results and Discussion

nitial microstructure of the as-received alloys were evaluated to contrast the microstructural lifferences between I/M casting and spray forming processes. Fig. 1 shows the typical nicrostructures in the as-received alloys. The as-cast I/M alloy exhibited a coarse dendritic tructure with a grain size of 200-440 µm. The constituent particles were ıonhomogeneously distributed to dendrite interstices and dendrite boundaries. They had lual morphology: nearly spherical with a diameter of 1-3 µm and nonspherical or longated with a size of 4 x 15 µm.

A unique feature of the spray formed alloys is the fine equiaxed microstructure with a ıniform distribution of constituent particles as shown in Fig. 1b-d. The grain size ranged etween 2 and 60 µm (Table 2). The average grain size was the finest (9 µm) for the N202 lloy followed by the LDC 2618 (14 µm) and Osprey 2618 (25 µm) alloys. The difference n grain size is related to the spray forming conditions which are in part dictated by the alloy hemistry.

'able 2. Porosity pore and grain sizes of the as-received alloys.

	Grain Size (µm)			
	Average	Range	Pore Size (µm)	Porosity (%)
518/Osprey	25	10-60	20-450	0.06-2.61
518/LDC	14	2-47	10-100	0.8
202	9	6-17	5-100	0.16-1.52
M 2618	300	200-440	NA	NA

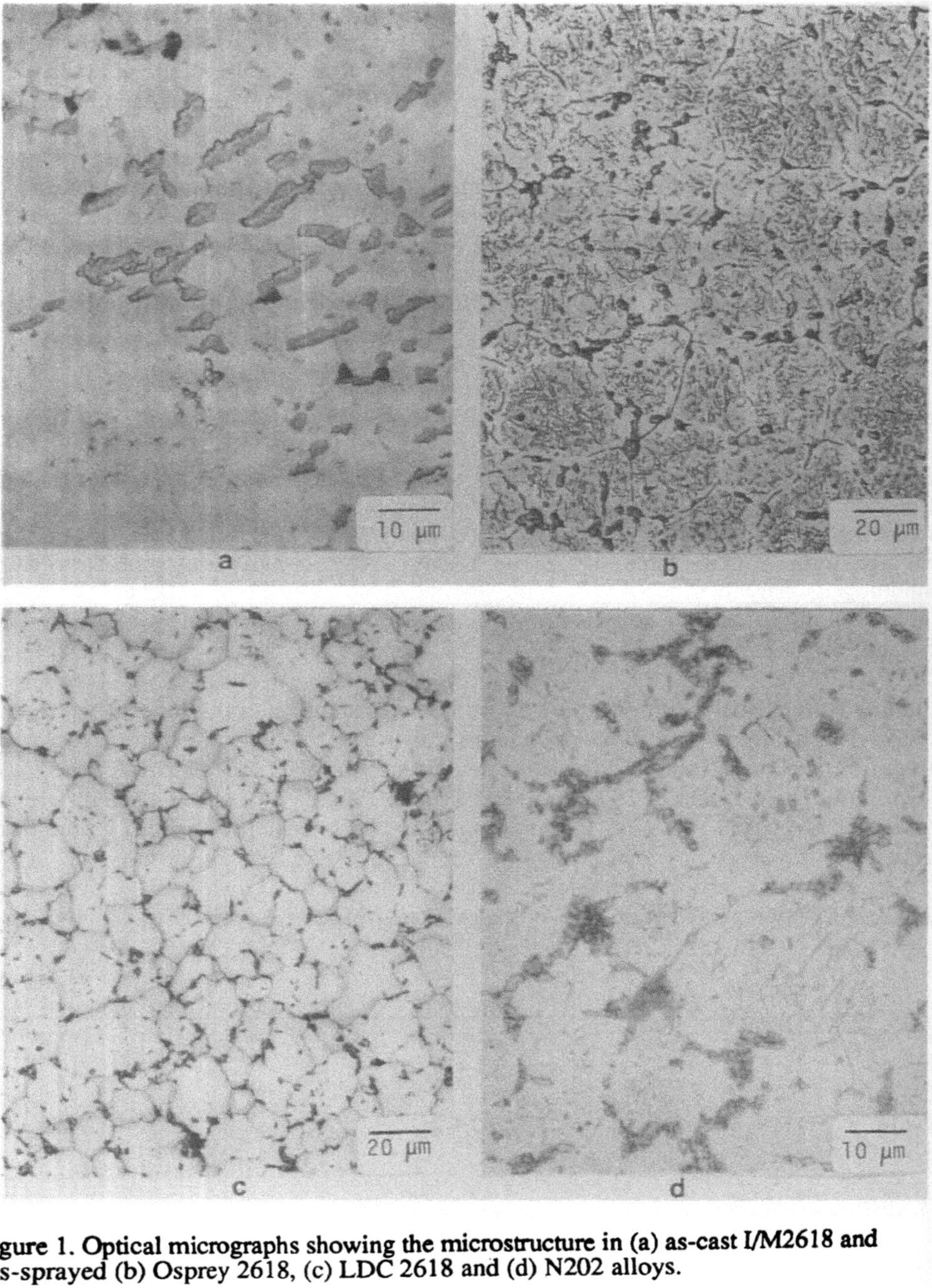

Figure 1. Optical micrographs showing the microstructure in (a) as-cast I/M2618 and as-sprayed (b) Osprey 2618, (c) LDC 2618 and (d) N202 alloys.

The LDC 2618 alloy contained irregularly shaped constituent particles which are mostly distributed at the grain boundaries as shown in Fig. 1c. The elongated particles had a thickness of up to 2 μm. The particles in grain interiors appear to form as a result of liquid entrapment between coarsening flanged cells. The constituent particles in the Osprey 2618 alloy appeared not only at the grain boundaries, but also grain interiors. The size of nearly spherical constituent particles was similar (1-3 μm) to that of the I/M2618 alloy, but the nonspherical particles were finer (2 x 8 μm) in the S/F alloy.

The Osprey 2618 alloy often showed signs of localized melting as shown in Fig. 1b. The remelting process was so progressed in some regions that the original grains appeared to be fragmented into subgrains and or smaller new grains. Partial remelting is believed to be a result of cyclic temperature variation on deposit surface due to billet rotation during spray deposition, thus causing the reheating of the newly solidified grains to above solidus temperatures with the incoming partially liquid droplets. In this case, elongated or rod-type particles formed and/or coarsened in solute rich liquid environment under particle impact deposition conditions. On the etched micrographs, the remelted regions showed little or no contrast, whereas the remnants of the original grains exhibited a contrast resembling an orange surface appearance. Based on this contrast difference, the liquid content of the mushy region on the deposit surface is estimated to be about 63 vol.%. This suggests that a significant fraction of the particles arrived at the deposit surface with a high liquid-to-solid ratio. This hot deposition condition is believed to be responsible for the large grain size observed in the Osprey 2618 alloy.

The microstructure of the N202 alloy contained constituent particles mostly on the grain boundaries. They are enriched especially in Mn and Cu, and some in Fe and Cr ($Al_{20}Cu_2Mn_3$-type phase). These particles were 2-5 μm, although some of them attained coarser size up to 10 μm. Zr-rich particles (Al_3Zr) were 2-10 μm in size and inhomogeneously distributed in the matrix.

In a typical spray deposited microstructure regime, the as-sprayed alloys contained up to 2.6 vol.% porosity (Table 3). The pores were nonuniformly distributed in the microstructure and their sizes were in the range of 5-450 μm. Larger pores were found in the Osprey 2618 alloy. Mostly spherical nature of the pores indicated that they formed as a consequence of gas entrapment during deposition. The porosity and pore size of the Osprey 2618 alloy decreased from center to the outer (side surface) section of the billet.

Except Cu, other alloying elements had a uniform distribution in the as-received Osprey 2618 billet. Cu concentration increased from 2.26 wt.% at the center to 2.55 wt.% at the outer region of the billet. This may be related to the size and thermal state of particles in the spray plume as well as the rotation of the billet during deposition. One possible scenario is that Cu-rich particles which were deposited or formed on the substrate were pushed outward with the mushy layer on the deposit surface by the centrifugal forces generated by the rotation of the billet. No chemical analysis was conducted on the N202 alloy, and the LDC 2618 alloy was small in size to determine the chemical homogeneity of the as-sprayed deposit from location to location.

Fig. 2 shows the microstructure of the alloys after hot deformation. The S/F alloys exhibited more uniform distribution of the particles in the matrix than the I/M618 alloy. Band regions consisting of predominately fine particles were found in the extruded I/M 2618 alloy as shown in Fig. 2d. The particles in the S/F2618 alloy were significantly finer than those in the I/M alloy. For example, the particle size was less than 6 μm in the S/F alloy compared to less than 10 μm in the I/M2618 alloy. In the N202 alloy, however, some particles were coarser than 10 μm and they were identified to be (Mn, Cu, Fe)-rich ($Al_{20}Mn_3Cu_2$-based) and Zr-rich (Al_3Zr-based) phases.

Hot deformation further refined the grain size of the 2618 alloys. The extruded Osprey 2618 alloy had a uniform grain size of about 4 μm as shown in Fig. 3. The I/M 2618 alloy, however, exhibited a duplex grain structure: the grain size was about 15 μm in the regions of coarse constituent particles and 3-50+ μm in coarse constituent-free regions (Fig. 4b). The generation of fine grain size requires a duplex particle size: particles larger than 1 μm to generate strain concentrations and promote nucleation of the recrystallized grains, and small particles to hinder the grain growth after recrystallization.[13] High stored energy and particle simulated nucleation facilitated the recrystallization in the regions of constituent particles. Some of the recrystallized grains rapidly coarsened because of the lack of fine particles to suppress the grain growth in the microstructure. Fine grain structure developed in the band regions where extensive small particles were present in the as-extruded I/M2618 alloy.

In the solution treated condition, the grain size of 2618 alloys varied between 5 and 125 μm. As shown in Fig. 4, extensive grain growth occurred in the Osprey 2618 alloy despite homogeneous distribution of the constituent particles. The average grain size was about 30 μm for the Osprey alloy and 50 μm for the I/M alloy. The present results suggest that a bimodal particle size is required to retain the fine grain size in the spray formed materials.

MECHANICAL PROPERTIES OF THE ALLOYS

Fig. 5 shows the yield strength and elongation of the alloys as a function of aging. Fracture toughness/yield stress relationship is given in Fig. 6. The yield strength of the alloys increased with aging time in the underaged regime. The S/F 2618 alloys exhibited

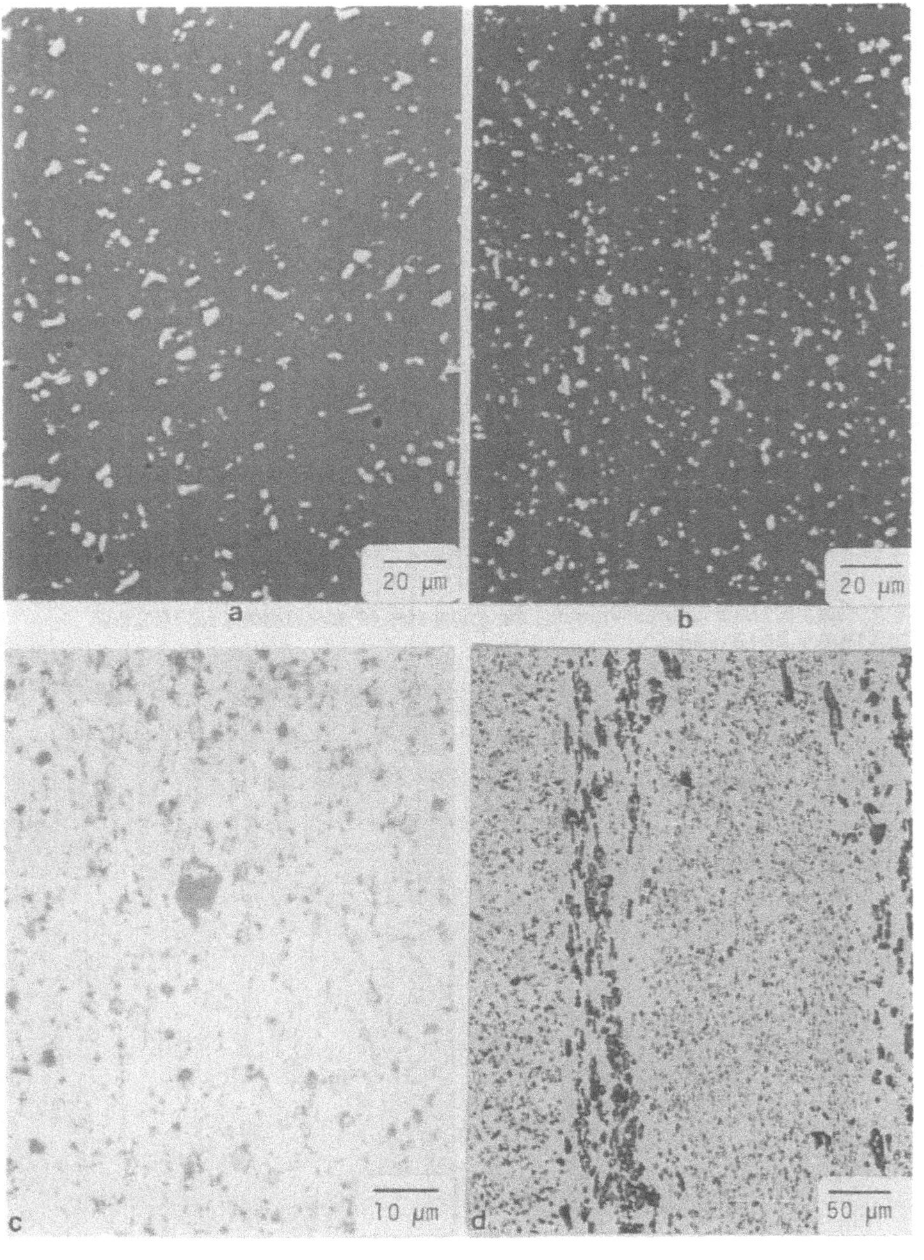

Figure 2. Back scattered electron images showing the particle distribution in as-extruded (a) I/M2618, (b) Osprey 2618 and (c) as-forged N202 alloys. (d) Optical micrograph showing bimodal particle distribution in the extruded I/M 2618 alloy.

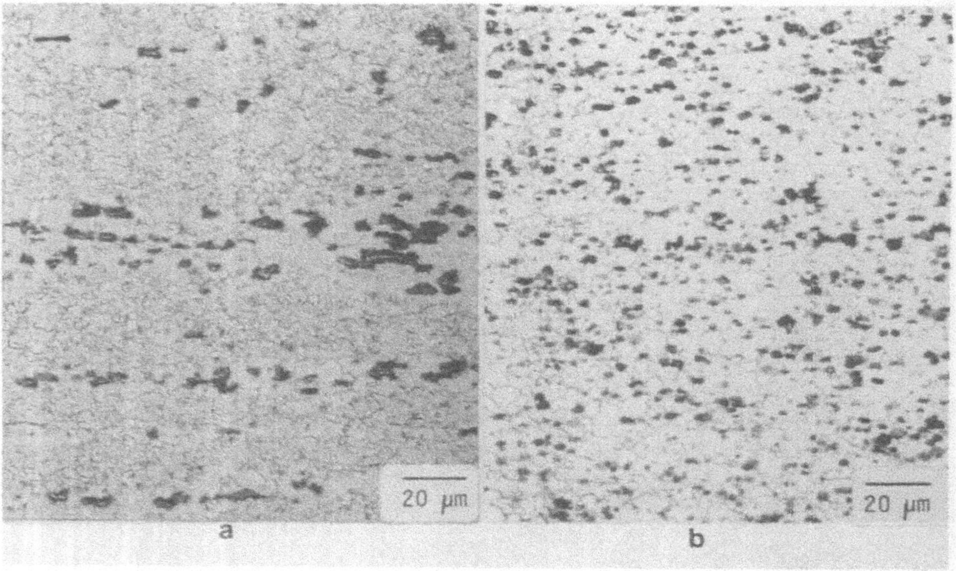

Figure 3. Optical micrographs showing the grain size of as-extruded (a) I/M2618 and (b) Osprey 2618 alloys.

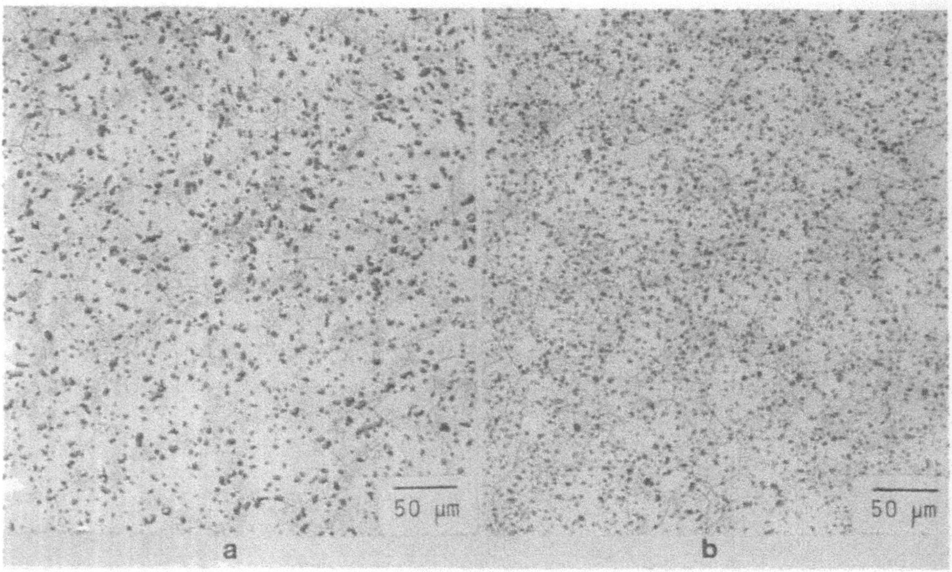

Figure 4. Optical micrographs showing the grain structure in T6 heat treated (980°F/4 hrs/WQ/390°F/20 hrs) (a) I/M2618 and (b) Osprey 2618 alloys.

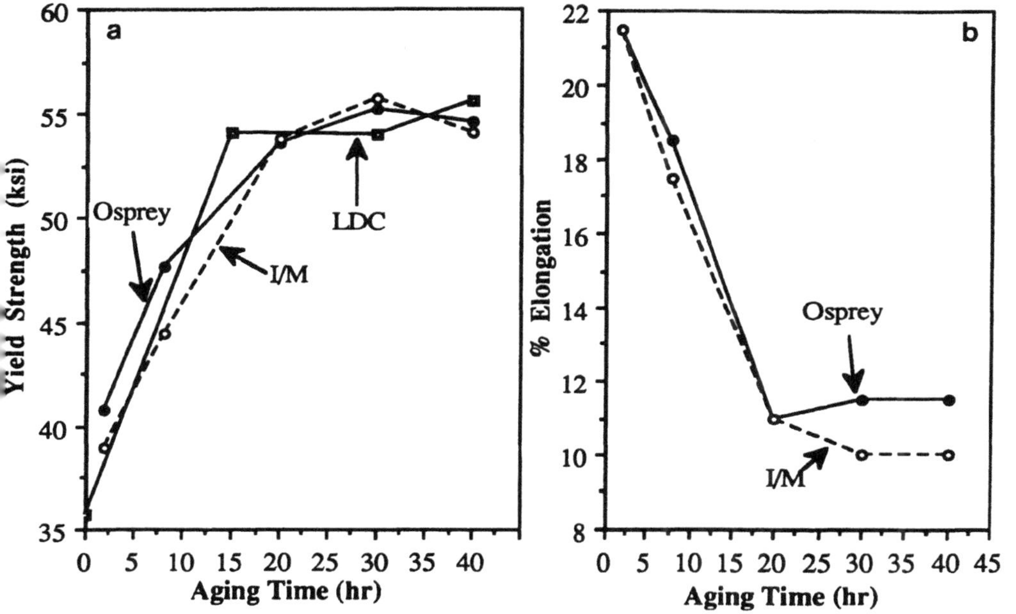

Figure 5. Tensile yield strength (a) and ductility (b) of the S/F and I/M 2618 alloys for various T6 conditions.

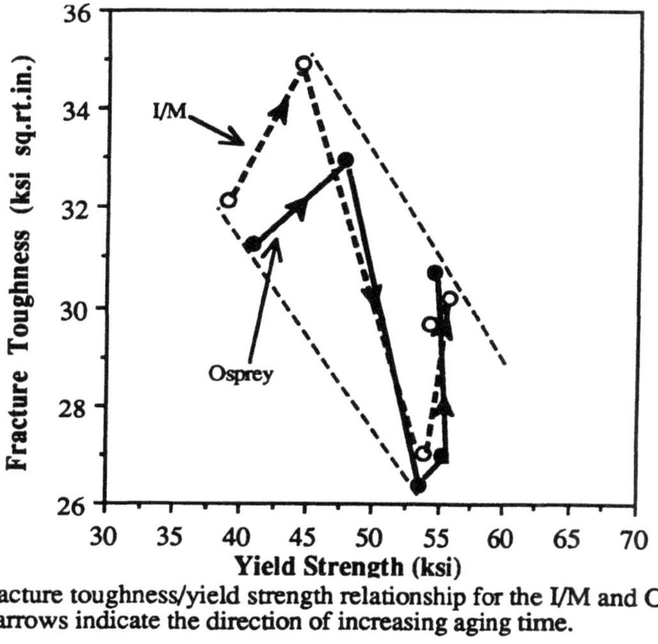

Figure 6. Fracture toughness/yield strength relationship for the I/M and Osprey 2618 alloys. The arrows indicate the direction of increasing aging time.

slightly higher yield strength than the I/M alloy, but the ductilities of the alloys were very comparable for this regime. The yield strength of the 2618 alloys attained 55.5 ksi for the Osprey, 55.6 ksi for the LDC and 57.5 for the I/M processes at the peak-aged condition, and then remained constant or decreased with further aging. As expected, the ductilities of the alloys decreased with aging and remained constant after peak-aged condition (Fig. 5b). The Osprey 2618 alloy had slightly higher (1.5%) elongation than the I/M alloy. The observed increase in strength is related to the formation of S' and S phases during aging.[14]

The plane strain fracture toughness (L-T) values for the S/F were slightly lower than those of the I/M alloy despite the fact that constituents in the S/F alloy were significantly finer and more uniformly distributed than that in the I/M alloy. The toughness increased with yield strength early in the underaged regime and then decreased until the maximum yield strength was reached. During overaging, the toughness again increased without significant loss in the yield strength. The best toughness-strength combination is achieved for the overaged condition. The recovery in toughness at the same strength level for the overaged condition differs from that commonly observed for the toughness-strength behavior of the aluminum alloys. Overaging beyond the peak aging condition generally results in improvement in the toughness with an accompanying decrease in strength.[15] The extent of recovery in the toughness depends on which precipitate microstructure (matrix vs. grain boundary) controls the fracture behavior. Significant recovery in the toughness is achieved when matrix microstructural changes dominate the fracture process, whereas very little or no chance in toughness is observed when grain boundary precipitates control the fracture behavior.[15]

The primary alloying constituents for the precipitate formation in the 2618 alloys are Cu and Mg, and one possible precipitation sequence is outlined as:

Supersaturated α ----> GPB zones---->S'---->S(Al_2CuMg)

After solution treatment and quenching, the first solid state reaction is expected to be GPB zone formation. No S' or S precipitates were reported in naturally aged 2618 alloy.[14,16] S' and S precipitates were found in peak aged condition without any precipitate free zones.[14] S' precipitation occurs uniformly through the matrix, whereas S precipitates initially form in mainly on the grain boundaries and the Al_9FeNi/matrix interfaces, and grain interiors with further aging.[14]

The strain concentrations are expected to develop at the constituent particle/matrix interfaces during deformation, leading to void nucleation and subsequent linkage through the planar slip bands when coherent phases are present in the microstructure, i.e., GP zones. Initial precipitation of S' phase is believed to improve the strain distribution since S' precipitates in 2XXX alloys suppress strain localization and thus promote homogeneous slip.[17] This results in more crack branching leading to higher toughness and yield strength as observed early in the underaged regime.

The toughness decrease in the underaged regime is attributed to the competing effects of continuous S' phase precipitation in the matrix and S phase precipitation on the grain boundaries and the Al_9FeNi/matrix interfaces. Increased amounts of incoherent S phase precipitation made grain boundaries and the particle interfaces susceptible to early void nucleation through embrittling effects of the incoherent grain boundary precipitates.[18] This is confirmed by the increasing amount of intergranular fracture as shown in Figs. 9-12. The recovery in toughness in the overaged condition is believed to be associated with the homogeneous deformation produced by the matrix precipitates of the S phase.[19]

The dependence of fatigue crack growth rate (da/dN) of the alloys on the stress intensity factor (ΔK) is shown in Fig. 7. Significant features of the test results are: relatively high stress intensity range required to propagate the cracks at low crack growth rates ($<10^{-6}$ in./cycle), and higher crack growth rate for the S/F alloy at FCGR higher than 10^{-6} in./cycle. A ΔK of approximately 9 ksi sq in. for the S/F alloy and of 8 ksi sq in. for the I/M alloy is required to grow the cracks at a rate of about 10^{-7} in./cycle. High ΔK values at slow crack growth regime are attributed to the residual stresses in the specimens. When this effect is taken into account, the resultant $\Delta K(\Delta K_{effective})$ is found to be in the range of 3-4 ksi sq in. for the alloys. The presence of residual stresses in the samples was revealed by a high level of crack closure during tests and by the change in the height of the samples around the fatigue notch before and after machining of the notch. Displacement of 0.0035 and 0.0025 in. occurred in the notched dimension of the S/F and I/M alloys, respectively. Higher displacement and the resultant higher ΔK value for the S/F alloy indicates the presence of higher residual stress in this alloy.

Fatigue performance of forged N202-T6 alloy is compared in Fig. 8 with 2014-T6 and extruded N202 alloys. The forged N202 shows improved fatigue performance with a fatigue strength as high as about 55 ksi compared to the extruded N202 alloy. In the high stress/low cycle regime, its performance is similar to that of the conventional 2014-T6, whereas in the low stress/high cycle regime, it tends to have better fatigue behavior than the 2014-T6. SEM examination of the fatigue fractured samples of the forged N202 alloy indicated that crack initiation takes place at the surface with the cracking of large constituent particles or interfacial separation of agglomerated constituent particles as shown in Fig. 15. Large particles lead to stress concentration for early crack nucleation and thus, lower fatigue strength. A large fraction of the higher fatigue strength of the N202 alloy can be attributed to the overall smaller constituent and grain sizes.

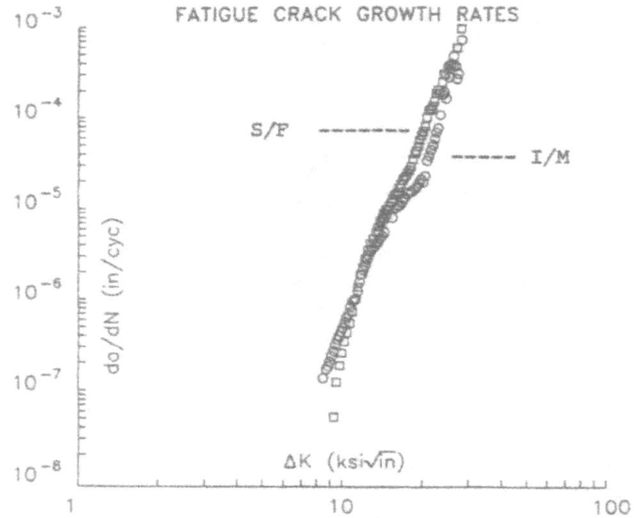

Figure 7. Fatigue crack growth rates(da/dN) of the I/M and Osprey 2618 alloys. Heat treatment: 980°F/4 hrs/WQ/390°F/20 hrs.

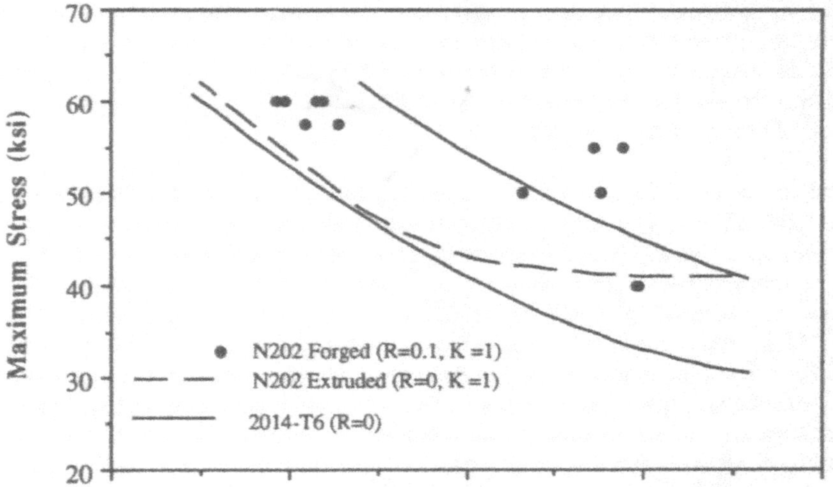

Figure 8. S-N curves for the N202 and 2014 alloys. Data for the extruded N202 alloy were taken from reference 20.

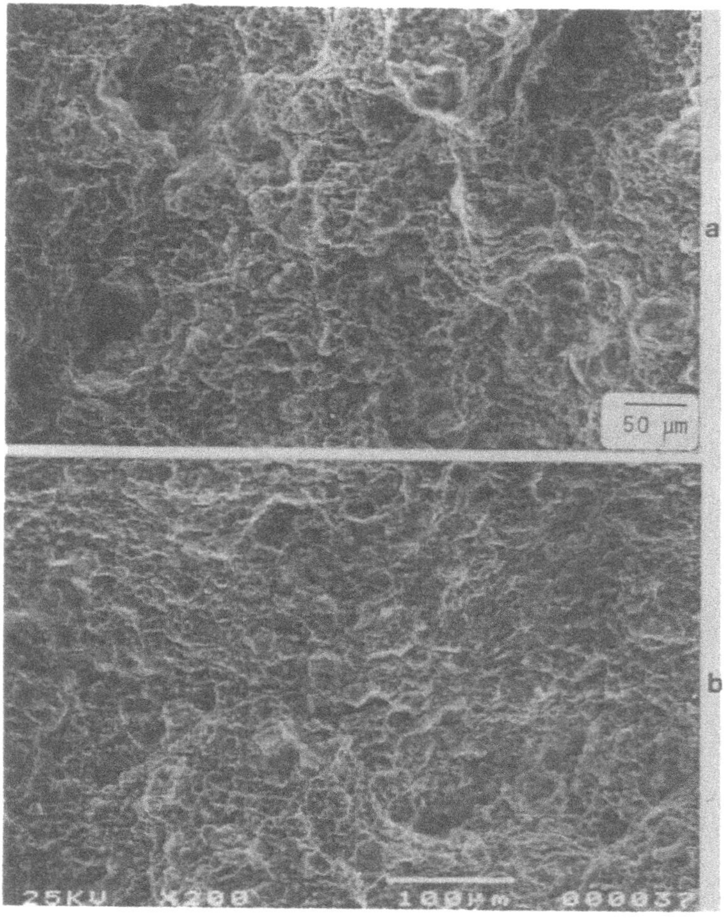

Figure 9. SEM micrographs showing the tensile fracture surface of the Osprey 2618 alloy. The microstructure shown in (a) had 4.5 ksi lower yield strength than that in (b). Heat treatment: 980°F/4 hrs/WQ/390°F/8 hrs.

FRACTOGRAPHIC EXAMINATION

Figs. 9-15 show the fracture surface of the alloys. The fracture surface of the tensile and fracture toughness samples consisted of a mixture of smooth intergranular fracture morphology surrounded by a transgranular fracture morphology with well-defined dimple structure. In the transgranular regime, the dimples were associated with the second phase (mostly Al_9FeNi) particles. This indicates that microvoids nucleated at the particle/matrix interface and subsequent growth and coalescence of these voids with further straining

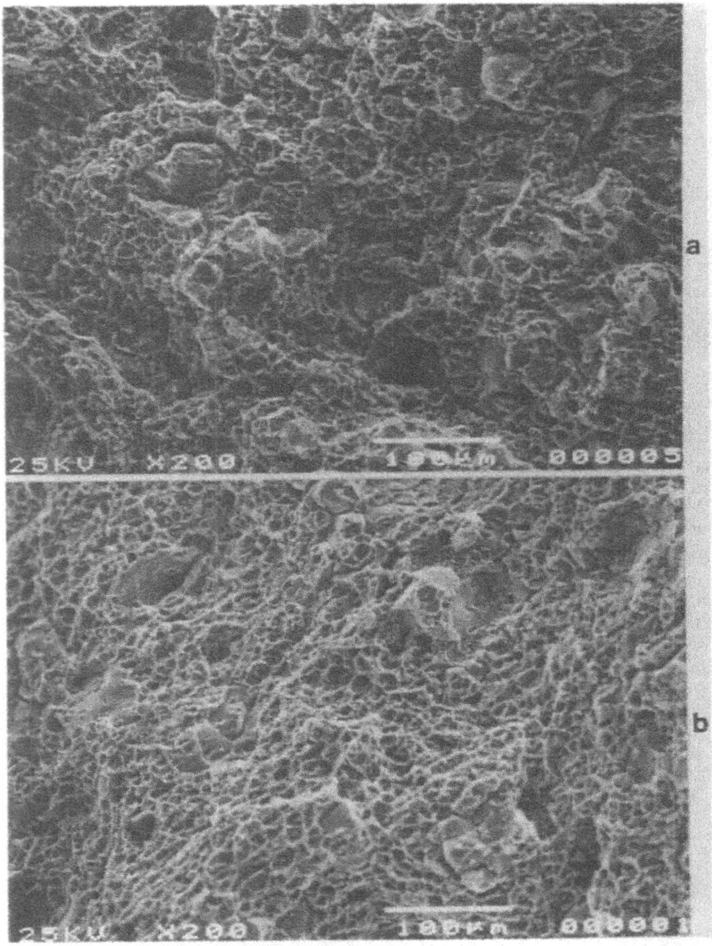

Figure 10. SEM micrographs showing the tensile fracture surface of the I/M 2618 alloy. The microstructure shown in (a) had about 3 ksi lower yield strength than that in (b).Heat treatment: 980°F/4 hrs/WQ/390°F/30 hrs.

resulted in the observed transgranular dimpled fracture behavior. As shown in Fig. 13, some dimple contained cracked particles or cluster of particles. Weak interfacial strength between the particles in a cluster provided sites for void nucleation by interfacial separation and easy path for crack propagation. The particles in the I/M alloy exhibited more cracking than those in the Osprey 2618 alloy. This may be related to the elongated particle morphology which appears to be incapable of accommodating plastic strains during deformation.

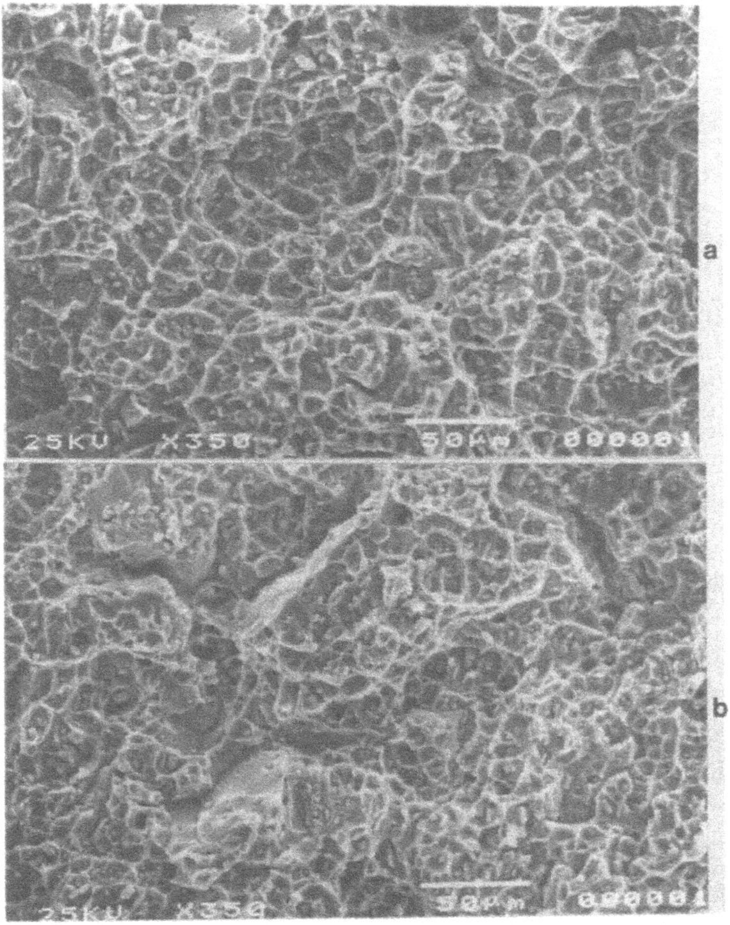

Figure 11. SEM micrographs showing the fracture surface of the I/M 2618 fracture toughness specimen for (a) 8 hr and (b) 20 hr aged. The microstructure shown in (a) had 6-10 ksi sq.in. higher fracture toughness than that in (b).

Some grain boundaries failed by interfacial decohesion and accommodated very little plastic strain leaving less tortured fracture surface as shown in Fig. 14. Duplicate test samples with notable strength difference were examined to establish the effect of intergranular fracture on the mechanical properties of the 2618 alloys. For example, the Osprey 2618 microstructure shown in Fig. 9a had 4.5 ksi lower yield strength than that in Fig. 9b. Similarly for the I/M alloy, the microstructure in Fig. 10a had about 3 ksi lower yield strength than that in Fig. 10b. Similar observations were also made on the fracture toughness specimens. For example, the I/M 2618 microstructure as shown in Fig. 11a

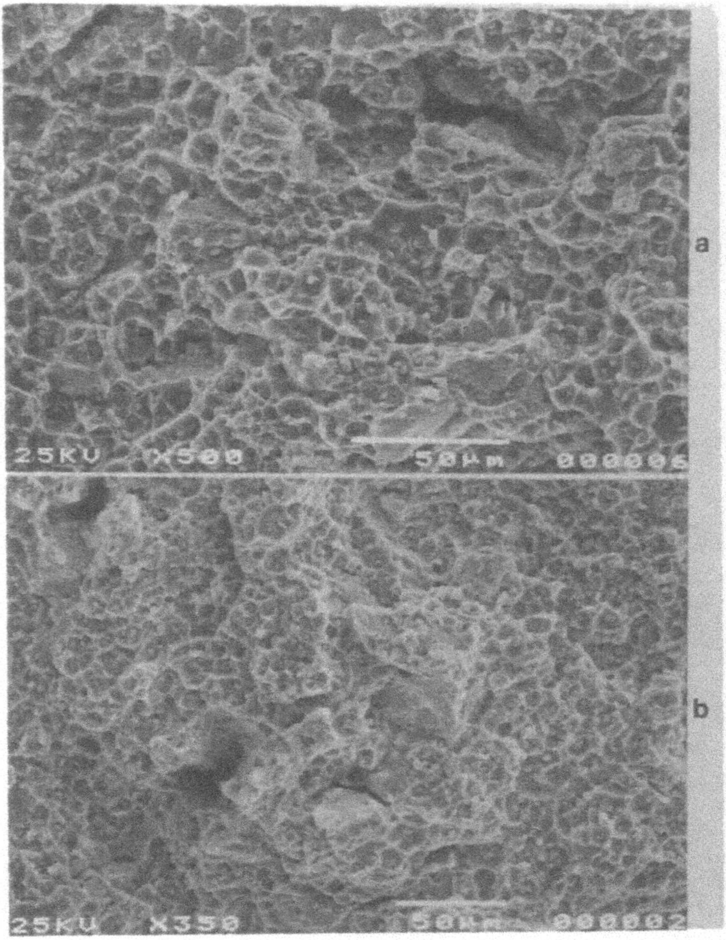

Figure 12. SEM micrographs showing the fracture surface of the Osprey 2618 fracture toughness specimen for (a) 8 hr and (b) 20 hr aged condition. The microstructure shown in (a) had about 7 ksi sq.in. higher fracture toughness than that in (b).

(8 hr aging) had 6-10 ksi sq in. higher fracture toughness than that in Fig. 11b (20 hr aging). The Osprey 2618 microstructure in Fig. 12a (8 hr aging) had about 7 ksi sq in. higher fracture toughness than that in Fig. 12b. The increase in fracture toughness or yield strength correlates well with the extent of the intergranular fracture behavior. For all the above cases, the higher yield strength or fracture toughness microstructure has less intergranular fracture as shown in Figs. 9-12. Overall fracture surface appearance of the FCG samples was similar to those observed on the tensile and fracture toughness specimens.

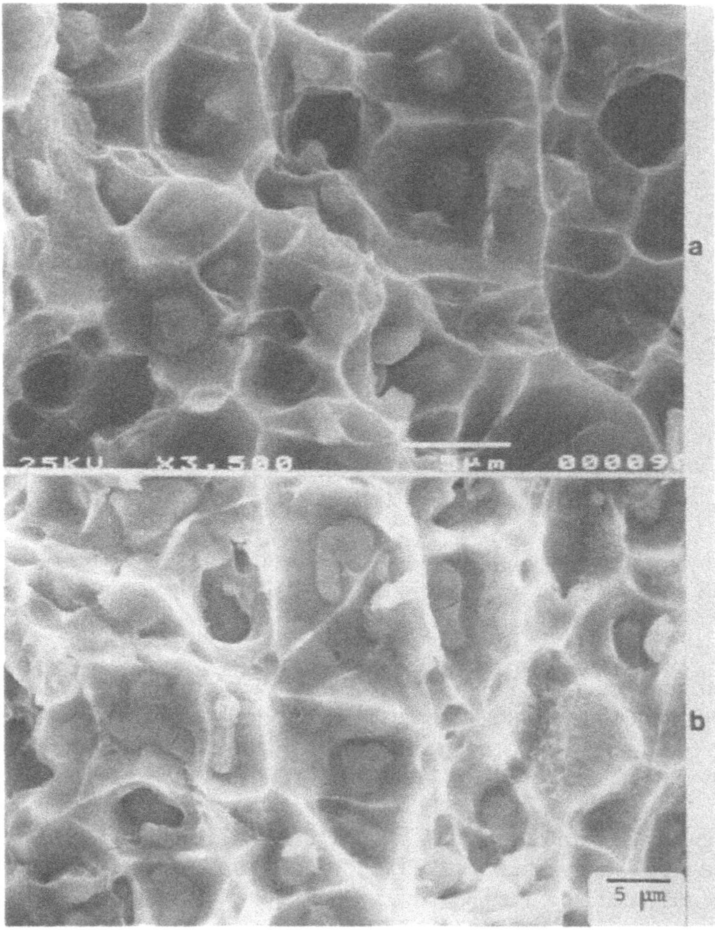

Figure 13. Details of tensile ductile fracture region of (a) the Osprey and (b) I/M 2618 alloys for 8 hr aged condition. Note the particle/particle interfacial separation in (a) and particle fracture in (b).

Conclusions

1. Porosity content of the as-sprayed alloys was in the range of about 0.06%-2.61% and the pores were mostly spherical in shape ranging from 5 to 450 μm in size. Larger pores were found in the Osprey 2618 alloy. The grain size was finer in the N202 alloy (9 μm) followed by LDC 2618 (14 μm) and Osprey 2618 (25 μm) alloys.

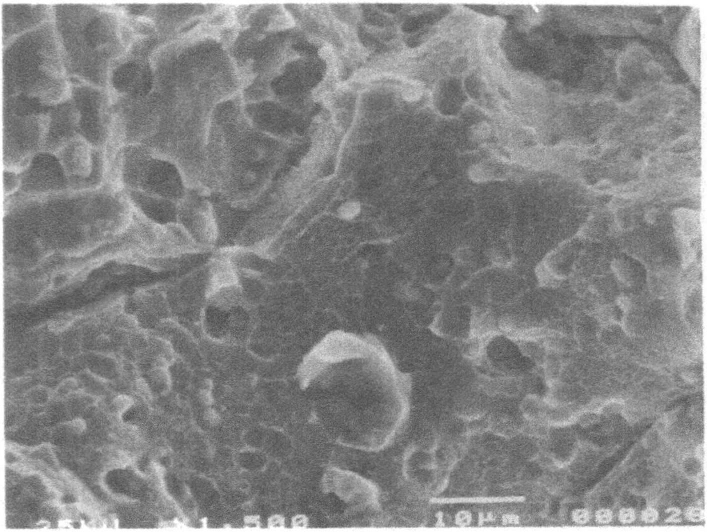

Figure 14. SEM micrographs showing the tensile intergranular fracture surface of the Osprey 2618 alloy for 8 hr aged condition.

2. Particle distribution and size are much more uniform in the hot deformed S/F alloys because of the initial finer particle size and more uniform distribution of the particles in the as-sprayed alloys. Significant grain growth occurred in the Osprey 2618 alloy during solution treatment, but overall grain size was finer in the spray formed alloys.

3. Despite much higher extrusion ratio of the I/M alloy, the mechanical properties of the S/F alloy are very comparable to that of the I/M 2618 alloy. Compared to the 2014-T6 alloy, the forged N202 alloy seemed to have better high cycle fatigue property.

4. Failure is initiated by void nucleation at particle matrix interfaces, cracking of large constituent particles and particle/particle interfacial separation. Crack propagation occurred by transgranular and intergranular fracture processes by the growth and coalescence of voids and by grain boundary separation. Intergranular fracture significantly affected the mechanical properties of the 2618 alloys.

Acknowledgments

The author expresses his appreciation to Prof. N. J. Grant and Dr. I. Ucok of MIT for spray forming of the LDC 2618 alloy, and to P. R. Chvala, C. P. Husack and G. Gray of Alcoa for sample preparation and microstructural evaluations.

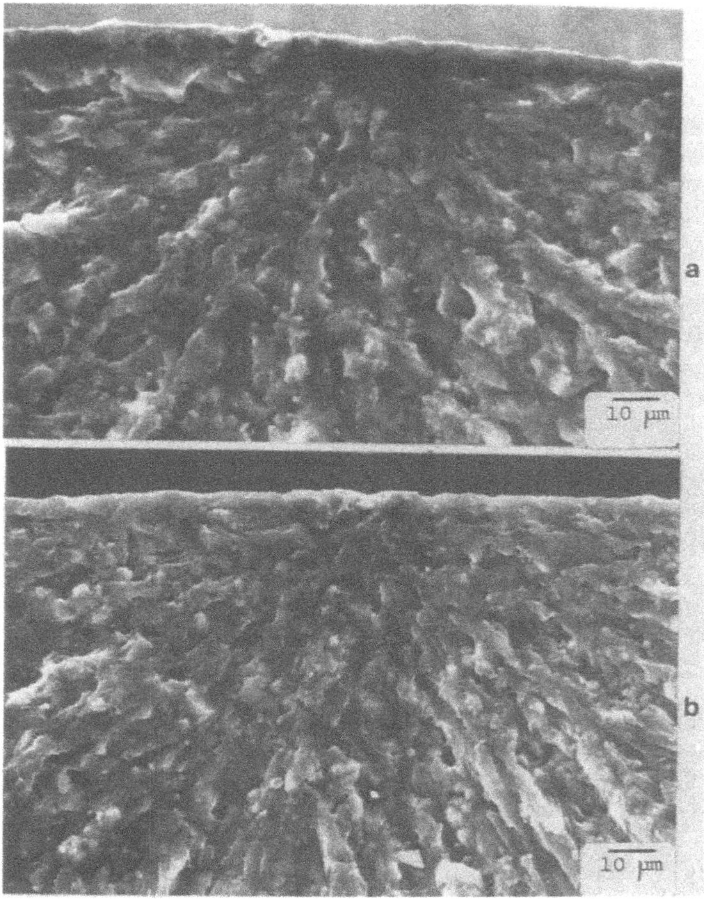

Figure 15. SEM micrographs showing the fatigue crack initiation sites in the forged N202-T6 alloy.

References

1. A. G. Leatham and Y. Kawaskima, "The Past, Present and Future Developments of the Osprey Preform Process" Proc. 2nd Japan International SAMPE Symposium (1991), pp. 369-376.

2. O. Beffort, P. J. Uggowitzer and M. O. Speidel, "Spray Formed Aluminum Alloys for Use at Elevated Temperatures" Proc. 3rd Int. Conf. on Aluminum Alloys, Vol. I, Trondheim, Norway (1992), pp. 46-51.

3. A. R. E. Singer, "Recent Developments in the Spray Forming of Metals" in New Perspectives in Powder Metallurgy, Vol. 8 - Powder Metallurgy for Full Density Products, K. M. Kulkarni (ed.), MPIF (1987), pp. 513-533.

4. E. J. Lavernia and N. J. Grant, "Structures and Properties of a Modified 7075 Aluminum Alloy Produced by Liquid Dynamic Compaction," Int. J. Rapid Solidification, Vol. 2, pp. 93-106 (1986).

5. A. Lawley, "Atomization: The Production of Metal Powders," MPIF (1992), p. 32.

6. M. J. Couper and R. F. Singer, Proc. 5th Int. Conf. on Rapidly Quenched Metals (Wurzburg 1984).

7. V. Anand, Cabot Corporation International Powder Metallurgy Conference (June 17-22, 1984, Toronto).

8. A. D. Styrkas and R. A. Oganyan, "Production and Properties of Tin Powders Produced by Ultrasonic Atomization and Solid State Transformation," Powder Metallurgy, Vol. 35, No. 2, pp. 117-119 (1992).

9. T. W. Clyne, R. A. Ricks and P. J. Goodhew, The Production of RS-Aluminum Powder by USGA, Part II, Int. Journal of Rapid Solidification Vol. 1, pp. 85-101, 1984-85.

10. H. Mohmann and S. Jonsson, Production of Fine Metal Powder by Conventional and Ultrasonic Gas Atomization, in Special Melting and Processing Techniques, edited by G. K. Bhat, Proc. 9th Int. Vacuum Metal. Conf. on Special Melting, San Diego, CA (1988) pp. 946-951.

11. A. Lawley, P. Mathur, D. Apelian and S. Annavarapu, "Process Modeling and Fundamentals in Spray Forming" in Spray Forming: Science, Technology and Applications, PMIF, 1992, pp. 3-31.

12. O. Beffort, P. J. Uggowitzer and M. O. Speidel, "Spray Deposited Aluminum Alloys for Use at Elevated Temperatures," Proc. 3rd Int. Conf. on Aluminum Alloys: Their Physical and Mechanical Properties, Vol. 1 (1992), pp. 46-51.

13. M. A. Zaidi and J. A. Wert, "Thermomechanical Processing of Aluminum Alloys" in Aluminum Alloys-Contemporary Research and Applications, A. K. Vasudevan and R. D. Doherty (eds.), Academic Press, Inc., San Diego, CA (1989) pp. 137-170.

14. D. L. Zhang and B. Cantor, "TEM Characterization of 2618/SiC Composites," Powder Metallurgy: An Overview, I. Jenkins and J. W. Woods (eds.), Inst. of Metals, London (1991), pp. 197-207.

15. A. K. Vasudevan, R. D. Doherty and S. Suresh, "Fracture and Fatigue Characteristics in Aluminum Alloys" in Aluminum Alloys-Contemporary Research and Applications, A. K. Vasudevan and R. D. Doherty (eds.), Academic Press, Inc., San Diego, CA (1989) pp. 445-462.

16. J. L. Lorca and P. Poza, "Fracture Toughness of Al/SiC Composites in the Temperature Range -136° to 190°C," Scripta Metallurgica et Materialia, Vol. 29/2 (1993) pp. 261-266.

17. R. E. Crooks and E. A. Starke Jr., Metal. Trans. A, 15A (1984) 1367.

18. A. K. Vasudevan, R. D. Doherty, Acta. Met. 35, 1193 (1987).

19. Gregson and H. M. Flower, Acta. Metal. Vol. 33 (1985), pp. 527-537.

20. W. Kahl and B. Zettler, Alusuisse-Lonza Brochure on Spray Deposited High Performance Materials-N202 (1990).

SELF-DIFFUSION AND RELAXATION PROCESSES IN AMORPHOUS METALLIC ALLOYS

H. KRONMÜLLER, W. FRANK, and P. SCHARWAECHTER
Max-Planck-Institut für Metallforschung
Institut für Physik
Postfach 80 06 65
D-70506 Stuttgart
Germany

ABSTRACT. Diffusion and relaxation experiments have been performed for a large number of binary, ternary and quarternary alloys including metal-metal as well as metal-metalloid compositions. Self-diffusion data were obtained by the ratio-tracer technique, and the activation parameters of structural relaxation were derived from studies of magnetic relaxation and induced magnetic anisotropy. A review of the experimental results obtained by these techniques will be given and the present state of our theoretical understanding of diffusion in amorphous and heterogeneous materials will be discussed. In particular it will be shown that the diffusion processes in as-quenched amorphous alloys are governed by the migration and agglomeration of quasi-vacancies and in the case of relaxed amorphous alloys by direct collective mechanisms where thermally activated displacement chains play a major role. Accordingly, in contrast to crystalline alloys, in relaxed amorphous alloys, self-diffusion does not occur indirectly with the aid of diffusion vehicles (e.g. quasi-vacancies). Since the self-diffusion enthalpies are smaller by 0.5 - 1.0 eV as compared with crystalline alloys, this may indicate that diffusion enthalpies correspond to migration energies but not to the sums of formation and migration energies.

For a theoretical description of diffusion in amorphous alloys the effective medium approximation has been successfully applied to determine the temperature dependence of the diffusion coefficient in the case of a random diffusion potential. Describing the random walk of a particle by a Markoffian master equation the diffusion coefficient has been determined for three types of potentials: a) The random barrier model. b) The random trap model. c) The mixed model with randomness of barriers and traps. The diffusion coefficients have been determined for short and long times and for one-, two- and three-dimensional models. The implications to the interpretation of self-diffusion and relaxation phenomena in amorphous alloys will be discussed.

1. Introduction

Diffusion in amorphous alloys has been a matter of controversy concerning the diffusion mechanism as well as the interpretation of activation parameters as for example activation enthalpies, pre-exponential factors and the validity of Arrhenius equations [1-4]. In crystalline solids self-diffusion may be controlled by vacancies in thermal equilibrium (noble metals, refractory metals) [5], by interstitial atoms (silicon) [6] or by direct atomic exchange (sodium) [7]. Non of these mechanisms, however, seem to apply directly to amorphous alloys.

Amorphous alloys correspond to metastable structures of a certain composition of alloying components which by diffusion processes tend to transform and separate into more stable

M. A. Otooni (ed.), Science and Technology of Rapid Solidification and Processing, 271–289.

crystalline phases. The underlying diffusion processes in general correspond to irreversible phenomena because the structural changes on an atomic scale cannot be removed by a cyclic process. In crystalline, stable phases the self-diffusion coefficient correspond to an intrinsic material property only dependent on the thermodynamic variables of state, temperature T, and pressure p. This behaviour is due to the fact that the diffusion mechanisms in crystalline materials are controlled by atomic defects which exist in thermodynamic equilibrium depending only on T and p. In contrast amorphous materials are neither in thermodynamic equilibrium nor do these materials have a well defined defect structure. Therefore, diffusion data of amorphous alloys often are controverse depending on the pre-treatment and the measuring technique. Crystalline materials are characterized by unique diffusion parameters, as e.g. a unique activation enthalpy, whereas in amorphous materials it turns out that the measured diffusion parameters are effective ones depending on the temperature and the time interval within which they were determined. An equivalent point of view of this behaviour is that of short-range and long-range diffusion processes. Therefore measurements at low temperatures or short times lead to different results than measurements at high temperature or long times. In the first of these two cases small activation enthalpies are measured and in the latter case larger activation enthalpies play a role.

In the following experimental results obtained by the radioactive tracer diffusion are presented together with results derived from magnetic relaxation experiments. Furthermore, results for the in radiation-enhanced diffusion will be discussed in relation to the type of diffusion mechanism in amorphous alloys.

2. Experimental Techniques

Measurements of structural relaxation and of self-diffusion in amorphous alloys in general lead to a wide spectrum of activation enthalpies, thus indicating a large variety of possible atomic displacements. The observed ranges in activation enthalpies for the same material may vary from 0.5 eV up to 2.3 eV. These differences have a number of sources:

1. Different pre-treatments (as-quenched or pre-annealed materials).
2. Different temperature ranges where the measuremetns have been performed.
3. Different time intervals over which the measurements extended.

2.1. DIFFUSION IN MICROMETER AND MESOSCOPIC RANGES

Self-diffusion measurements in general are performed by the radiotracer technique. After carefully smoothing the surface of the specimen a thin layer of about 1 nm of the radioactive element, the diffusion of which has to be studied (e.g. ^{59}Fe or ^{95}Zr), is deposited electrochemically on the surface. By an annealing treatment the radiotracers diffuse into the bulk of the specimen producing a diffusion profile where the concentration obeys a simple Gaussian law [8,9,10],

$$c(x) \propto \exp\left[-\frac{x^2}{4Dt}\right],$$

(D = diffusion coefficient, t = annealing time, x = depth where the concentration c(x) of the radiotracer concentration is measured). The diffusion profiles are determined by micro-sectioning the specimen applying the ion-beam sputtering technique and subsequently measuring the radioactivity of the sputtered-off section. In this way the sectioning achieved a precision of ± 1 nm thus allowing the determination of D with an accuracy of half-an-order-of-magnitude.

Self-diffusion experiments in amorphous alloys are performed in the temperature range form 500 K - 800 K in time intervals from 300 s to 10^5 s. The diffusion profiles extend over a depth of 100 nm - 1000 nm.

2.2. DIFFUSION ON MESOSCOPIC AND NANOMETER RANGES

Induced magnetic anisotropy measurements are another tool for the investigation of thermally activated diffusion processes. In amorphous alloys which are magnetically isotropic in the as-quenched state, a long-range magnetic anisotropy may be induced by annealing in a magnetic field below the Curie temperature [11,12]. As in crystalline alloys the induced anisotropy in amorphous alloys is due to an atom pair ordering by thermally activated atomic jumps which lead to a preferred orientation of magnetic atom pairs with respect to the orientation of the spontaneous magnetization. In order to constitute such an atom pair ordering only few jumps of the magnetic ions involved are required. Therefore, this effect may be investigated at lower temperatures than the self-diffusion experiments. Usual temperatures and annealing times are 500 K up to 10^5 s.

2.3. DIFFUSION IN NANOMETER AND ATOMIC RANGES

Magnetic after-effect measurements are performed by investigating the time and temperature dependence of the initial susceptibility after a demagnetization process. The mobility of domain walls which determines the susceptibility is sensitively influenced by atomic rearrangements and therefore may be used as a tool for investigating local rearrangements of magnetic ions on an atomic scale. The average temperature ranges and measuring times are 4 K - 400 K and 1 s - 10^5 s [13,14].

Whereas in the case of the self-diffusion measurements form plots

$$\ln c(x) \quad \text{vs.} \quad x^2 / t \qquad (1)$$

an effective temperature dependent diffusion coefficient is determined from the slopes of straight lines, in the case of the induced anisotropy and the magnetic relaxation from the measured isotherms spectra of activation enthalpy may be determined. In these cases in general it is assumed that a distribution function $p(\tau)$ of Arrhenius type relaxation times

$$\tau = \tau_0 \, e^{H/kT} \qquad (2)$$

exists, form which the distribution function p(H) of activation enthalpies may be derived.

Fig. 1 gives a summary on the activation enthalpies as obtained for an amorphous $Co_{58}Ni_{10}Fe_5Si_{11}B_{16}$ alloy by the different techniques described above [11-15]. According to the different regimes of measuring times and measuring temperatures these different techniques monitor different parts of the activation enthalpy spectra of diffusion and structural relaxation. The abrupt cut-off observed in the case of the magnetic after-effect and the induced magnetic anisotropy at the large enthalpy regime is due to the fact that in these cases short-time and medium-time experiments have been performed. Therefore large relaxation times or low jump frequencies characterized by large activation enthalpies could not be observed. The meaning of the discrete activation enthalpies measured for the self-diffusion will be discussed in section 5.2.

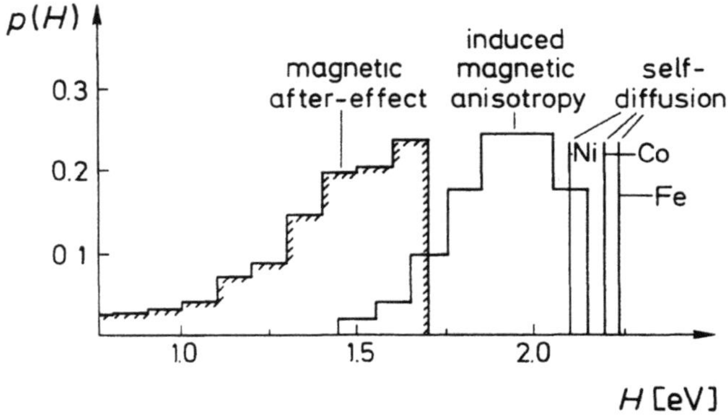

Fig. 1. Self-diffusion enthalpies of Fe, Co, Ni in $Co_{58}Ni_{10}Fe_5Si_{11}B_{16}$ as obtained by radiotracer experiments and spectra of activation enthalpies derived from isothermal relaxation curves of magnetic properties. The average temperatures and measuring times were 400 K/10^2 s, 500 K/10^5 s and 700 K/10^5 s for the magnetic after-effect, the induced magnetic anisotropy and the radiotracer experiment, respectively [4].

3. Non-Equilibirum and Quasi-Equilibrium of Diffusional Properties

As-quenched amorphous alloys show irreversible structural changes and consequently in this state the diffusional properties are not thermodynamic reversible properties. This may be observed as changes of the diffusion coefficient which primarily are due to changes of the spectrum of activation enthalpies. Accordingly, in as-quenched alloys the diffusion coefficient is not only a function of the thermodynamic variables T and p but also of the pre-treatment. In the as-quenched state free volumes or quasi-vacancies exist in supersaturation which at temperatures above RT enable a rearrangement of the atomic structure until a metastable quasi-equilibirum is established. The existing quasi-vacancies lead to an enhancement of the diffusion coefficient as in crystalline materials if vacancies are produced by fast particle irradiation. The diffusion coefficient may be written as [4,16]

$$D(t,T) = D_R^T(T) + \delta D^T(t,T) \quad , \tag{3}$$

where $D_R^T(T)$ represents the tracer diffusion coefficient of the relaxed state, whereas $\delta D^T(t,T)$ corresponds to the diffusivity enchancement due to the exess free volume present in the as-quenched state. The effect of an annealing treatment at 360°C on the diffusion coefficient of an as-quenched $Fe_{91}Zr_9$ is shown in fig. 2 where the tracer atom profiles (measured at 360°C) are plotted as a function of x^2/t for different annealing times [16]. It is obvious that the negative slopes of the straight lines increase with increasing annealing times thus demonstrating the decrease of D_R^T, i.e., a decrease of δD_R^T with annealing time. After an annealing time of 26 h the diffusion coefficient approaches a "stable" value since the structure achieves a metastable equilibrium. The existence of such a state where the physical properties become reversible properties is demonstrated by fig. 3 where the change of D_R in a cyclic temperature-time process is investigated. Fig. 3 proves that the saturation value (equilibrium value) of D depends only on T and not on the diffusion time: After an 18 h annealing treatment at 633 K (state 1) an $Fe_{91}Zr_9$ specimen was coated with a radioactive ^{59}Fe monolayer and than diffusion annealed at 673 K (state 2 in fig. 3). The state (2) also was achieved after a diffusion treatment of a non-preannealed specimen (state 3 in fig. 3). Annealing the specimen of state (3) at 633 K for 1 h led to state (4) which corresponds to the saturation value of state (1). This rather time-consuming experiment clearly proves that in amorphous alloys which were stabilized by an annealing treatment the material properties obey the reversible thermodynamics as in stable crystalline solids.

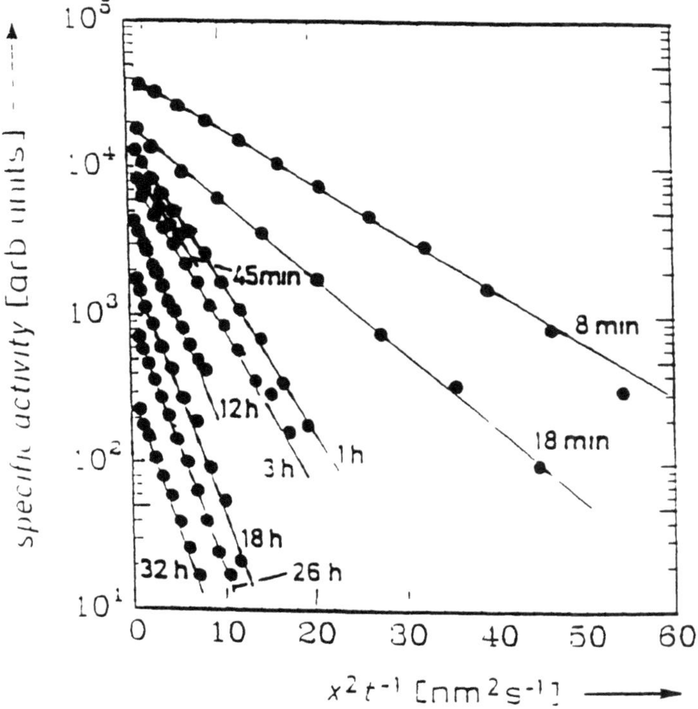

ig. 2. Diffusion profiles of ^{59}Fe radiotracers at 360°C in amorphous $Fe_{91}Zr_9$. x = penetration depth, t = annealing time for the development of the profile) [16].

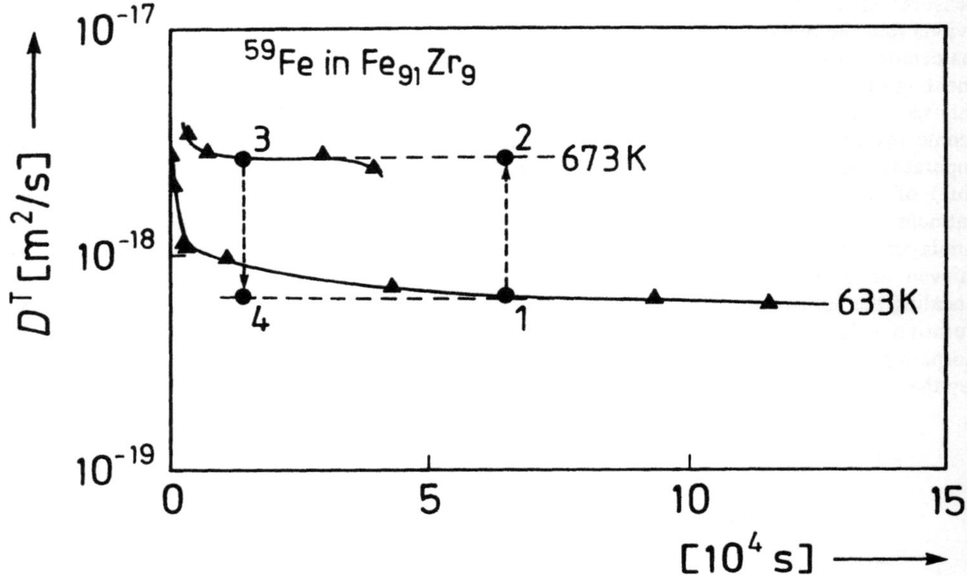

Fig. 3. **Prove of the existence of metastable states of equilibrium depending only on T in the pre-annealed state (see text) [16].**

Fig. 4 shows the dependence of the effective spectra of activation enthalpies, p(H), on the measuring time t for a pre-annealed ($T_A = 623$ K, 90 min) $Co_{58}Ni_{10}Fe_5Si_{11}B_{16}$ alloy as derived from isothermal magnetic after-effect measurements [14]. Comparing the results for the measuring times t = 10 s and t = 750.000 s the average activation enthalpy increases from 1.47 eV to 2.03 eV. Accordingly, in the case of the long-time experiment the activation enthalpy measured by the magnetic after-effect technique approaches the activation enthalpies obtained from radio-tracer experiments (see fig. 1). This fig. 4 clearly demonstrates that the effective activation enthalpies measured in amorphous alloys depend in a well-defined manner on the temperature range and the time interval choosen.

4. Review of Experimental Results

One of the most surprising experimental result concerning the self-diffusion in amorphous alloys is the fact that in relaxed amorphous alloys the self-diffusion coefficient D_R shows the features of the usually observed Arrhenius type behaviour

$$D_R^T(T) = D_{R,0}^T \exp\left[-H_R / kT\right] \quad , \tag{4}$$

where the pre-exponential factor $D_{R,0}^T$ may be expressed as

$$D_{R,0}^{T} = D_{00} \exp\left[S_R / k\right] \tag{5}$$

with S_R denoting the diffusion entropy.

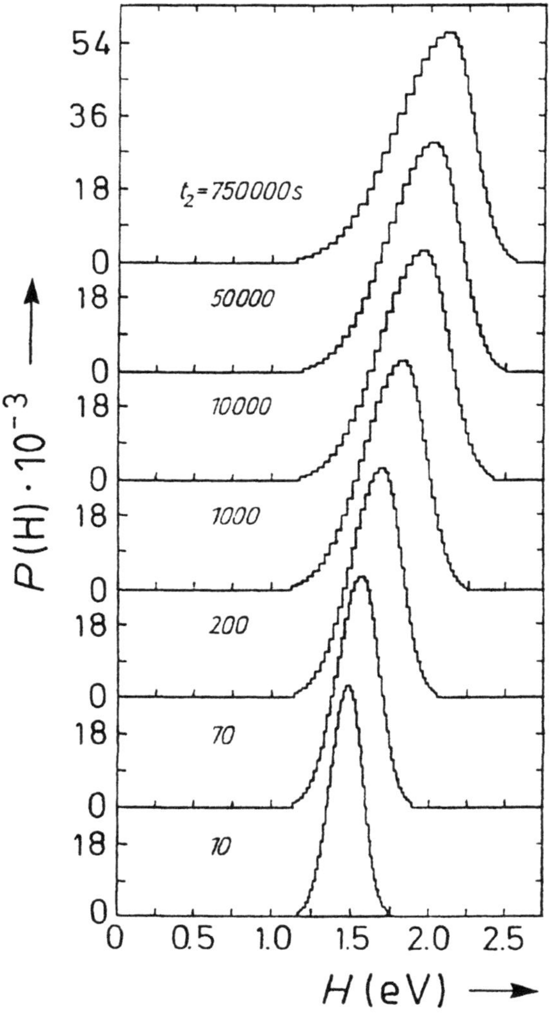

Fig. 4. Spectra of activation enthalpies as derived from isothermal magnetic after-effect mesurements using different time intervals t_2. The role of large activation enthalpies with increasing measuring time is demonstrated [14].

Fig. 5 gives a summary of a large number of Arrhenius plots of radioactive tracer diffusion coefficients for a series of amorphous alloys. In the Fe-rich as well as the Zr-rich FeZr-alloys the diffusion of Fe and Zr has been studied. Large variations in the activation enthalpies and pre-exponential factors can be observed, e.g., $H_R = 1.5$ eV, $D^T_{R,0} = 10^{-7}$ m²/s for Fe-diffusion in $Fe_{91}Zr_9$ and $H_R = 3.2$ eV, $D_{R,0} = 10^7$ m²/s for Zr-diffusion in $Zr_{76}Fe_{24}$. Here it is obvious that the diffusion enthalpy depends on the size of the diffusing element as well as on the matrix within which this atom diffuses.

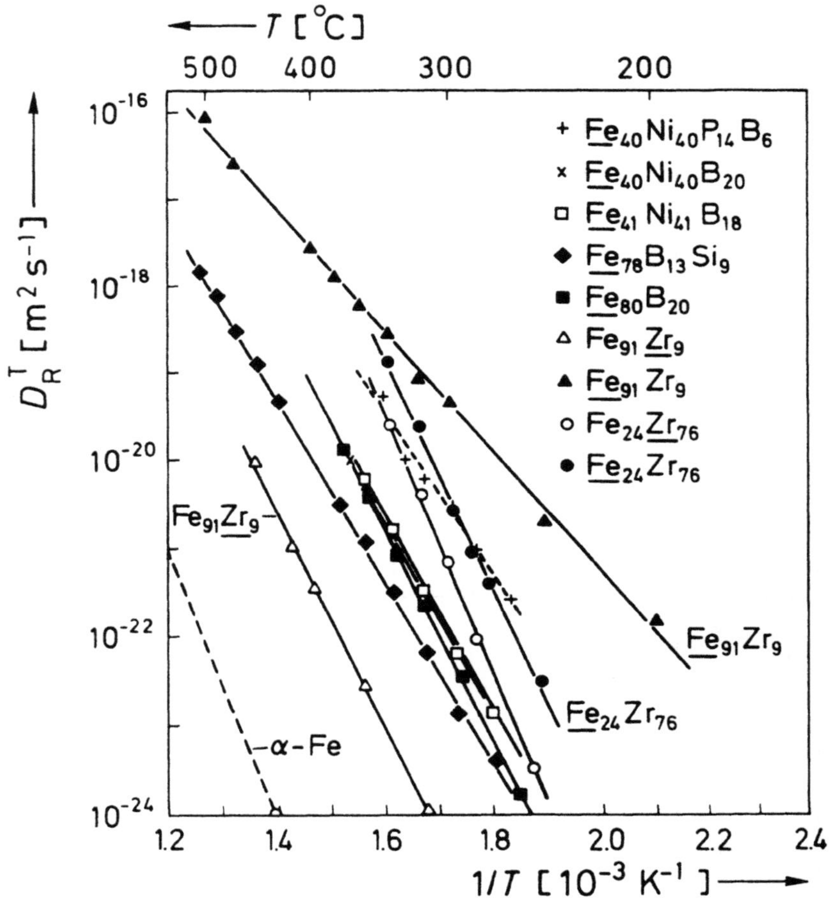

Fig. 5. Arrhenius plots of radiotracer self-diffusion coefficients, D^T_R, in relaxed amorphous alloys. The diffusing elements are underlined. For comparison the self-diffusion coefficient of α-Fe has been included [4].

In the special case of the high-permeability alloy $Co_{58}Ni_{10}Fe_5Si_{11}B_{16}$ the self-diffusion coefficients of all three transition metals, Co, Ni, Fe, have been determined showing that only small variations of H_R and $D_{R,0}$ take place as demonstrated by fig. 6 [15].

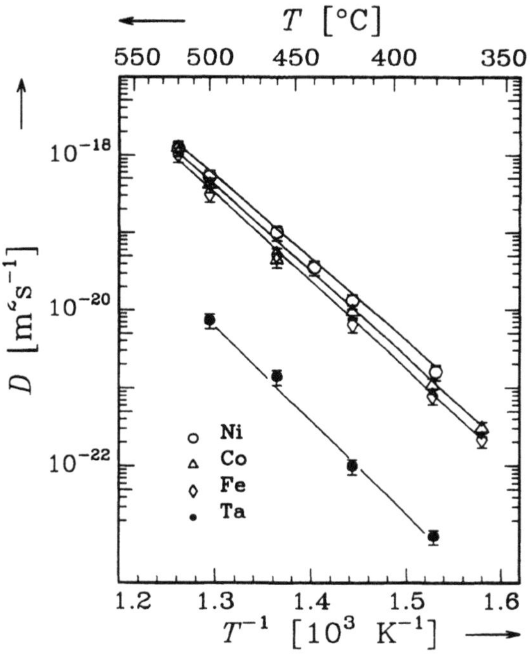

Fig. 6. Arrhenius plots of the radiotracer self-diffusion coefficients of Fe, Co, Ni and Ta in relaxed $Co_{58}Ni_{10}Fe_5Si_{11}B_{16}$ [16] (○ Ni, Δ Co, ◊ Fe, ● Ta).

According to fig. 7 it is obvious that in amorphous alloys the pre-exponential factor $D^T_{R,0}$ increases with increasing activation enthalpy and for the FeZr alloys an empirical relationship

$$\ln D^T_{R,0} = A(H_R - B) \tag{6}$$

exists with $A = 8\ (eV)^{-1}$ and $B = 2.4$ eV.

In contrast the pre-exponential factors of crystalline metals range only from approximately 10^{-5} m²/s to 10^{-4} m²s⁻¹ slightly decreasing with increasing activation enthalpy as demonstrated in fig. 7 for the refractory metals and Al [5]. These experimental results also clearly show that in amorphous alloys the diffusion mechanism is different from that of crystalline materials.

A very important experiment concerning the diffusion mechanism is the investigation of the irradiation-enhanced diffusion. From crystalline materials it is well known [17] that the increase of the density of atomic defects by irradiation leads to a decrease of the effective activation energy. In the case where we deal with an **indirect** diffusion mechanism where diffusion vehicles are required, e.g., self-diffusion by vacancies in thermal equilibirum, the irradiation-induced vacancies reduce the effective self-diffusion enthalpy because these vacancies do not require the energy consumption of the formation enthalpy. On the other hand no decrease of activation enthalpy is expected if we deal with a **direct** diffusion mechanism because in this case no diffusion vehicles have to be produced by thermal activation. The increase of the density of diffusing defect configurations by irradiation in this case only increases the pre-exponential factor leaving H_R constant. Both effects mentioned so far have been observed.

A decreasing activation enthalpy under Kr^+ irradiation has been observed in amorphous $Ni_{50}Zr_{50}$ [18] and in amorphous FeNiB under Ni^+ irradiation [19]. It should be noted, however, that a decrease of the activation enthalpy is not a necessary prove to an indirect diffusion mechanism in the unirradiated material because this also may be due to the introduction of a new species of diffusing vehicles.

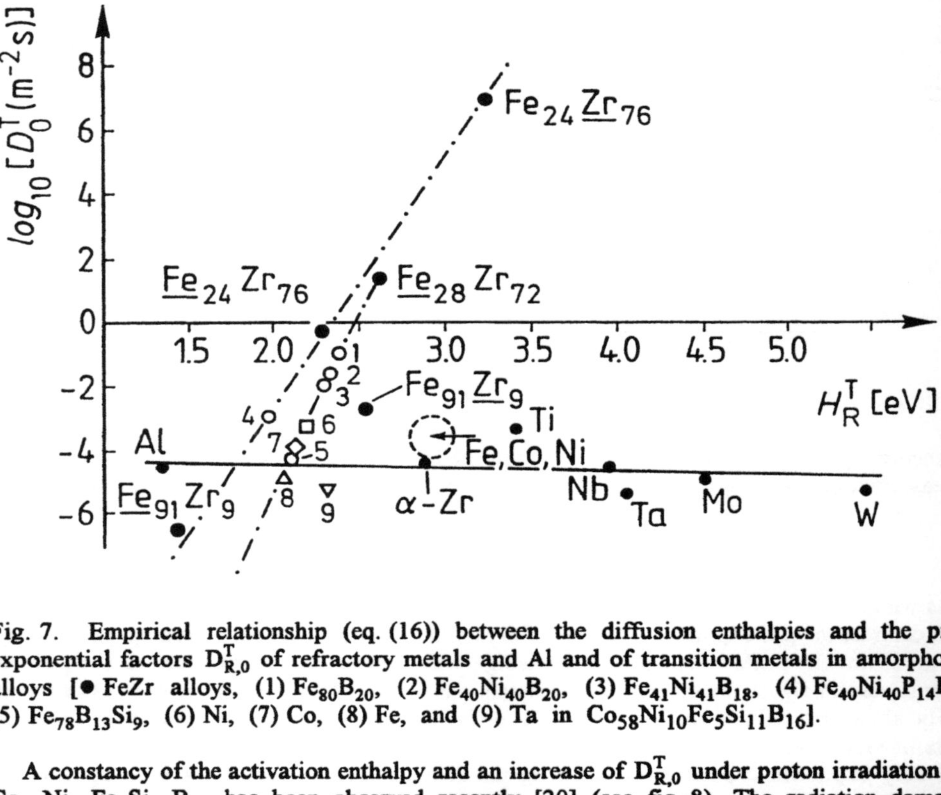

Fig. 7. Empirical relationship (eq. (16)) between the diffusion enthalpies and the pre-exponential factors $D^T_{R,0}$ of refractory metals and Al and of transition metals in amorphous alloys [● FeZr alloys, (1) $Fe_{80}B_{20}$, (2) $Fe_{40}Ni_{40}B_{20}$, (3) $Fe_{41}Ni_{41}B_{18}$, (4) $Fe_{40}Ni_{40}P_{14}B_6$, (5) $Fe_{78}B_{13}Si_9$, (6) Ni, (7) Co, (8) Fe, and (9) Ta in $Co_{58}Ni_{10}Fe_5Si_{11}B_{16}$].

A constancy of the activation enthalpy and an increase of $D^T_{R,0}$ under proton irradiation of $Co_{58}Ni_{10}Fe_5Si_{11}B_{16}$ has been observed recently [20] (see fig. 8). The radiation damage produced by 400 keV protons used in this experiment consists mainly in close quasi-Frenkel pairs being composed of a quasi-vacancy and a quasi-interstitial atom. At the temperature of irradiation of 360°C - 460°C these quasi-Frenkel pairs have a finite life time, because their configurations are smeared out. Thus during proton irradiation a steady-state concentration of smeared-out quasi-vacancies of the quasi-Frenkel pairs is established. This also is supported by the increase of the diffusion coefficient with increasing radiation flux $\dot{\phi}$. At a diffusion temperature of 420 K according to fig. 9 the diffusion coefficient increases by a factor of 10 for an increase of the flux by a factor of 10 [20]. Measurements performed at 380 K also show a similar increase of D with flux. Also this result is compatible with a direct diffusion mechanism and the increase of D with flux has to be attributed to the increase of the steady-state concentration of smeared out quasi-vacancies for a large flux.

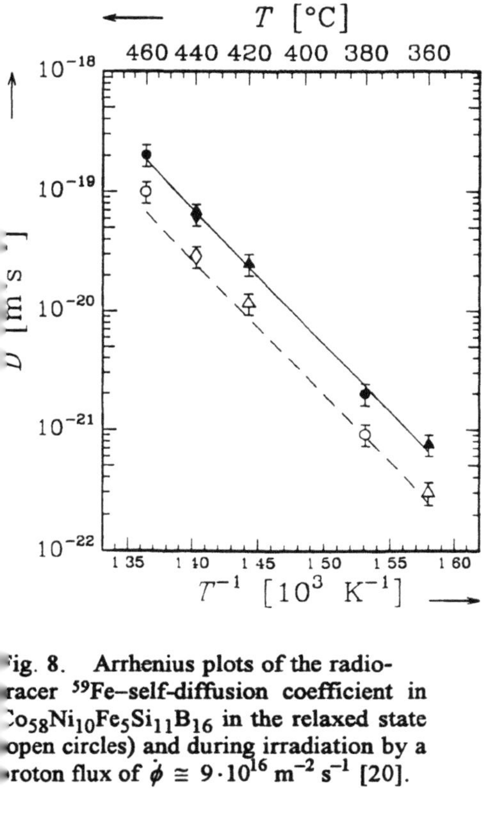

Fig. 8. Arrhenius plots of the radiotracer ^{59}Fe–self-diffusion coefficient in $Co_{58}Ni_{10}Fe_5Si_{11}B_{16}$ in the relaxed state (open circles) and during irradiation by a proton flux of $\dot{\phi} \cong 9 \cdot 10^{16}\ m^{-2}\ s^{-1}$ [20].

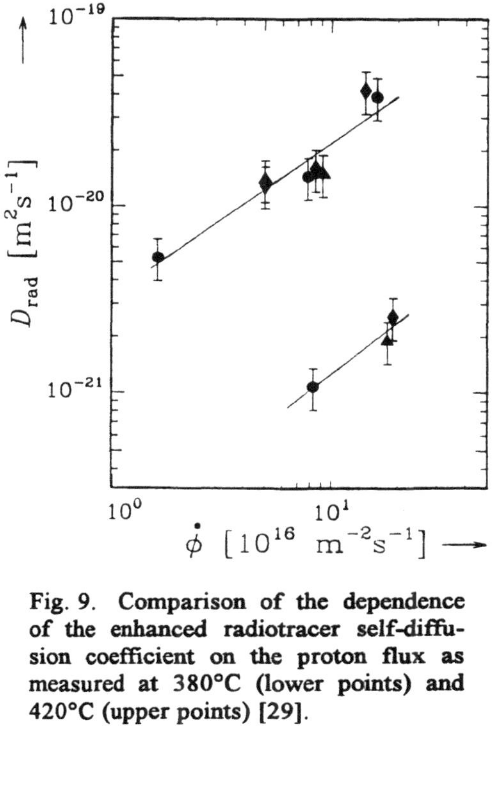

Fig. 9. Comparison of the dependence of the enhanced radiotracer self-diffusion coefficient on the proton flux as measured at 380°C (lower points) and 420°C (upper points) [29].

5. Theory of Diffusion in Disordered Media

5.1. THE "EFFECTIVE-MEDIUM APPROXIMATION" (EMA)

The description of diffusion in amorphous alloys in general is based on the following assumption: The jump frequency of an atom from a site i to a neighbouring site j obeys an Arrhenius law

$$\nu_{ij} = \nu_{ij,0} \exp\left[-\left(h_{ij} - h_i\right)/kT\right] , \tag{7}$$

where the attempt frequency, $\nu_{ij,0}$, contains the migration-entropy term and h_i denotes the enthalpy at site i and h_{ij} the enthalpy at the saddle point between i and j, respectively. The quantity to be determined is the diffusion coefficient which in general depends on time, t, and is defined as

$$D(t) = \frac{1}{2d} \frac{\partial}{\partial t} \langle \underline{R}^2 \rangle = \frac{1}{2d} \frac{\partial}{\partial t} \sum_i \underline{R}_i^2 \, P_i(t) \quad . \tag{8}$$

Here $\underline{R}_i$ are the site vectors and d corresponds to the dimensionality of the diffusion process. $P_i(t)$ denotes the conditional probability of occupation of site i at time t after starting at t = 0 at the origin. The motion of a particle is governed by a Markoffian master equation for the occupation probability

$$\frac{d}{dt} P_i(t) = \sum_j^{n.n.} \left[\nu_{ij} \, P_j(t) - \nu_{ij} \, P_i(t) \right] \quad , \tag{9}$$

where the summation extends over the nearest neighbours of site i or site j.

The solution of eq. (9) is well-known for periodic crystals where the atomic jump frequency is characterized by a unique activation enthalpy $h_{ij} - h_i = \Delta H$ and a unique pre-exponential factor $\nu_{0,ij} = \nu_0$ and gives a time independent diffusion coefficient

$$D = \frac{z}{2d} a^2 \, \nu_0 \, \exp\left[- \Delta H / k T \right] \quad , \tag{10}$$

where a corresponds to the distance between nearest neighbour sites and z to the number of nearest neighbours. The situation is far more complex in the case of amorphous alloys. Exact solution of eq. (9) are known for the random-trap model (RTM) [21,22] for d = 1,2,3 and for the random-barrier model (RBM) for d = 1 [23,24].

For all other cases, in particular for the generalized random-trap–random-barrier model (mixed model (MM)) suitable approximations have to be applied. A very powerful approximation, which approaches the exact solutions in the limiting cases of short and large diffusion times, is the so-called **effective-medium approximation** (EMA). A general description of this theory allowing a treatment of the random-trap–random-barrier model has been given by Haus and Kehr [25]. Starting from eq. (9) all jump frequencies depend on site i and nearest neighbour sites j and also depend on the direction of the transition. Introducing Laplace transforms eq. (9) leads to a set of coupled linear equation which in principle could be solved using extensive computer facilities. In the actual case we do not want to have the solution for a special site i but just the configuration average $G_i(t) = \langle P_i(t) \rangle$ which obeys the following generalized master equation

$$\frac{d}{dt} G_i(t) = \int_0^t \nu_c(t-t') \sum_j \left[G_j(t') - G_i(t') \right] dt' \quad , \tag{11}$$

where $G_i(t)$ corresponds to the conditional probability to find a particle after a time t at position i in the effective medium. Here $\nu_c(t)$ denotes an associated effective jump frequency of an effective medium describing the averaged probabilities of the disordered medium. In order to obtain the connection between the local jump frequencies, ν_{ij}, and ν_c we consider a cluster of bonds with transition rates corresponding to the original configuration and embedding it into the effective medium at position i'. All other bonds are described by eq. (11). Eq. (9) than writes (i ≠ i')

$$\frac{d}{dt} P_i(t) = \sum_j^{n.n.(i')} \left[\nu_{ji'} P_j(t) - \nu_{i'j} P_{i'}(t)\right] + \int_0^t \nu_c(t-t') \left[\langle P_j(t') - P_i(t')\rangle\right] dt' \quad , \qquad (12)$$

where the first term on the right side just describes the cluster and the second term the time averaged effective medium. Eq. (12) can be solved by a Laplace transformation of the time variable: The final set of linear homogeneous equations now can be solved by algebraic methods. The effective medium frequency ν_c than is determined from the self-consistency condition that the averaging over the jump frequencies of the cluster should give the same expression for $\tilde{G}_i(s)$ as given by the solution of eq. (11). This leads to the following condition:

$$\tilde{P}_i(s) = \int_0^\infty P_i(t) \exp(-st) \, dt \qquad (13)$$

(the tilt indicates the Laplace transform).

$$\left\langle \tilde{P}_i(s) \right\rangle_{\{\nu_{i'j}\}} = \tilde{G}_i(s, \nu_c) \qquad (14)$$

where $\{\nu_{i'j}\}$ denotes averaging over the cluster jump frequencies.

A solution of this self-consistency equation allows a calculation of the time dependent diffusion coefficient D(t). Under the assumption that the tracer-atom resides on all sites with equal probability the Laplace transformed diffusion coefficient is given by

$$\tilde{D}(s) = \frac{z}{2d} a^2 \tilde{\nu}_c(s) / s \quad . \qquad (15)$$

In the limit of short times this gives

$$D_0 = \lim_{t \to 0} D(t) = \frac{z}{2d} a^2 \tilde{\nu}_c(\infty) \quad . \qquad (16)$$

and for the long time limit we obtain

$$D_\infty = \lim_{t \to \infty} D(t) = \frac{z}{2d} a^2 \tilde{\nu}_c(0) \quad . \qquad (17)$$

5.2. APPLICATIONS OF THE "EFFECTIVE-MEDIUM APPROXIMATION"

The above outlined diffusion theory of disordered media has been applied to three types of diffusion potentials shown in fig. 10a-c [26,27]. 1) The random barrier model (RBM) with h_i independent of i and a Gaussian distribution of h_{ij} with a variance h_{RB}^2. 2) The random trap model (RTM) with h_{ij} independent of i and j and a Gaussian distributions $p(h_i)$ with a variance h_{RB}^2. 3) The mixed model (MM) with independent Gaussian distribution $p(h_i)$, $p(h_{ij})$ and variances $h_{RB}^2 = h_{RT}^2$. The temperature dependence of the diffusion coefficients of these three types of potentials have been determined for the limiting cases of short times (st) and long times (lt) and for a topological structure of a three-dimensional fcc-like disordered structure. Furthermore, the averaging indicated in eq. (14) has been performed within the so-called single-site–single-bond model where only on equilibrium site and one saddle point (bond) has been varied acording to Gaussian distributions. In fig. 10 the broken lines correspond to the Arrhenius plots for the average activation enthalpy $H_0 = \langle h_{ij} - h_i \rangle$. In the RBM model both the Arrhenius plots for the st-and the lt-case lie above the Arrhenius line of the average activation enthalpy since in the three-dimensional diffusion the high barriers can be avoided. The st effective diffusion enthalpy is given by $H_{eff} = \langle h_{ij} \rangle - h_{RB}^2 / kT$. In the RTM model D(T) is time independent and lies below the straight average enthalpy Arrhenius plot because the trapping positions are unavoidable. The lt effective diffusion enthalpy is given by $H_{eff} = \langle h_{ij} \rangle + h_{RT}^2 / kT$. In the MM model the st-limit coincides with the Arrhenius plot for H_0 since the fluctuations in the barrier heights and the trap depths just cancels for $h_{RB}^2 = h_{RT}^2$. In the case of the lt-limit D(T) lies below the H_0-Arrhenius line because of the dominance of the strong traps at long times.

It is evident that the three-dimensional mixed model is most suitable for describing diffusion in amorphous alloys. Therefore, time averaged Arrhenius plots have been determined for different measuring times which are shown in fig. 11. The st line is identical with the Arrhenius plot for the average activation enthalpy $H_0 = 1.65$ eV and $\nu_{0,ij} = 10^{12}$ s^{-1}. For the calculation of the Arrhenius plots, for the measuring times $t = 10^2$ s (curve 2) and $t = 10^5$ s (curve 1) and for the lt-limit the variances $h_{RB} = h_{RT} = 0.22$ eV have been used. Curve 2, with $t = 10^2$ s, is typical for the magnetic after-effect measurements and curve 1 with $t = 10^5$ s is typical for the radio tracer experiments and the induced magnetic anisotropy measurements. The hatched regions in fig. 11 indicate the temperature ranges where the different techniques probe different ranges of the spectrum of activation enthalpies. In agreement with the experimental results shown in fig. 1 for the three experimental techniques the largest effective activation enthalpy

$$H_{eff}(t) = -\left[\frac{\partial}{\partial(1/kT)}\right] \ln D(t,T) \tag{18}$$

is obtained for the radio tracer diffusion followed by those of the induced magnetic anisotropy and of the magnetic after-effect.

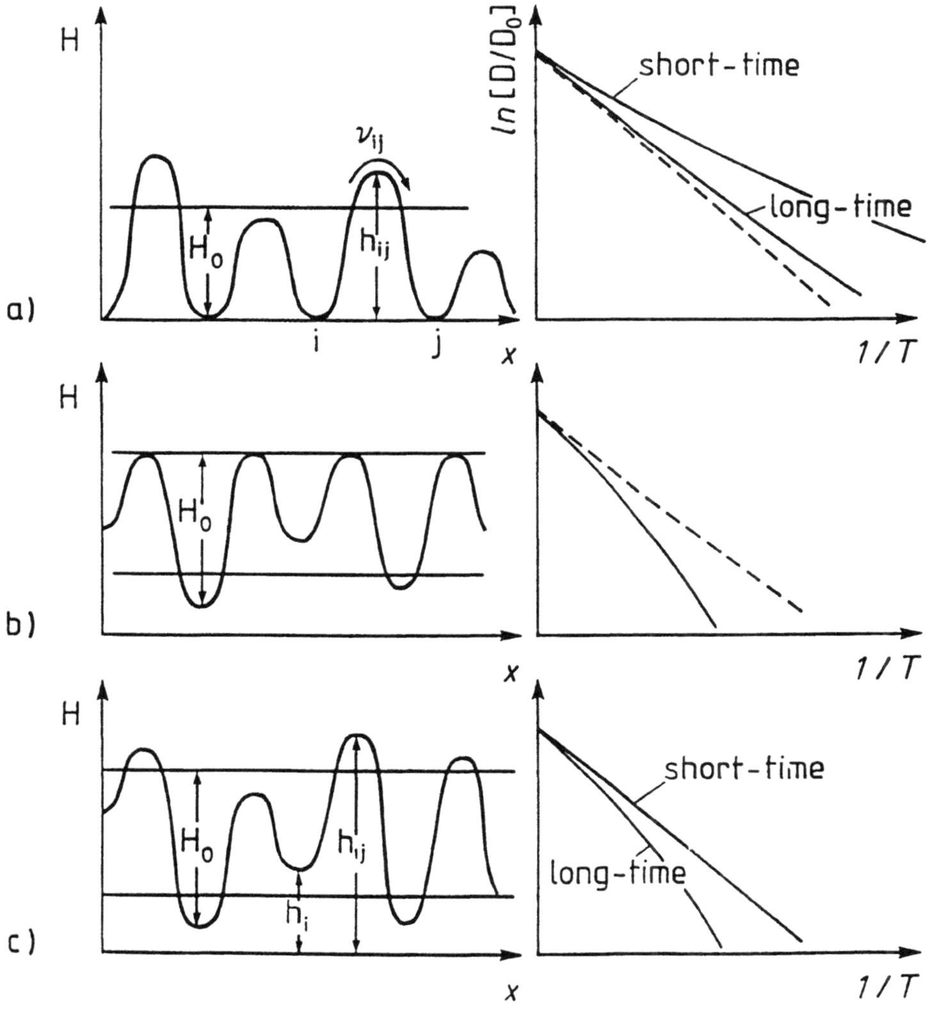

Fig. 10. Schematic representation of different diffusion potentials and of the corresponding Arrhenius plots of self-diffusion coefficients. (a) Random barrier model, (b) random trap model, (c) mixed model. The horizontal lines in the H-x diagrams correspond to $H = 0$, $H = \langle h_i \rangle$, and $H = \langle h_{ij} \rangle$, respectively (see text) [27].

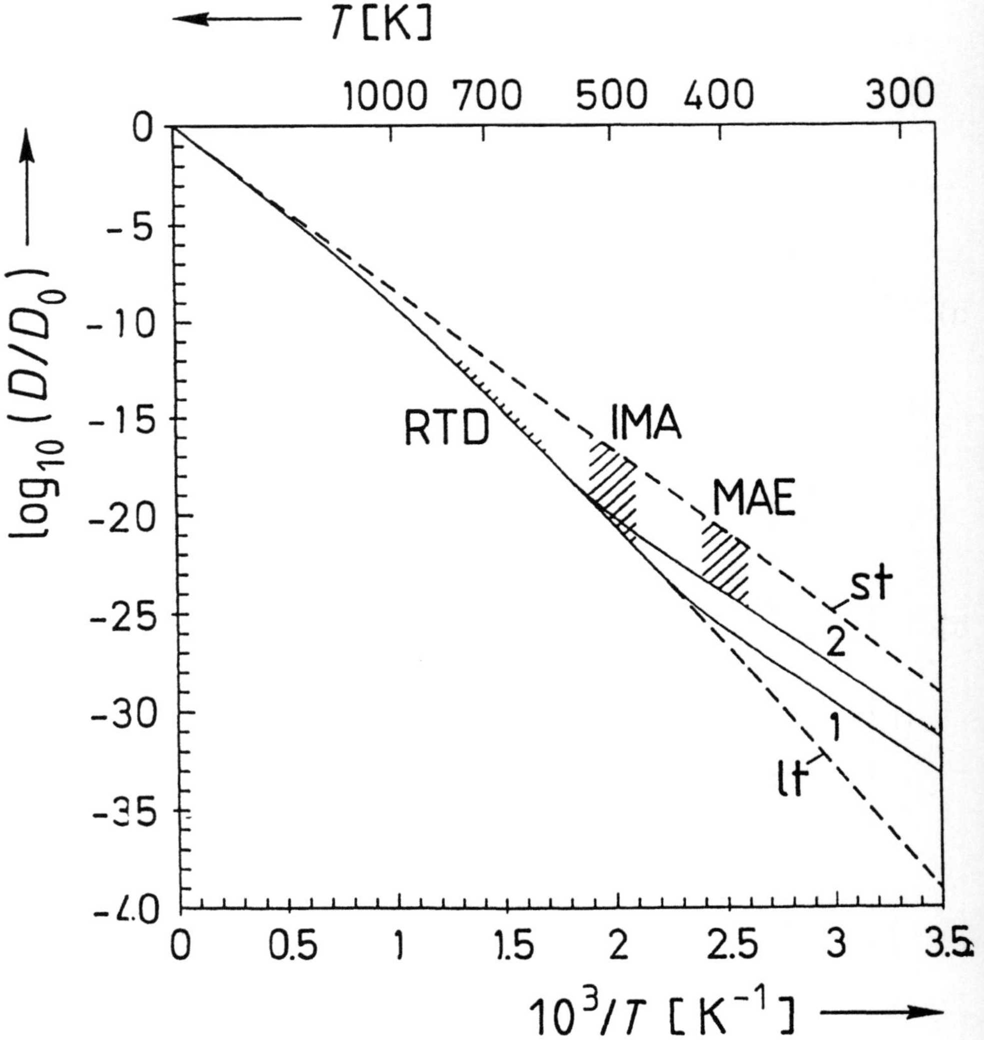

Fig. 11. Arrhenius representation of time averaged diffusion coefficients according to the mixed model for different diffusion times. The curves "st" and "lt" correspond to the limiting cases $t \rightarrow 0$ and $t \rightarrow \infty$. The hatched regions indicate the regimes where radiotracer (RTD), induced magnetic anisotropy (IMA), and magnetic after-effect experiments (MAE) have been performed [4].

The existence of rigorous Arrhenius equations for the diffusion coefficient in the long-time limit has been shown by Schirmacher [28] and Hörner [29]. Within the framework of the RBM and the single-site–single-barrier approximation the following relations hold for the long-time limit

$$D(T) = \left(\frac{z}{2d}\right) a^2 \nu_c(T) \quad , \tag{19}$$

where ν_c for $z > 2$ obeys the self-consistency condition

$$\left\langle \frac{\nu_{i',j} - \nu_c}{\nu_{i',j} + \left(\frac{z}{2} - 1\right) \nu_c} \right\rangle = 0 \quad . \tag{20}$$

Introducing the distribution function $\rho\,(\nu_{i',j})$ for the transition frequencies and $\nu_{i',j} = \nu_0 \cdot \exp\,[- h_{i',j} / kT]$ eq. (20) writes

$$2/z = \int_0^\infty d\,h_{i',j}\;\rho\left(h_{i',j}\right)\;\left\{1 - \exp\left[\left(h_{i',j} - \mu\right) / kT\right]\right\}^{-1} \quad . \tag{21}$$

Here the bracket in (21) corresponds to the Fermi function with a Fermi energy

$$\mu = -kT\,\ln\left[\left(z/2 - 1\right)\,\nu_c / \nu_0\right] \quad . \tag{22}$$

μ has the meaning fo an effective barrier height defining

$$\nu_c(T) = \frac{\nu_0}{z/2 - 1}\,\exp\left[-\mu / kT\right] \quad . \tag{23}$$

Here it is of interest to note that for $T \rightarrow 0$ the Fermi function can be replaced by a step function and μ is determined by

$$2/z = \int_0^\mu \rho\left(h_{i',j}\right) \cdot d\,h_{i',j} \quad . \tag{24}$$

Accordingly, this consideration yields an Arrhenius law for the diffusivity independent of the barrier distribution.

6. Conclusions

Self-diffusion data of amorphous alloys show the following characteristic features:

1. The self-diffusion coefficients of transition metals obey Arrhenius laws in the temperature range 200°C - 500°C and are a factor of 10^4 - 10^6 larger than in crystalline alloys of comparable compositions.
2. The effective activation enthalpies, H, in relaxed alloys throughout are smaller by 0.5 - 1.0 eV than in comparable crystalline alloys.
3. The pre-exponential factors, D_0, vary between 10^{-7} m²/s and 10^7 m²/s, whereas in crystalline alloys D_0 is nearly constant around $10^{-4\pm1}$ m²/s.
4. Relaxation phenomena due to short range diffusional processes have to be described by broad spectra of activation enthalpies which approach the self-diffusion enthalpies in the long time limit.
5. The diffusion processes in as-quenched, unrelaxed amorphous alloys are governed by migration and agglomeration of quasi-vacancies.
6. In the case of relaxed amorphous alloys self-diffusion occurs by a direct collective mechanism, where quasi-vacancies are involved.
7. Radiation enhanced diffusion under proton irradiation is governed by an increase of D_0 at constant H thus confirming a direct diffusion mechanism.
8. The effective medium approximation may be successfully applied to analyse the temperature and time dependence of diffusion enthalpies and spectra of diffusion enthalpies. The application of EMA to the "random barrier" model, the "random trap model" and the "mixed model" give a very informative insight into the role of random diffusion potentials.

References

[1] Y. Limoge, G. Brebéc and Y. Adda, in: Diffusion and Defects, F.J. Kedeves and D.L. Beke (Eds.), Trans. Tech. Publ.; Aedermannsdorf 1983, p. 285.

[2] B. Cantor, in: Rapidly Quenched Metals, S. Steeb and H. Warlimont (Eds.), Elsevier Amsterdam, 1985, p. 595.

[3] M.R.J. Gibbs, J.E. Evetts and J.A.J. Leake, J. Mater. Sci. Eng. **18** (1993) 278.

[4] H. Kronmüller and W. Frank, Rad. Eff. & Def. in Solids **108** (1989) 81; W. Frank, A. Hörner, P. Scharwaechter and H. Kronmüller, J. Mater. Sci. Eng. (Proc. 8th Intern. Conf. Rapidly Quenched and Metastable Materials, Sendai 1993), to be published.

[5] K. Maier, M. Peo, B. Saile, H.-E. Schaefer and A. Seeger, Phil. Mag. A **40** (1979) 701.

[6] A. Seeger and K.P. Chik, phys. stat. sol. **29** (1968) 455.

[7] A. Seeger, Defect and Diffusion Forum, Trans. Tech. Publ. Zürich, Switzerland, **95-98** (1993) 147.

[8] J. Horvath, J. Ott, K. Pfahler and W. Ulfert, J. Mater. Sci. Eng. **97** (1988) 409.

[9] W. Frank, J. Horvath and H. Kronmüller, Mater. Sci. Eng. **97** (1988) 415.

[10] J. Horvath and H. Mehrer, Cryst. Latt. Def. and Amorph. Mater. **13** (1986) 1.

[11] H. Kronmüller, H.-Q. Guo, W. Fernengel, A. Hofmann and N. Moser, Cryst. Latt. Def. and Amorph. Mater. **11** (1985) 135.

[12] W. Chambron and A. Chamberod, J. de Physique **42-C5** (1981) 511.
[13] H. Kronmüller, J. Magn. Magn. Mat. **41** (1984) 366.
[14] M. Bourrous and H. Kronmüller, phys. stat. sol. (a) **113** (1989) 169.
[15] P. Scharwaechter, Diploma Thesis, Universität Stuttgart, 1989.
[16] K. Pfahler, J. Horvath, W. Frank and H. Mehrer, in: Rapidly Quenched Metals, S. Steeb and H. Warlimont (Eds.), Elsevier, Amsterdam, 1985, p. 755; K. Pfahler, J. Horvath and W. Frank, Cryst. Latt. Def. and Amorph. Mater. **17** (1987) 249.
[17] R. Sizmann, J. Nucl. Mater. **69-70** (1968) 386.
[18] R.S. Averback and H. Hahn, Mater. Sci. Forum **37** (1989) 245.
[19] A.K. Tyagi, M.-P. Macht and V. Naundorf, J. Nucl. Mater. **179-181** (1991) 1026.
[20] P. Scharwaechter, Dr. rer. nat. Thesis, Universität Stuttgart, 1993.
[21] I. Webmann and J. Klafter, Phys. Rev. B **26** (1982) 5950.
[22] J.W. Haus, K.W. Kehr and J.W. Lyklema, Phys. Rev. **25** (1982) 2905.
[23] R. Zwanzig, J. Stat. Phys. **28** (1982) 127.
[24] P.J.H. Denteneer and M.H. Ernst, Phys. Rev. B **29** (1984) 1775.
[25] J.W. Haus and K.W. Kehr, Phys. Reports **150** (1987) 263.
[26] A. Hörner, Dr. rer. nat. Thesis, Universität Stuttgart, 1993.
[27] H. Kronmüller, W. Frank and A. Hörner, Mater. Sci. Eng. A **133** (1991) 410.
[28] W. Schirmacher, Solid State Ionics **28** (1988) 129.

MICROSCOPIC CRITERIA FOR GLASS FORMATION

T. EGAMI
Department of Materials Science and Engineering,
and Laboratory for Research on the Structure of Matter,
University of Pennsylvania, Philadelphia, PA 19104-6272, USA

ABSTRACT. Factors that contribute to the formability of metallic glasses are discussed from a microscopic point of view. We focus in particular on the atomic size factor and its effect on the topology of the structure. A universal local topological instability criterion is proposed to explain melting, solid state amorphization, glass transition, as well as composition range for glass formation by rapid cooling of the melt.

1. Introduction

The process of a liquid transforming into crystals upon cooling is a first order transition, involving a substantial interfacial energy between the liquid and solid phases. It proceeds with the nucleation and growth process, and its kinetics is strongly temperature dependent. Since the rate of transformation approaches zero at the melting temperature it is always possible to supercool the liquid, and upon rapid cooling it is sometimes possible to suppress the crystallization altogether to obtain a glass. In fact in theory, any liquid will become a glass when it is cooled sufficiently fast through the glass transition, as was pointed out by D. Turnbull sometime ago [1]. However, in reality only certain alloy systems within limited compositional ranges can actually be vitrified by currently available rapid quenching methods.

The criteria for glass formation can be discussed at two levels, at a macroscopic thermodynamical level and a microscopic or atomic level. In this paper we discuss the question of glass formability for metallic alloys from a microscopic point of view, for various paths to the glassy state, i.e., vitrification from liquid through the glass transition, amorphization by radiation damage, and solid state amorphization by diffusion or mechanical alloying. For metallic alloy systems in which atomic bonding is not strongly directional, two main factors contributing to the glass formability can be identified. They are the chemical association energy and the atomic size factor. The latter contributes to the instability of the crystalline phase through the atomic level stresses. We show that the local instability condition for the coordination cage corresponds to the instability condition for the solid structure and hence generally to the conditions for glass transition, melting, and glass formation. In the case when the chemical short range order does not change much upon vitrification, the composition range for glass formation is determined by the size factor alone.

M. A. Otooni (ed.), Science and Technology of Rapid Solidification and Processing, 291–307.

2. Thermodynamic Criteria

Soon after the discovery of Au-Si metallic glass by Duwez and his students [2], Cohen and Turnbull suggested that the ease of glass formation must be related to the presence of a deep eutectic in the phase diagram of Au-Si [3]. Indeed all the binary glasses discovered soon after including Pd-Si were characterized by a deep eutectic phase diagram [4]. Deep eutectic implies that the liquid is stable down to low temperatures, therefore it is quite reasonable to assume that the glass is also relatively stable. This point was later quantified by Marcus and Turnbull [5].

In the eutectic system, even below the eutectic temperature the free energy of the liquid phase remains lower than those of other crystalline phases over a certain composition range. Thus if phase separation is suppressed by rapid cooling, crystallization will not take place, and the system remains a liquid which will freeze into a glass at the glass transition temperature. If we define the T_0 line as the locus of the temperature and composition at which the free energies of the crystalline phase and the liquid or glass become equal, in an idealized condition of quenching in which no phase separation occurs, the boundary between the solid solution and liquid or glass may be given by the T_0 line. Therefore the glass forming composition may be found in-between the T_0 lines, as pointed out by Massalski [6]. This argument implies an exclusive relation between a solid solution and a liquid or glass. This fact was recognized early by Giessen et al. who applied the traditional approach for the question of solid solubility to defining the glass forming alloy systems [7-9].

3. Solubility Limit in Alloys and Glass Formability

The question of a solubility limit in crystalline alloys, or the composition limit for the formation of a solid solution, is a very old question in physical metallurgy, and has been discussed by a number of authors. Early on in the 1930's, Hume-Rothery and coworkers [10] proposed that three factors contribute to the formation of a solid solution:

1. Atomic size ratio between the solute atom and the solvent atom.
2. Electronegativity.
3. Relative valence.

In particular with regard to the first factor they pointed out that if the atomic sizes are different by more than 15 % a solid solution will not form. The second factor was then quantified by Pauling [11]. Darken and Gurry put the first two factors together in the well-known Darken-Gurry plot [12]. The second and third factors are often expressed together in terms of the enthalpy of mixing, ΔH [13]. The role of the enthalpy of mixing seems to be simple; if it is strongly positive the elements will not mix and phase-separate. Thus it has to be negative or only weakly positive. As for the atomic size factor Eshelby explained the 15 % rule by equating the local elastic energy associated with the solute to $kT_m/2$, where T_m is the melting temperature [14]. The idea is that if the local elastic energy is too large the lattice will locally "melt", i.e. the atomic configuration will be locally rearranged, and the phase becomes unstable.

In the case of the Hume-Rothery rules when a solid solution is not formed the alloy will phase-separate into a two phase material. On the other hand if such a material is melted and rapidly cooled such that the phase separation is suppressed, since a solid solution is unstable there is an excellent chance that the material remains glassy. Thus as we mentioned earlier an exclusive relation exists between the solid solubility and glass formability. Giessen et al. used the plot of the atomic size ratio vs. ΔH in order to define the glass forming systems [7-9]. It appears that the role of ΔH is the same for solid solution and glass, that it needs to be negative or weakly positive. Thus the size factor seems to be the one that really separates glasses from solid solutions. In the following we focus on the role of the size factor in glass formation. However, we should first discuss the crucial difference in the effect of the atomic size on the structures of crystals and glasses.

4. Atomic Size and Structure

4.1 METALLIC GLASS

The first realistic model of the structure of a liquid is the well-known dense random packing of hard spheres (DRP-HS) model of Bernal and Scott [15,16]. The DRP-HS structure contained local clusters such as icosahedra which are rarely seen in crystalline structures but were predicted by Frank to exist in liquids [17]. It also had cavernous holes (Bernal holes) of various topologies [18]. Polk proposed that the stability of the alloy glass is obtained by filling these holes with smaller elements such as metalloids [19].

Indeed many metallic glass systems discovered earlier were alloys of transition metals and metalloids, so that Polk's argument appeared to be sound. However, soon metallic glasses such as Cu-Zr which do not fit this description became known. Furthermore studies of computer-built soft sphere models revealed that large Bernal holes are a consequence of unyielding hard spheres, and disappear when the potential is softened. In the soft potential model which is certainly more realistic than the hard sphere model only tetrahedral and octahedral holes remain [20]. These holes are too small to be fitted with any element except for hydrogen.

Underlying in Polk's argument is an intuitive argument that by mixing spheres of two different sizes the overall packing is improved. Even though it sounds perfectly reasonable, however, this argument does not hold water, unless the two sphere sizes are drastically different. As shown in the model calculation by Cargill [21], for instance, the average coordination number of the two-component DRP structure remains independent of composition. Also as shown in Fig. 1 [22] in the absence of chemical affinity the energy of the two-component DRP structure is independent of both the atomic size ratio and the composition. Thus unless there is chemical affinity, ***mixing two components does not contribute to the stability of a glass***, since it does not reduce the ground state energy of the DRP structure. As we will discuss later, upon alloying glass formation becomes easy, ***not because the glass phase is stabilized, but because the crystalline phase is destabilized***.

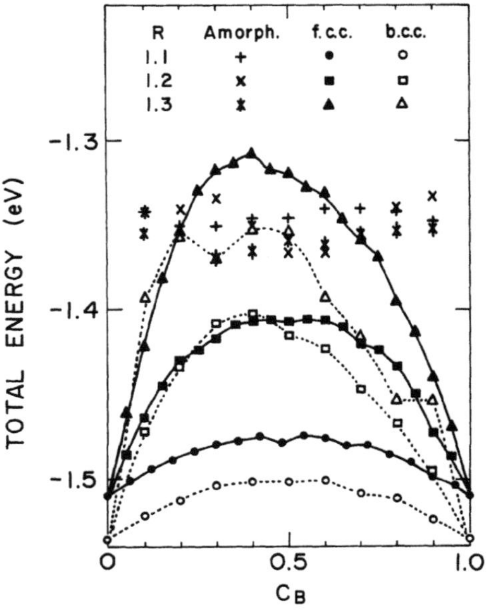

Fig. 1. Energy of binary (AB) alloy in the amorphous state and in the crystalline solid solution (f.c.c. and b.c.c.) against the composition, calculated by computer simulation. Note that the energy of the amorphous state is almost independent of composition, while that of a crystalline solid solution increases rapidly with the solute concentration [22].

Another demonstration of the fact that the size difference does not change packing efficiency is given by the calculation of the coordination number of an impurity atom in a metallic glass [22]. Fig. 2 shows the average coordination number of an impurity atom A embedded in the B metallic glass as a function of the ratio of the radii, R = r_A/r_B. Also the coordination number of an interstitial site in regular polyhedra is plotted. It is clear that the number of nearest neighbors increases with the size of the atom, as expected. The data show an excellent fit to the solid curve given by an equation

$$N_{AB} = \frac{4\pi(1-\frac{\sqrt{3}}{2})}{1-\frac{\sqrt{R(R+2)}}{R+1}} \tag{1}$$

which was derived assuming the 2-dimensional packing fraction, or the total solid angle to view the nearest neighbor from the central atom, to be independent of the size and composition [22]. Note that when $r_A = r_B$, $N_{AB} = 4\pi$.

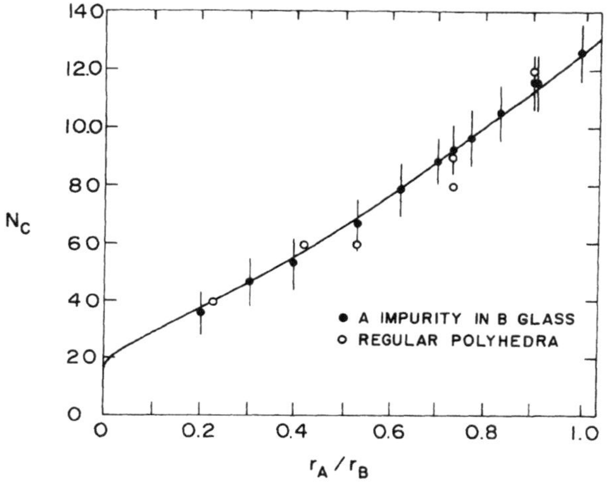

Fig. 2. Coordination number of an A impurity atom in a B glass as a function of the atomic size ratio. Also data for regular polyhedra such as icosahedron are shown. The solid curve is equation (1) [22].

The example above illustrates the crucial difference in the ways with which the structures of crystals and glasses are formed. *The crystal structure is topologically fixed, so that a substitutional or interstitial impurity atom has a fixed coordination number, and creates a strain field around it. On the other hand glasses are made by quenching the liquids, so that the local stresses or strains around an atom due to atomic substitution are completely relaxed in the liquid state by changing the local coordination number.* Liquids, thus glasses, are very accommodating. In fact a notion such as substitutional or interstitial defects has no meaning in glasses and liquids, except in the chemical sense, since they are not accompanied by long range stresses. In general it is dangerous to extend the concept developed for crystalline solids directly to glasses without carefully examining the premises.

4.2 CRYSTALLINE SOLID SOLUTION

An impurity atom in the crystalline matrix develops a stress and strain field around it, if there is a size mismatch as we just mentioned. The atomic level stresses can be rigorously defined as long as the two-body force can be defined [23-25], and the atomic level strain can then be defined by dividing through the elastic moduli. This stress field can be readily calculated in a continuum approximation [14,24,26]. Atoms are considered as elastic bodies, and the misfit strain is given by the strain field when an elastic sphere is inserted into a hole with a different size in an elastic matrix. This may appear to be too crude an approximation, but computer simulations show that it is sufficiently accurate for the purpose of the discussion [22]. Eshelby equated the elastic energy associated with this stress field to a half of the thermal energy at the melting point, kT_m, to deduce the size limit for solubility [14]. The local elastic energy is given by

$$E_{el} = E_0 \delta^2 c(1-c) \qquad (2)$$

where E_0 is the term involving the elastic constant as given in Ref. 14, c is the concentration of the impurity, and $\delta = R - 1$. Thus unlike liquids and glasses, the energy of a crystalline solid solution increases with the solute concentration, as shown also in Fig. 1. This increase destabilizes the solid solution. On the other hand the energy of a glass is independent of the solute composition. Thus there is a cross-over of the energies of the two states, if the size ratio $R = r_A/r_B$ is far from unity, and this cross-over point gives the composition at which $T_0 = 0$ K.

5. Glass Formation Criteria

5.1 VITRIFICATION BY MELT QUENCHING

According to the results in Fig. 1, the cross-over occurs only when R is greater than 1.25 ~ 1.3, larger than expected for the 15 % rule. Therefore the enthalpy cross-over at T = 0 K does not explain the solubility limit. Also since the local elastic energy due to impurity is given by equation (2), the cross-over condition at T = 0 K is given by

$$\delta^2 c_{crit}(1 - c_{crit}) = const. \qquad (3)$$

where c_{crit} is the cross-over concentration, thus the critical concentration for glass formation. For a small concentration of impurities therefore

$$c_{crit} \propto \delta^{1/2} \qquad (4)$$

However, experimentally c_{crit} is proportional to δ^{-1}, not $\delta^{-1/2}$ [26]. This again gives a serious doubt about the use of the enthalpy cross-over condition at T = 0 K as the glass formation criterion.

But this is not surprising, since whether the alloy forms a solid solution or a glass is determined during cooling from the liquid state, not by the energy balance at T = 0 K. In fact the relevant criterion is the cross-over of the ***free energies*** at the effective glass transition temperature in the time scale of cooling, or the fictive temperature. Since the liquid has a higher entropy than the crystal, by about $k_B \ln 2$ [27], the free energy of the glass/liquid rises more slowly than that of the crystal, and they cross each other at the melting temperature as shown schematically in Fig. 3. During rapid cooling the liquid becomes thermally arrested when $q(d\tau/dT) \approx 2$, where q is the cooling rate, $-dT/dt$, and τ is the structural relaxation rate [28]. For $Pd_{84}Si_{16}$ from the measured temperature dependence of the viscosity [29] this condition gives the fictive temperature T_f of about 750 K at a rate attainable with the standard melt-quenching machine, i.e. 10^6 K/sec, which also agrees with the estimate in Ref. 29 based upon the position of the nose in the time-temperature-transformation (T-T-T) diagram. Thus we expect T_f/T_m to be about 0.6, and therefore a cross-over can occur for an alloy with the size factor R greater than 1.15 or so, in agreement with the Hume-Rothery rule and the estimate by Eshelby.

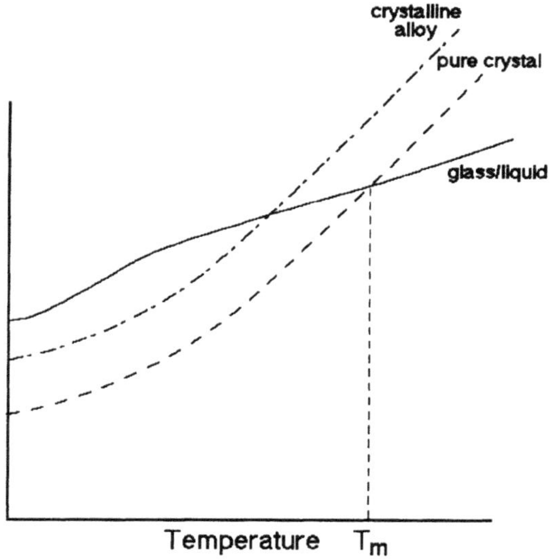

Fig. 3. Schematic diagram of the temperature dependence of the free energy for glass/liquid (solid curve), pure crystal (dashed curve) and crystalline solid solution (chained curve).

On the other hand the criterion adopted in Ref. 26 is the matrix instability or catastrophe condition, based upon the average stress in the matrix. If $r_A > r_B$, the impurity atom A is under compression, since the vacancy in the B matrix in which the A atom is fitted in is smaller than the A atom. Then the B atoms in the matrix must be under tension, since the total stress of the sample is zero. Or,

$$p_B = \frac{c_A}{1-c_A}\frac{\Omega_A}{\Omega_B}p_A \tag{5}$$

where p_A and p_B are the averages of the local pressure on the A and B atoms, and Ω_A and Ω_B are the atomic volume of A and B, respectively. The catastrophe condition is that when the local volume strain in the matrix

$$\varepsilon_B = \frac{p_B}{B_B} \tag{6}$$

where B_B is the atomic bulk modulus of the B atom, exceeds the critical value the crystal falls apart. Since p_A is proportional to the volume mismatch, $\Delta\Omega = \Omega_A/\Omega_B$ -1

[24,26], and thus the size mismatch δ, for small values of c_A the critical concentration is given by

$$c_{crit} \propto \Delta\Omega^{-1} \propto \delta^{-1} \tag{7}$$

which agrees with the observation [26,30],

$$c_{crit} \approx \frac{0.1}{|\Delta\Omega|} \quad . \tag{8}$$

More accurately, the critical solute concentration for the instability of solid solution, or glass formation by quenching, is given by

$$c_A^{\min} = \frac{1}{1 - y + \dfrac{y}{\varepsilon_v^T}\left[1 - \left(\dfrac{2}{1+R}\right)^3\right]} \tag{9}$$

where

$$y = R^3\left[\frac{2(1-2\nu)}{3(1-\nu)}\left(\frac{2}{1+R}\right)^3 + \frac{1+\nu}{3(1-\nu)}\right] \tag{10}$$

and ν is the Poisson's ratio (ν = 1/4 for a central force potential). The critical value of the strain, ε_v^{crit}, is estimated by fitting the predicted concentration dependence to the experimental data. This strain is the volume strain necessary to fit a B atom in the co-ordination cage of the matrix defined by the nearest neighbor atoms without causing the strain around it, or the transformation volume strain as defined by Eshelby [14]. This value comes to ε_{crit} = 0.054 [26], which is very close to the expected value as we discuss below.

5.2 SOLID STATE AMORPHIZATION BY RADIATION DAMAGE

Glasses can also be formed by accumulated radiation damage [31]. Irradiation of a crystal by high energy particles produces Frenkel pairs, or pairs of vacancy and interstitial. When the density of the Frenkel pairs is increased the crystalline lattice gradually becomes disintegrated, and finally becomes totally glassy. Since this process resembles melting, interpretations of this phenomena as a special form of melting invoking Lindeman's criterion for melting have been proposed [31-33]. Although the analogy is quite reasonable, this alone is not sufficiently enlightening, if we cannot explain the Lindeman's criterion itself sufficiently well. In addition, melting is a thermodynamic equilibrium phenomenon which occurs as a consequence of the cross-over of the free energies of the solid and the liquid, while amorphization by irradiation is a non-equilibrium kinetic phenomenon relating to the destabilization of the damaged lattice in the solid state. The equivalence of these two phenomena is not necessarily obvious, and needs to be explained as below.

On the other hand Kulp et al. [34] observed by molecular dynamics simulation that as the number of Frenkel pairs is increased the r.m.s. value of the atomic level stresses increases, and the crystal lattice becomes gradually disintegrated because it cannot sustain high levels of stress. It is important to note that during this process there is no sign of stress overshooting, and the average stress increases monotonocally until it saturates to a value that was observed in the glass quenched from the liquid, or in terms of volume strain, 0.065 [22,24]. Since this residual strain in glasses is related to glass transition temperature as discussed in Ref. 24, if solid state amorphization is equivalent to melting, glass transition can also be considered to be some form of melting, as we discuss below.

The fact that no stress overshooting was observed implies that the threshold stress or strain for structural instability is the same for crystalline and amorphous phases. In a crystalline solid well below the glass transition temperature if the stress exceeds this threshold the structure will rearrange itself, or collapse, into a glass, since there is not enough thermal energy to create a new ordered crystalline structure. In a glass if the stress exceeds the threshold the structure will also be rearranged, but the product still is essentially the same glassy structure. Unlike the crystalline state there is a very high degree of degeneracy in the glass structure.

The r.m.s. value of the volume strain in the glass, thus the threshold strain for the structural instability in the solid state, 0.065, corresponds to the transformation strain of 0.146. This value is more than twice the critical value of transformation strain for the instability of the solid solution, 0.054. Thus these two phenomena, solid state amorphization and glass formation by quenching the liquid, apparently have different size criteria. This point will be discussed further later on. Also not all crystals can be amorphized by radiation. Various criteria for irradiation amorphization, in particular the chemical destabilization, have been extensively discussed [35], but the size effect has not been clearly resolved.

5.3 SOLID STATE AMORPHIZATION BY DIFFUSION

Another path of solid state amorphization is diffusive reaction either by low temperature annealing or by mechanical alloying [36]. In this case the criterion is most likely the T_0 criterion, or the free energy cross-over, as we discussed above [36,37]. While a strongly negative enthalpy of mixing is necessary for the reaction to occur at all, the borderline between a glass and a solid solution must be again determined by the size effect. However, no systematic analysis has been made on this subject. The condition of shear instability used in identifying the amorphization condition [37,38] and its relation to melting, again via Lindeman's criterion [39], suggest a strong link to the local structural instability discussed below.

6. Universal Microscopic Criterion

Surveys above show that the atomic size effect on glass formation may be different for each path of glass formation, vitrification through glass transition, radiation damage, and solid state reaction by diffusion. On the other hand a strong similarity exists among amorphization, glass transition and melting phenomena. Furthermore it is not likely that the principle behind the size effect be so much different for each path to the glassy state. In this chapter we show that the condition for glass formation as well as glass transition and melting can indeed be elucidated by a single principle of local structural instability.

6.1 MAXIMUM HOOP STRESS

Let us examine equation (1) again. This equation describes the number of B atoms which form a cavity of the size r_A. Simply put, it takes a larger number of atoms to form a larger hole. By differentiating eq. (1) by R and evaluating it at R = 1 one obtains

$$\frac{\partial R}{\partial N_{AB}} = \frac{\sqrt{3}}{\pi}\left(1-\frac{\sqrt{3}}{2}\right) = 0.0739 \qquad (11)$$

Thus by changing the value of R by this much the coordination number is changed by one. The physical meaning of this result is easy to see as follows: If we consider the cage made of the nearest neighbor atoms, with the radius $r_A + r_B$, and define the surface strain of the cage, ε_s^T, which will bring it to the size with R = 1,

$$\varepsilon_s^T = \left(\frac{R+1}{2}\right)^2 - 1 \qquad (12)$$

then the surface strain due to changing the coordination number by one is

$$\left.\frac{\partial \varepsilon_s^T}{\partial N_{AB}}\right|_{R=1} = \frac{1}{N_{AB}} = \frac{1}{4\pi} = 0.0796 \qquad (13)$$

Since

$$\frac{\partial R}{\partial N_{AB}} = \frac{\partial \varepsilon_s^T}{\partial N_{AB}} \qquad (14)$$

the result in eq. (13) is very close to that in eq. (11), suggesting that the eq. (11) describes a very reasonable result.

Now, if the coordination cage has this much of a surface strain, then by rearranging the atoms and changing the local coordination number by one the strain is wiped out, reducing the local elastic energy. Thus the system will not be able to sustain the level of strain as large as in eq. (11). In fact we do not have to have the full amount of the strain before the change in the coordination number to happen, half as much is enough, since if the strain is larger than a half the value in eq. (11), by changing the coordination number by one the sign of the strain is reversed, and the magnitude of the strain is reduced. This defines the critical surface, or hoop, strain,

$$\varepsilon_s^{crit} = \frac{\sqrt{3}}{2\pi}\left(1 - \frac{\sqrt{3}}{2}\right) = 0.0369 \tag{15}$$

or in terms of the volume strain,

$$\varepsilon_v^{crit} = \frac{3}{2}\varepsilon_s^{crit} = \frac{3\sqrt{3}}{4\pi}\left(1 - \frac{\sqrt{3}}{2}\right) = 0.0554 \tag{16}$$

When the volume strain exceeds this value the nearest neighbor cage structure becomes unstable and some atomic rearrangement occurs which would change the local coordination by one.

6.2 CONDITION FOR TOPOLOGICAL INSTABILITY

The maximum volume strain given in eq. (16) represents the topological instability point for the close packed structure. However, the way this instability condition applies is critically different in the liquid state and in the solid state. In the solid state the local size mismatch produces a static long range stress field, both in glasses and in crystals [24], while in the liquid state the temporarily fluctuating atomic level stress is confined to the nearest neighbors, and the long range stress is almost immediately relaxed. Therefore the instability condition is different when the critical point is approached from the liquid and from the solid.

In the liquid state since the local stress fluctuation produces no stress field beyond the nearest neighbor cage, the relevant strain is the one that defines the size of the coordination cage, or the so-called transformation strain in Eshelby's definition, ε^T. If, due to thermal fluctuation, the size of the cage becomes too big, then there will be some atomic rearrangement to take in an atom to the nearest neighbor cage to increase the local coordination number. If it becomes too small, one atom will be ejected from the coordination cage. Because of high fluidity in the liquid the structural perturbation due to this change in the local coordination number is relaxed within the Debye time, usually of the order of a pico-second, and does not produce any stress field around it.

On the other hand in the solid state the coordination cage is static, and is embedded in the elastic matrix. Consequently the stress is partially relaxed. Therefore the relevant strain is not the transformation strain but the static partially relaxed strain which is related to the transformation strain by

$$\varepsilon^R = \frac{2(1-2\nu)}{3(1-\nu)}\varepsilon^T = \frac{4}{9}\varepsilon^T \tag{17}$$

The transformation strain corresponding to the partially relaxd strain being critical is

$$\varepsilon_v^{T,crit} = \frac{9}{4}\varepsilon_v^{R,crit} = 0.125 \tag{18}$$

When the local strain exceeds this value there will be spontaneous atomic rearrangements even in the solid state.

6.3 STRUCTURAL INSTABILITY AND MELTING

The structural instability condition eq. (16) should apply equally well to a pure crystalline solid with the close packed structure, if we neglect a small difference in the coordination number, N_c , between the f.c.c. lattice ($N_c = 12$) and the DRP structure ($N_c = 4\pi = 12.56$ [23]). Therefore it is possible that this condition corresponds to the critical dilatation for melting. Indeed the dilatation at T_m,

$$\varepsilon_v(T_m) = \frac{V(T_m) - V(0)}{V(0)} \tag{19}$$

where V(T) is the volume at temperature T, is close to the value in eq. (16) [38], corresponding to the linear strain,

$$\left.\frac{\Delta a}{a}\right|_{crit} = \frac{1}{3}\varepsilon_v^{crit} = 0.0185 \tag{20}$$

This instability condition explains Lindeman's melting condition which was derived empirically.

6.4 SOLID STATE AMORPHIZATION

In the case of melting of a pure elemental metal, the macroscopic strain, eq. (19), is equal to the local strain. However, during the solid state amorphization by irradiation the local stress is highly inhomogeneous. Therefore, we have to consider the atomic level local strain,

$$\varepsilon_v^R = \frac{\langle p^2 \rangle^{1/2}}{B} \tag{21}$$

When this strain reaches the critical value given in eq. (16) amorphization should take place. In other words when the relaxed strain around the atom in the solid state exceeds the critical value the structure collapses. As we mentioned above this is in agreement with the results of computer simulation. Therefore amorphization indeed is melting in an extended sense as proposed by Okamoto and Meshii [31], and by Wolf et al. [32]. However, the critical parameter is not the macroscopic volume as they initially proposed. In the case of melting the macroscopic volume strain is equal to the microscopic volume strain as we mentioned, but in the case of radiation damage the macroscopic dilatation reflects the local strains only through anharmonicity. Therefore the dilatation itself is NOT the parameter of structural instability. Indeed the result of computer simulation indicates that the volume itself is not a reliable indicator [33,34]. Instead the local strain must be the critical parameter of structural instability.

The standard deviation of the distribution in the locally averaged nearest neighbors, σ_N, used by Devanathan et al. [33] is a good measure of the local stress in an elemental system, since it is related to the local atomic volume, thus to the local pressure. However, in the case of alloys or compounds the partial pressure of each element is not necessarily zero, and in such a case σ is not a good measure of pressure. In the case of the result in Ref. 33 it is possible that just before the amorphization chemical order was almost completely lost, thus making the partial pressure on each element zero. Indeed, since radiation damage destroys chemical order, ordered compounds or alloys will be destabilized, and this contributes to glass formation as we discuss below [35].

During the solid state amorphization by diffusion when the concentration of the solute becomes high enough for the strain in the matrix, eq. (6), to exceed the critical value, eq. (16), amorphization should occur. In this sense, this phenomenon is also the melting in an extended sense [37]. Indeed the shear instability condition suggested by Fecht and Johnson [39] implies that the T_0 line is determined by the structural instability. The increase in the r.m.s. displacement due to alloying [40] in relation to the general melting criterion [41] also suggests that the equivalence of the local strain instability and melting. Then the composition limit for solid state amorphization can be predicted by equation (9), but using the value of transformation strain given in eq. (18). Thus the composition range for amorphization should be considerably narrower than that for glass formation by liquid quenching. This prediction remains to be tested.

6.5 GLASS TRANSITION

Now this instability condition should apply to the glassy structure as well. The inverse of the logic behind eq. (16) is that if the local strain is below the critical value the structure is stable. Thus the critical strain in eq. (16) should define the level of strain in the amorphous state at $T = 0$ K. Indeed the value of the real strain in the DRP structure built by a computer is close to this value, being equal to about 0.065 [22,24] as we mentioned before, slightly higher than the calculated value. This small difference is most likely the effect of structural relaxation [42]. The DRP structure created by a computer is not likely to be completely relaxed, since the structural relaxation is a very time consuming process. Furthermore this level of the strain defines the glass transition temperature along the line of the mechanism described in Ref. 24, although the details were corrected later as a result of a molecular dynamics study [43,44].

6.6 GLASS FORMATION BY QUENCHING

The phenomena we discussed above are the cases of solids being destabilized by local stresses, resulting in melting. Thus the direction of approach is from the side of the solid. On the other hand in the case of melt-quenching we approach from the side of the liquid, and consequently the criterion eq. (16) has to be applied ***in the liquid state***. During the rapid quenching of a molten alloy, crystallization occurs very rapidly if nuclei of the crystalline solid solution can be easily formed in the liquid state, resulting in copious nucleation. Otherwise atomic rearrangements are required at the crystal/liquid interface for solidification, which slows down the crystallization rate. The crystalline nuclei are in a dynamic state with plentiful thermal energy, thus ***the criterion for instability should be that the transformation strain exceeds the critical value***. Indeed the critical value of transformation, 0.054 [26], is virtually identical to the critical value, 0.055, given in eq. (16). In Ref. 26 we used a simpler criterion similar to eq. (13), i.e. the critical strain of 0.0625. The eq. (16) is more accurate, and indeed shows better agreement with the experimental results.

Thus a single criterion, eq. (16), can explain the melting, glass transition [24], vitrification by rapid cooling, and solid state amorphization by irradiation and by diffusion. It also makes it clear that while the solid state amorphization phenomena are understood as equivalent to melting by crossing the T_0 line so that the composition range for glass formation may be defined by the T_0 lines, the stability of solid solution during rapid quenching is NOT defined by the T_0 line, so that the glass forming composition cannot be defined by the T_0 lines, but by the equation (9). Even though they share the same local instability condition, eq. (16), the way the condition is applied is crucially different for glass formation by rapid quenching and by solid state amorphization. Consequently the composition range for glass formation by quenching is considerably wider than that by solid state amorphization.

6.7 EFFECT OF CHEMICAL AFFINITY

In the preceding argument we focused only on the atomic size effect, whereas other factors such as chemical affinity and the electron density have been suggested to contribute to glass formability [45]. The argument by Nagel and Tauc on electron density [45] is an extension of the Mott-Jones explanation of the Hume-Rothery rule for compound formation [46], and states that when $2k_F$, where k_F is the fermi momentum, is equal to Q_p, the position of the first peak in the structure factor, the glass phase is stabilized. When $2k_F$ is equal to Q_p the minimum of the Friedel oscillation agrees with the nearest neighbor distance, so that the potential minimum is deeper and the solid is more stable [47]. However, this stabilization applies almost equally for glasses as well as for crystals, since crystals also usually have strong Bragg peaks near Q_p. Thus while this stabilization contributes to raise the glass transition temperature and make it more difficult for a glass to crystallize once the glass is obtained, it may not contribute to the ease of producing the glass.

The same argument applies to the general role of chemical affinity, or the negative heat of mixing, ΔH. Chemical affinity increases the glass transition temperature, but does not affect the ease of glass formation, unless there is an appreciable difference in the degree of chemical ordering between the glass and crystalline phase. In fact chemical order is often higher in crystalline compounds than in glasses, thus this effect makes it more difficult to produce a glass by liquid-quenching, but easier by irradiation. By destroying chemical order by particle irradiation a crystalline compound or alloy can be destabilized, resulting in formation of a glass [35]. In general these chemical effects can be incorporated in the interatomic potential. In case the competeing solid phase has a similar degree of chemical order or disorder, the critical factor which plays a decisive role in glass formability is the atomic size factor.

7. Conclusions

In this article we have discussed the microscopic criteria of glass formation focusing on the effect of atomic size. The argument is purely geometric, which may give rise to a very crude impression. However, other factors such as electron density and chemical affinity contribute almost equally to the stability of glasses as well as that of crystals. While they increase the absolute stability of glasses, they do not contribute to the relative stability against crystallization during quenching. The atomic size factor still is the dominant factor that influence the ease of glass formation. In this paper we extended the concept of the critical strain for the instability of the coordination cage first introduced in our theory of glass forming composition by melt-quenching [26], to the general universal structural instability condition including the glass forming condition by solid state amorphization, Lindeman's criterion for melting, and the instability of the glass structure itself which determines the glass transition. The simplicity of the concept as well as its wide applicability suggest the validity of this concept.

Acknowledgment

This work was supported by the National Science Foundation DMR93-00728. The author benefited greatly from discussions with D. E. Luzzi and V. Vitek.

References:

1. D. Turnbull, Contemp. Phys., **10**, 471 (1969).
2. W. Klement, Jr., R. H. Willens and Pol Duwez, Nature, **187**, 869 (1960).
3. M. H. Cohen and D. Turnbull, Nature, **189**, 131 (1961).
4. Pol Duwez, R. H. Willens and R. C. Crewdson, J. Appl. Phys., **36**, 2267 (1965).
5. M. Marcus and D. Turnbull, Mater. Sci. Eng., **23**, 211 (1976).
6. T. B. Massalski, *Proc. 4th Int. Conf. Rapidly Quenched Metals*, eds. T. Masumoto and K. Suzuki (The Japan Inst. Metals, Sendai, 1982) p. 203.
7. D. E. Polk and B. C. Giessen, in *Metallic Glasses,* eds. J. J. Gilman and H. J. Leamy (ASM, Metals Park, OH, 1978) p. 1.
8. B. C. Giessen and S. Whang, J. de Phys. **41**, C8-95 (1980).
9. B. C. Giessen, Proc. *4th Int. Conf. Rapidly Quenched Metals,* eds. T. Masumoto and K. Suzuki (The Japan Inst. Metals, Sendai, 1982) p. 213.
10. W. Hume-Rothery, G. W. Mabbott and K. M. Channel-Evans, Phil. Trans. Royal Soc., London, **A233**, 1 (1934).
11. L. Pauling, *The Nature of the Chemical Bond* (Cornell Univ. Press, Ithaca, 1939).

12. L. S. Darken and R. W. Gurry, *Physical Chemistry of Metals* (McGraw-Hill, New York, 1953).
13. A. R. Miedema and P. F. de Châtel, *Theory of Alloy Phase Formation,* ed. L. H. Bennett (TMS-AIME, Warrendale, PA, 1980) p. 344
14. J. D. Eshelby, Solid State Physics, **3**, 79 (1956).
15. J. D. Bernal, Nature, **188,** 410 (1960).
16. G. D. Scott, Nature, **188,** 408 (1960).
17. F. C. Frank, Proc. Roy. Soc., **A215**, 3 (1952).
18. J. D. Bernal, Proc. Roy. Soc., **A284,** 299 (1964).
19. D. E. Polk, Script. Metal., **4**, 117 (1970).
20. J. L. Finney and J. Wallace, J. Non-Cryst. Solids, **43,** 165 (1981).
21. G. S. Cargill, III, *Amorphous Materials:Modeling of Structure and Properties,* ed. V. Vitek (TMS-AIME, Warrendale, PA, 1983) p. 15.
22. T. Egami and S. Aur, J. Non-Cryst. Solids, **89,** 60 (1987).
23. T. Egami, K. Maeda and V. Vitek, Phil. Mag., **A41,** 993 (1980).
24. T. Egami and D. Srolovitz, J. Phys. F, **12,** 2141 (1982).
25. V. Vitek and T. Egami, phys. stat. sol., (b) **144,** 145 (1987).
26. T. Egami and Y. Waseda, J. Non-Cryst. Solids, **64,** 113 (1984).
27. R. M. J. Cotterill and J. U. Madsen, *Amorphous Materials:Modeling of Structure and Properties,* ed. V. Vitek (TMS-AIME, Warrendale, PA, 1983) p. 179.
28. G. W. Scherer, *Relaxation in Glass and Composites* (John Wiley, New York, 1986).
29. N. Naka, Y. Nishi and T. Masumoto, *Rapidly Quenched Metals III,* ed. B. Cantor (The Metals Society, London, 1978) p. 231.
30. S. H. Liou and C. L. Chien, Phys. Rev., **B35**, 2443 (1987).
31. P. R. Okamoto and M. Meshii, *Science of Advanced Materials,* eds. H. Wiedersich and M. Meshii (ASM International, Materials Park, OH, 1990) p. 33.
32. M. J. Sabochick and N. Q. Lam, Phys. Rev., **B43,** 5243 (1991).
33. R. Devanathan, N. Q. Lam, P. R. Okamoto and M. Meshii, Phys. Rev., **B48**, 42 (1993).
34. D. T. Kulp, T. Egami, D. E. Luzzi and V. Vitek, J. Non-Cryst. Solids, **156-158,** 510 (1993).
35. D. E. Luzzi, H. Mori, H. Fujita and M. Meshii, Acta Metall., **34,** 629 (1986).
36. R. B. Schwarz, J. B. Rubin and T. J. Tiainen, *Science of Advanced Materials,* eds. H. Wiedersich and M. Meshii (ASM International, Materials Park, OH, 1990) p. 1.
37. W. L. Johnson, Mo Li and C. E. Krill III, J. Non-Cryst. Solids, **156-158,** 481 (1993).
38. J. L. Tallon, Nature, **342,** 658 (1989).
39. H. J. Fecht and W. L. Johnson, Nature, **334**, 50 (1989).
40. C. E. Krill III, J. Li, C. Ettl, K. Samwer, W. B. Yelon and W. L. Johnson, J. Non-Cryst. Solids, **156-158**, 506 (1993).
41. A. Voronel, S. Rabinotich, A. Kisliuk, V. Steinberg and T. Sverbilova, Phys. Rev. Lett., **60,** 2402 (1988).
42. D. Srolovitz, T. Egami and V. Vitek, Phys. Rev., **B24**, 6936 (1981).

43. S. -P. Chen, T. Egami and V. Vitek, Phys. Rev., **B37,** 2440 (1988).
44. T. Egami, T. Tomida, D. Kulp and V. Vitek, J. Non-Cryst. Solids, **156-158,** 63 (1993).
45. S. R. Nagel and J. Tauc, Phys. Rev. Lett., **35**, 380 (1975).
46. N. F. Mott and J. Jones, *The Theory of the Properties of Metals and Alloys* (Oxford University Press, Oxford, 1931).
47. J. Hafner, *From Hamiltonians to Phase Diagrams* (Springer-Verlag, Berlin, 1987).

MAGNETIC BEHAVIOR IN NANOCRYSTALLINE Fe-Cu-Nb-Si-B ALLOYS AT ELEVATED TEMPERATURES

I. ŠKORVÁNEK*, C.K. KIM, AND R.C. O'HANDLEY
Department of Materials Science and Engineering
Massachusetts Institute of Technology, Cambridge, MA 02139

ABSTRACT. Magnetic behavior in nanocrystalline $Fe_{72}Cu_1Nb_{4.5}Si_{13.5}B_9$ alloy is investigated by measurements of low and high field magnetization and coercivity over the temperature range from 300 K to 800 K. When the measuring temperatures approach the Curie temperature of intergranular amorphous phase T_c (am), the nanocrystalline materials consists of ferromagnetic Fe-rich nanocrystals embedded in a weakly magnetic or non-magnetic matrix. Samples with different volume fractions of crystalline phase are used to vary the nature and strength of inter-particle interactions between the nanocrystallites. It is shown that at temperatures above T_c (am), the specimens with low fraction of nanocrystalline particles exhibit superparamagnetic behavior. In the case of specimens containing larger fraction of crystalline phase, the magnetic interaction between particles prevents superparamagnetic relaxation and superferromagnetic ordering of the particle magnetic moments above T_c (am) is observed.

1. Introduction

Nanocrystalline materials have drawn much attention in recent years for both scientific and technological reasons. Samples of material in the nanocrystalline state exhibit properties that can differ considerably from those of polycrystalline specimens of the same composition. One example is nanocrystalline iron-rich FeSi alloys, with a small admixtures of Cu and Nb (typical composition is $Fe_{73.5}Cu_1Nb_3Si_{13.5}B_9$), prepared by controlled crystallization of initially amorphous ribbons. These materials, also known under the trade name FINEMET®, have excellent soft magnetic properties which are comparable to those of Co-based amorphous alloys and permalloys, but with a significantly higher saturation magnetization [1,2]. Their microstructure consists of very fine bcc Fe-Si grains embedded in residual Nb, Cu and B-enriched amorphous matrix phase [3,4].

In order to understand better the relationship between the microstructure and the observed soft magnetic properties of these materials, detailed investigations of the influence of various parameters such as modified composition, crystal size and the ratio of crystalline to amorphous phase have been performed by several methods [2, 5-8]. However, in contrast to the ambient and low temperature studies, much less attention has been devoted to the magnetic properties of these nanocrystalline systems at elevated temperatures.

When the measuring temperatures exceed the Curie temperature of the amorphous phase T_c (am), which is typically more than 200 K below that of nanocrystalline phase

* On leave from Institute of Experimental Physics, Košice, Slovakia

M. A. Otooni (ed.), Science and Technology of Rapid Solidification and Processing, 309–316.

[2], the ferromagnetic Fe-rich nanocrystals exist within a paramagnetic amorphous residue. This decreases the exchange coupling between the individual grains and a degradation of soft magnetic properties takes place, as observed experimentally [9]. If the size of the nanocrystals is small enough and the paramagnetic intergranular matrix is sufficiently thick to minimize the magnetic interactions between them, then superparamagnetic behavior of the ensemble of nanocrystalline particles can be expected. This was indeed recently confirmed for the $Fe_{66}Cr_8Cu_1Nb_3Si_{13}B_9$ alloy system containing a low volume fraction of nanocrystalline particles [10]. On the other hand, for the samples with larger fractions of nanocrystalline phase the inter-particle interactions will certainly play a significant role in determining the magnetic behavior at elevated temperatures. However, a more detailed investigation of the character and strength of the interaction effects in the above nanocrystalline systems is still missing.

The aim of the present work is to study the superparamagnetism and the interparticle interactions in nanocrystalline $Fe_{72}Cu_1Nb_{4.5}Si_{13.5}B_9$ alloy. The 50% higher Nb-content compared to the standard composition leads to a decrease of Curie temperature in the amorphous phase by more than 60 K, which results in widening of the temperature range where the above experiments can be performed.

2. Experimental

Samples of amorphous $Fe_{72}Cu_1Nb_{4.5}Si_{13.5}B_9$ alloy were prepared in ribbon form by the method of planar flow casting. Pieces of these ribbons were annealed under protective argon atmosphere for 1 hour at various temperatures in order to prepare nanocrystalline alloys with different ratios of amorphous to crystalline phase. The crystallization behavior of the samples was investigated by differential scanning calorimetry. Thermal data were measured with Perkin-Elmer DSC-7 unit at the heating rate of 10 K/min.

In order to obtain direct microstructural evidence of the various stages of crystallization in the investigated alloy, we used transmission electron microscopy (JEOL 200 CX and Akashi 002B operating at 200 keV). Several series of electron-transparent plan-view TEM specimens were prepared using a single or double gun Argon milling method. Bright and dark field diffraction contrast was used for the real space imaging; selected area electron diffraction was used for the reciprocal space information.

Measurements of the temperature and field dependence of magnetization were performed by using a DMS 880 A vibrating sample magnetometer in magnetic fields up to 800 kAm^{-1} and over the temperature range from 300 K to 800 K. The temperature of the magnetic measurements never exceeded the annealing temperature for any of the investigated samples.

3. Results and Discussion

3.1 Microstructure

The DSC-measurements have shown that the onset of the first exothermic peak (which corresponds to the formation of the bcc Fe-Si phase) appears at the temperature T_{x1} = 815 K, which is approximately 30 K higher than that for the standard (Nb_3) composition [6,7]. The onset of the second peak, (identified with the formation of other crystalline phases such as Fe-and/or Nb-borides) appears at T_{x2} = 961 K.

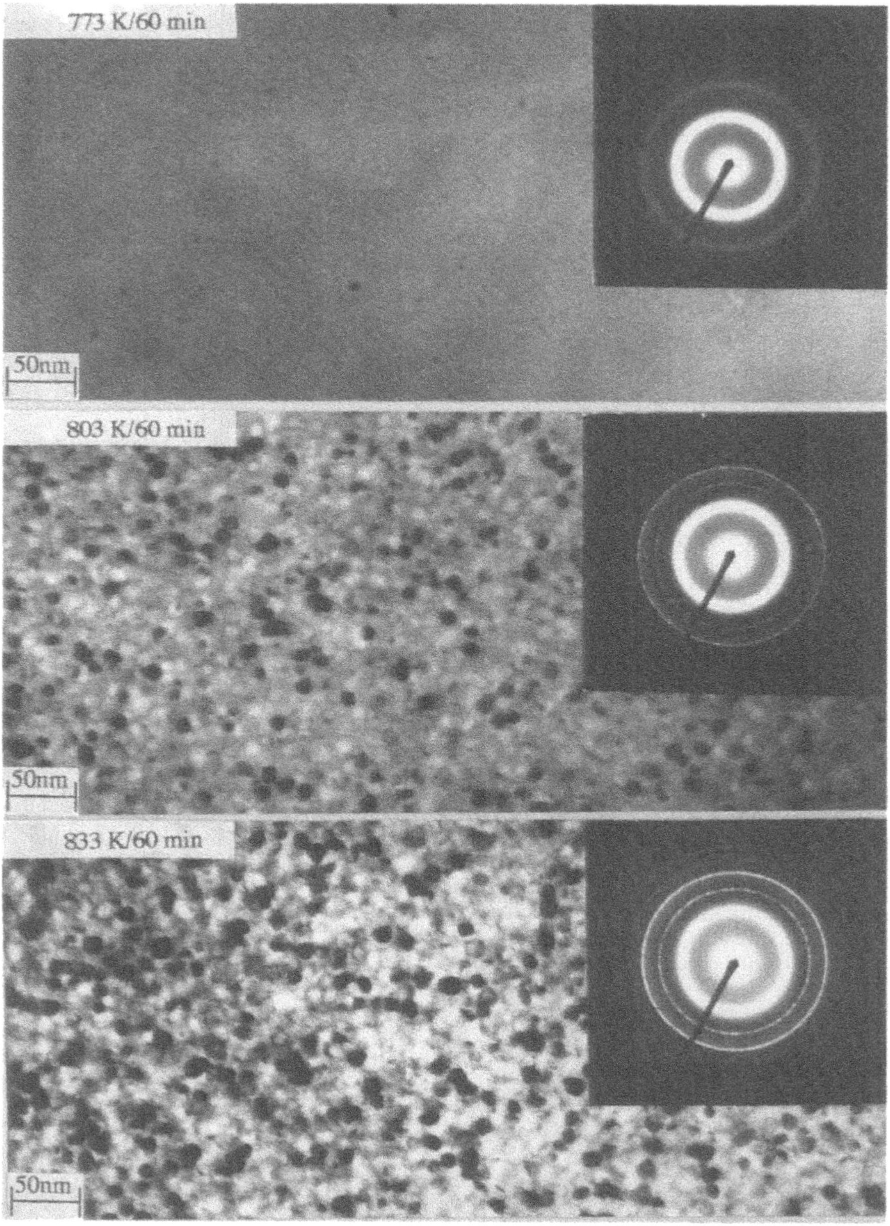

Fig. 1 TEM-micrographs and the selected area diffraction patterns of $Fe_{72}Cu_1Nb_{4.5}Si_{13.5}B_9$.

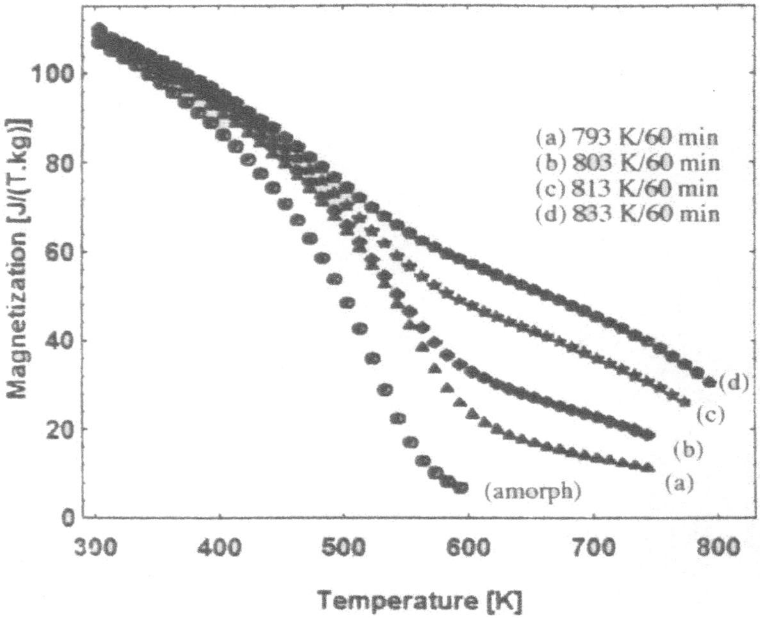

Fig. 2 Temperature dependences of magnetization for differently thermally treated $Fe_{72}Cu_1Nb_{4.5}Si_{13.5}B_9$ specimens.

Typical microstructures of the investigated alloy as observed by TEM are shown in Fig. 1. They correspond to the specimens prepared by annealing at different temperatures below or above T_{x1}. It can be seen that the sample annealed at 773 K for 1 hour exhibits traces of very early stage crystallization. Because the content of crystalline phase after such a thermal treatment is very low, the corresponding electron diffractogram (see insert Fig. 1a) shows a diffuse maximum typical for the amorphous structure. With an increase of the annealing temperature to 803 K (Fig. 1b) the homogeneous ultrafine grains of bcc Fe-Si measuring approximately 5-10 nm in diameter are formed. For an annealing temperature of 833 K we observe a further increase in volume fraction of crystalline phase and a slight increase in crystal size.

3.2 Magnetic behavior

Fig. 2 shows the temperature dependences of magnetization for thermally treated nanocrystalline samples measured at an applied field of 800 kAm^{-1}. The same dependence for the as-quenched amorphous sample is also shown in this figure. It can be seen that with increasing annealing temperature the fraction of nanocrystalline phase, that showing higher Curie temperature [2], increases. Consequently, samples with different ratios of crystalline to amorphous phase can be prepared. Above T_c (am), ferromagnetic nanocrystalline (bcc Fe-Si) particles exist within a paramagnetic amorphous matrix. A unique feature of such ultrafine magnetic particles is that at temperatures higher than the critical "blocking" temperature, T_B, the magnetization direction is not fixed as in large crystals, but fluctuates spontaneously. This behavior, called superparamagnetism, is due to the fact that magnetic anisotropy energy is proportional to the particle volume, and, for very small particles, it may therefore be comparable to the thermal energy even below room temperature. Assuming that nanocrystalline Fe-Si particles in the investigated alloys are non-interacting, we can determine their blocking temperature from the well known relation [11]

$$T_B = \frac{K \cdot V}{k\,(\ln \tau_m / \tau_o)} \tag{1}$$

where K and V are the anisotropy energy density and the volume of a particle, k is Boltzmann's constant, $1/\tau_o$ is a frequency factor estimated to be of the order of 10^9 s^{-1} and τ_m is the time of measurement (a typical value τ_m = 100s is chosen for dc measurements, and so, $\ln \tau_m/\tau_o \approx 25$). Using the value of $V = 524$ nm^3 (corresponding to particle dimension of 10 nm) and the $K = 8 \times 10^3$ $J{\cdot}m^{-3}$, which is the magnetocrystalline anisotropy constant for bcc Fe (20 at% Si) [12] and assuming that we have a system of isolated, non-interacting particles, we can estimate the value of T_B to be 12 K. Consequently, we expect the particle in our nanocrystalline alloys to be superparamagnetic when they do not interact. Ordinarily, we would expect their interactions to vanish for $T > T_c$ (am). However, we will show that for alloys with high fractional particle content, some magnetic interaction persists above T_c (am) and in those cases, superparamagnetic fluctuations are frozen out, not by low temperature, but by the magnetic interaction.

The results of the measurements of coercivity versus measuring temperature for the nanocrystalline samples with different fractions of crystalline phase are depicted in Fig. 3. For all samples the coercivity increases rapidly as T exceeds T_c (am). The reason for the observed degradation of soft magnetic properties is the decrease of exchange

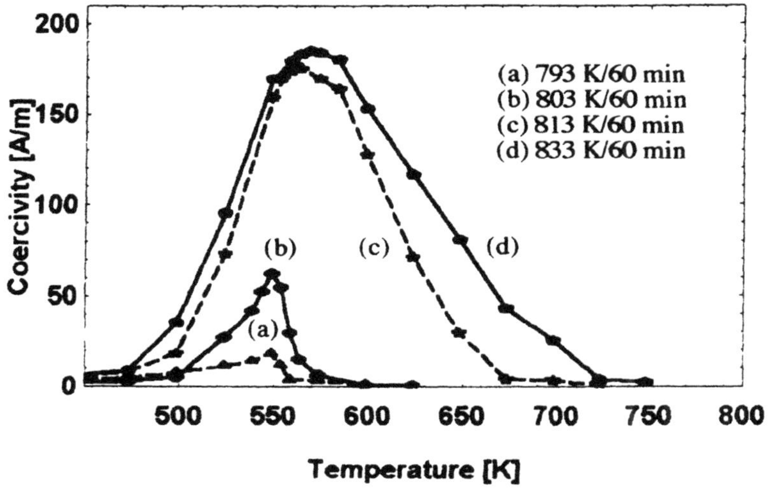

Fig. 3 Coercivity as a function of measuring temperature.

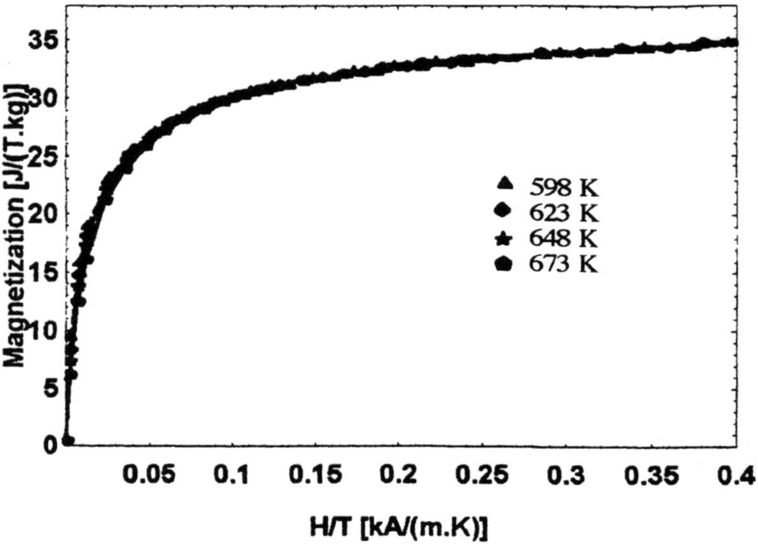

Fig. 4 H/T superposition of magnetization data for the $Fe_{72}Cu_1Nb_{4.5}Si_{13.5}B_9$ sample annealed at 803 K for 60 min.

coupling between individual grains through the amorphous matrix, which starts to lose its ferromagnetism. Samples with different volume fraction of nanocrystalline phase show different bevahior for $H_c(T)$. For the samples with lower fractions of nanocrystalline particles (Fig. 3 a, b), the coercivity drops sharply over a narrow temperature range. This drop suggests that these samples are superparamagnetic, above T_c (am). However, the zero coercivity is not enough to confirm superparamagnetism. The second criterion usually used for this purpose is that the magnetization curves taken at different temperatures must superimpose when plotted against H/T after correction for the temperature dependence of the spontaneous magnetization [11]. The set of such curves for the sample annealed at 803 K for 60 min (b) is plotted in Fig. 4. The good scaling behavior suggests that this sample is indeed superparamagnetic.

More puzzling behavior is observed for samples with higher crystalline content (see Fig. 3c, d). In this case the coercive behavior persists over a broad temperature interval well above T_c (am). As confirmed by TEM, the size of these particles increased only slightly in comparison to those of a and b, so their blocking temperature should not be significantly different. However, with the increase of crystalline volume fraction, the intergranular matrix is thinner, so the inter-particle interactions could play a more significant role in determining the magnetic behavior.

One of the most sensitive techniques for the investigation of particle interactions in fine particle systems is the measurement of low field susceptibility versus temperature [13]. The temperature dependences of dc-susceptibility measured in applied field of 80 Am^{-1} are depicted in Fig. 5. It can be seen that for the samples with low volume fraction of crystalline phase (a,b), susceptibility starts to decrease above T_c (am), where the ferromagnetic-superparamagnetic transition in the nanocrystalline phase takes place.

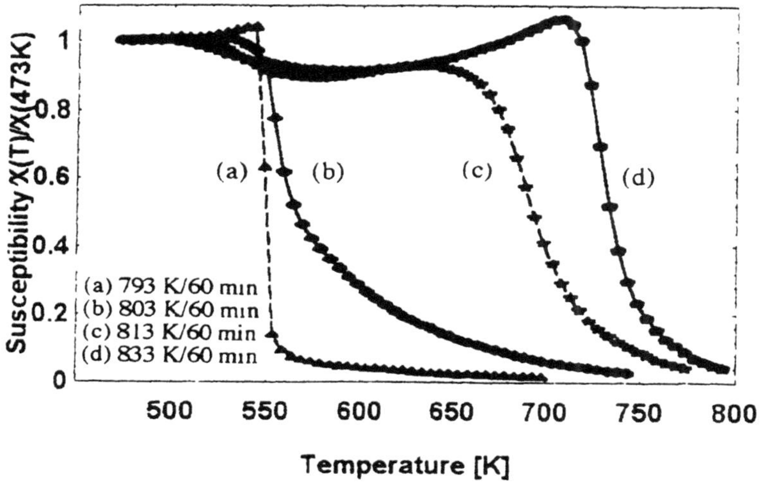

Fig. 5 Variation of low-field dc-susceptibility with measuring temperature.

For the samples with higher crystalline content (c,d), the χ versus T dependences exhibit a peak at temperatures well above T_C (am) and after reaching its peak, the susceptibility rapidly decrease. It can be seen that the susceptibility peak is shifted to higher temperature with increasing volume fraction of nanocrystalline phase and the temperature of peak susceptibility is very close to the temperature where the coercivity of the corresponding sample disappears. This suggests that the interparticle interactions in the samples with higher content of crystalline phase increase the height of the energy barrier for rotation of the magnetization in individual particles as was observed for some other fine particle systems, however at lower temperatures [13,14,15]. This results in the magnetic ordering of particles at temperatures above T_C (am). An analysis of the particle ordering process in systems in which the magnetic coupling between particles prevents superparamagnetic relaxation was performed by Morup et al [16,17], who named such a system "superferromagnetic." He showed that in the above coupling, both the exchange and dipole interactions between particles should be taken into consideration if the particles are close enough. On the other hand, the exchange interactions are expected to be insignificant if the particles are separated by more than 1-2 nm. At present, it is rather difficult to clarify the role of the above two coupling mechanisms in our system. However, as the ordering temperatures are rather high, it seems that the magnetic interaction between crystallites is strong, so, the exchange contribution can be significant.

4. Conclusions

In summary we have shown that FeCuNbSiB-type nanocrystalline materials offer an opportunity for studying interparticle interactions for a system of particles in a matrix whose magnetic properties can be controlled. Striking differences are observed and interpreted for the samples with low and high volume fractions of nanocrystalline particles as the matrix goes from ferromagnetic to paramagnetic state. It is shown that the samples with low volume fraction of crystalline phase exhibit superparamagnetic behavior. On the other hand, magnetic interactions between the nanocrystalline particles in the samples with higher crystalline content result in superferromagnetism (i.e. ordering of magnetic moments of particles in a paramagnetic matrix). Additional studies of the effects of fine-particle magnetism in these nanocrystalline alloys are currently in progress.

Acknowledgements

We are grateful to Dr. P. Duhaj, Institute of Physics SAS, Bratislava, for providing us with the samples of amorphous alloy. One of us, I.Š., acknowledges the support of the Fulbright Fellowship.

References

[1] Y. Yoshizawa, S. Oguma, and K. Yamauchi, J. Appl. Phys. 64, (1988) 6044.
[2]. G. Herzer, Mater. Sci. Eng. A133, (1991) 1.
[3]. K. Hono, K. Hiraga, Q. Wang, A. Inoue, and T. Suzuki, Acta metall. mater. 40, (1992) 2137.
[4] U. Köster, V. Schunemann, M. Blank-Bewersdorft, S. Brauer, M. Sutton, and G.B. Stephenson, Mater. Sci. Eng. A133, (1991) 616.
[5] M. Müller, N. Mattern, L. Illgen, Z. Metallkunde 82, (1991) 852.
[6] Y. Yoshizawa and K. Yamauchi, Mater. Sci. Engn. A133, (1991) 176.
[7] I. Škorvanek and R. Gerling, J. Appl. Phys. 72, (1992) 3417.
[8] G. Hampel, A. Pundt, and J. Hesse, J. Phys.: Cond. Matter 4, 3195 (1992).
[9] G. Herzer, IEEE Trans. Magn. 25, (1989) 3327.
[10] A. Slawska-Waniewska, M. Gutowski, H.K. Lachowicz, T. Kulik, H. Matyja, Phys. Rev. B46, (1992) 14594.
[11] C.P. Bean and J.D. Livingston, J. Appl. Phys. 30, (1959) 120S.
[12] H. Gengnagel and H. Wagner, Z. Angew. Physik 8, (1961) 174.
[13] K. O'Grady, M. El-Hilo and R.W. Chantrel, IEEE Trans. Magn. 29, (1993) 2608.
[14] G. Gittleman, B. Abelas and Bozowski, Phys. Rev. B9 (1974) 2891.
[15] R.W. Chantrell, M. El-Hilo and K. O'Grady, IEEE Trans. Magn. 27, (1991) 3570.
[16] S. Morup, M.B. Madsen, J. Franck, J. Villadsen, and C.J.W. Koch, J. Magn. Magn. Mat. 40, (1983) 163.
[17] S. Morup, P.H. Christensen, and B.S. Clausen, J. Magn. Magn. Mat. 68, (1987) 160.

Diffusion, Oxidation, and Nucleation of Crystalline Phases in the Glass- Forming System $Zr_{60}Ni_{25}Al_{15}$

S. Schneider, X. Sun, M.-A. Nicolet and W. L. Johnson
California Institute of Technology, Pasadena, California 91125

ABSTRACT

Oxidation kinetics and crystallization behavior of melt-quenched amorphous $Zr_{60}Ni_{25}Al_{15}$ samples were studied by MeV ^{4}He-backscattering spectrometry and transmission electron microscopy. The samples were annealed between 310°C and 430° C in dry oxygen. A planar diffusion-controlled growth of an amorphous ZrAl-oxide is observed, beginning at the surface. Ni atoms diffuse from the surface back into the sample in consequence of the oxidation. Up to the glass transition temperature the temperature dependence for the diffusivity of Ni in the amorphous matrix follows an Arrhenius law with an activation energy of 1.7eV. The pre-exponential factor is about $2.8 \cdot 10^{-5}$ m^2/s. Transmission electron microscopy studies of the amorphous alloy annealed in dry oxygen near the glass transition temperature T_g show the formation of nanocrystalline ZrO_2 particles at the interface between the amorphous metal alloy and the amorphous oxide. In the late stages, the crystallization of the amorphous metallic alloy into two intermetallic phases is observed as well.

INTRODUCTION

Recently, interesting amorphous metallic alloys with a wide supercooled liquid region and high glass forming ability have been found [1]. In contrast to binary amorphous metallic alloys like ZrNi or ZrCu [2] which crystallize very rapidly at the glass transition temperature T_g, this new family of glass-forming metallic alloys such as Zr-Ni-Al [3], La-Ni-Al [4], Zr-Ti-Ni-Cu-Be [5] show a high resistance against nucleation of crystalline phases in the undercooled liquid. For the first time it seems possible to study the atomic mobility of the different species above the glass transition temperature T_g. It is well known that far below the glass transition temperature the temperature dependence of the diffusivity D of the metallic atoms follows an Arrhenius law [6]. Diffusion data are important to prove or disprove the validity of the Stokes-Einstein relationship between the viscosity and the diffusivity in the undercooled liquid of the metallic alloy. Further, this study intends to examine the behavior of an amorphous metal surface in contact with an oxygen atmosphere. Previous investigations of amorphous ZrNi alloys have observed a crystalline ZrO_2 on $Zr_{36}Ni_{64}$ [7] and $Zr_{66}Ni_{34}$ surfaces [8]. Another aim of this work is to clarify the influence of Al and oxygen on the oxidation and crystallization behavior.

M. A. Otooni (ed.), Science and Technology of Rapid Solidification and Processing, 317–326.

EXPERIMENTAL PROCEDURES

$Zr_{60}Ni_{25}Al_{15}$ ingots are prepared by induction melting on a water cooled copper boat under a Ti-gettered argon atmosphere. Small pieces of about 60 mg of the initial ingot are levitation-melted in a high-frequency rf field. The samples are subsequently quenched into thin foils in a pure argon-filled twin-piston rapid quencher. The thickness of the as-splat quenched samples is about 50µm. The amorphous ternary splats are oxidized in an open-ended quartz tube furnace at temperatures between 310°C and 430°C. The oxygen gas flow was adjusted to 100 cm^3/min. X-ray diffraction analysis and 2MeV $^4He^{2+}$ backscattering spectrometry are carried out to obtain the structure and the concentration depth profiles before and after oxidation. Cross-sectional transmission electron microscopy and scanning transmission electron microscopy with energy dispersive X-ray spectroscopy are used to analyze the microstructure and to determine the elemental composition. The cross-sectional samples are embedded in epoxy and trimmed with a razor blade. For thin-sectioning microtomy the sections are cut with a diamond knife. The thickness of sections varies from 50nm to 80nm.

RESULTS AND DISCUSSION

Figure 1 shows a differential scanning calorimetric curve of a splatquenched amorphous $Zr_{60}Ni_{25}Al_{15}$ sample at a low heating rate of 5K/min. An endothermic heat

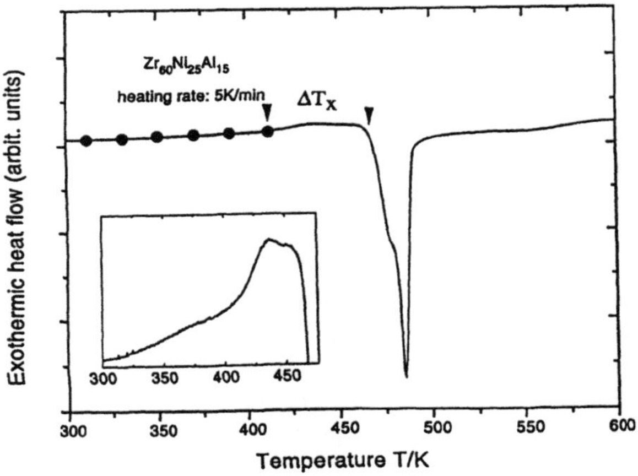

Fig. 1. Differential scanning calorimetric curve of the amorphous $Zr_{60}Ni_{25}Al_{15}$ alloy. The insert shows the low-temperature range of the curve with expanded ordinate.

flow effect which is related to the specific heat increase at the glass transition temperature is observed at 411°C. The insert in Figure 1 reveals the width of the undercooled liquid region ΔT_X on an expanded scale. The exothermic peak represents a single stage crystallization of the sample which starts at 463°C under Ar atmosphere. Both T_g and T_X increase logarithmically to higher temperatures with increasing heating rates. We find that the heating rate dependence of T_X is much stronger than that of T_g and the width of the undercooled region increases for higher heating rates too [9]. The arrows label the annealing temperatures in oxygen for the samples which are used for backscattering investigations. The X-ray diffraction patterns shown in Fig.2 are taken from the $Zr_{60}Ni_{25}Al_{15}$ sample which has been heated in a differential scanning calorimeter to 460 °C and 475°C at a heating rate of 5K/min. Figure 2(a) shows a broad maximum which is typical for a glassy structure. There is no evidence from the X-ray pattern that crystals have formed in the undercooled liquid below T_X. The X-ray diffraction pattern taken from samples heated above T_X reveals Bragg peaks from intermetallic compounds as shown in Fig.2(b). Most likely two intermetallic phases have appeared: the cubic $Zr_{50}Ni_{40}Al_{10}$ and the h.c.p. $Zr_{66}Ni_{11}Al_{22}$. From the peak positions the lattice constants determined for the h.c.p. phase are a = 0.792nm and c = 0.334nm [10] and for the cubic

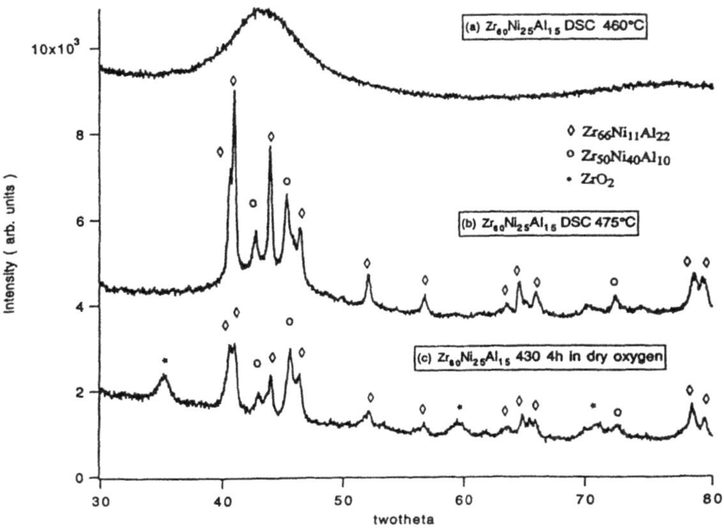

Fig. 2. X-ray diffraction pattern (Co K_α radiation) for different $Zr_{60}Ni_{25}Al_{15}$ samples heated in the differential scanning calorimeter up to 460°C (a) and up to 475°C (b), or annealed in dry oxygen at 410°C for 4h (c).

phase is a = 1.208nm. Exactly the same compounds form during annealing of the $Zr_{60}Ni_{25}Al_{15}$ foils above T_g at 430°C for 4h in dry oxygen atmosphere. This result is shown in Fig. 2(c). In addition, two more Bragg peaks appear. They belong to a cubic ZrO_2 phase with a lattice constant of a = 0.509nm [11]. A simple analysis based on the Debye Scherrer formula indicates that the average grain size of these oxide particles is only a few nanometers.

The backscattering spectra of an amorphous $Zr_{60}Ni_{25}Al_{15}$ sample annealed at 390°C for various times in dry oxygen are shown in Fig.3. The spectrum of the as-prepared foil reveals the elements Zr, Ni, and Al. The signal of the Al is poorly resolved in this typical thick-target spectrum because the yield of an element is proportional to the square of its atomic number. In addition there is only 15 at % Al in the alloy. After annealing in dry oxygen the yield of the $^4He^{2+}$ ions backscattered from the surface region decreases. The height of the Zr signal near 1800keV indicates the presence of an uniform and planar surface layer of ZrAl-oxide. Further, the shift in the front edge of the Ni signal gives evidence that the Ni concentration decreases much in the surface layer, while the increase in the yield at 1400keV indicates that Ni atoms are piling up behind the oxide layer. The thickness of the oxide layer can be derived from the shift in the high-energy edge of either the Zr or the Ni signal, as compared to their initial positions. For this

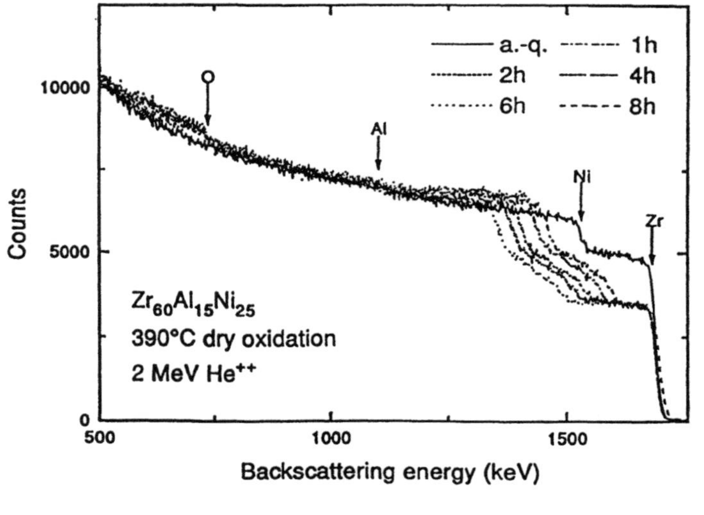

Fig. 3. 2 MeV $^4He^{2+}$ backscattering spectra of amorphous $Zr_{60}Ni_{25}Al_{15}$ samples before and after anealing in dry oxygen at 390°C for various times.

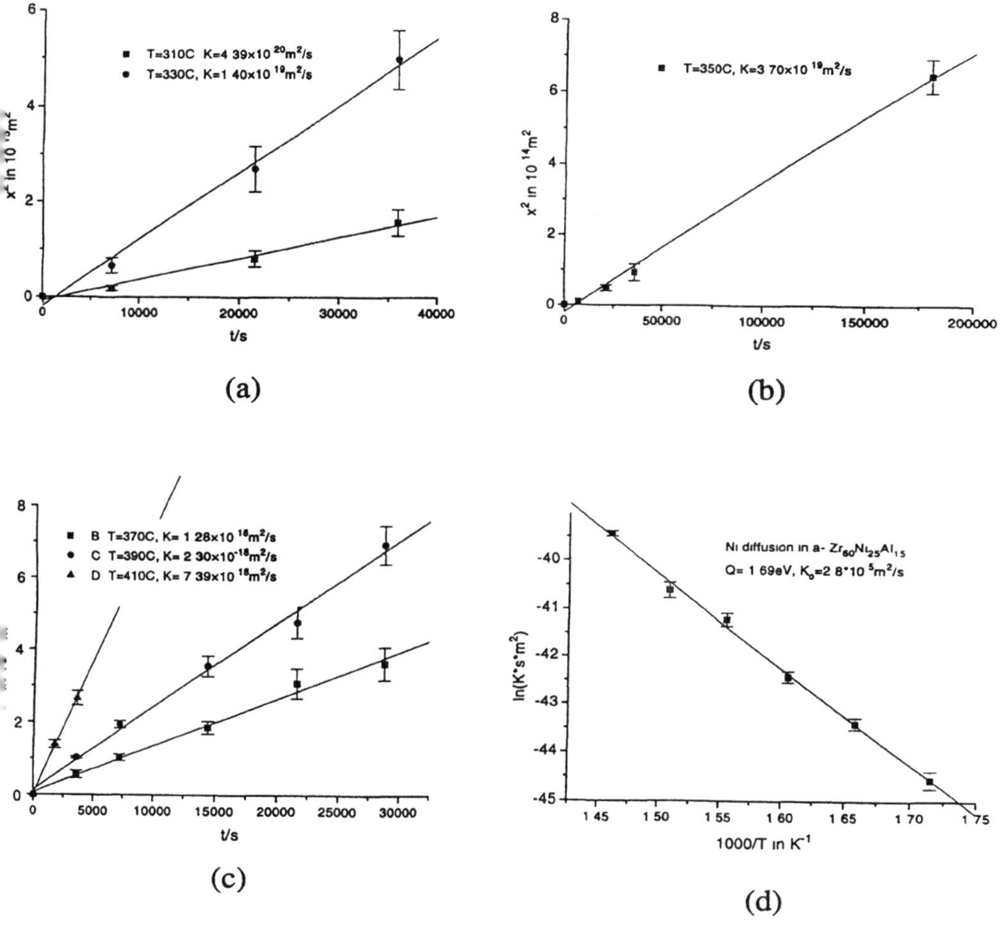

Fig. 4(a)-(c) The square of the oxide thickness formed versus the oxydation time for different temperatures. (d) The effective diffusion coefficient K for dry oxidation of amorphous $Zr_{60}Ni_{25}Al_{15}$ versus reciprocal oxidation temperature.

calculation we have taken the density of the amorphous ZrO_2 as $2.18 \cdot 10^{22}$ atoms/cm^3 [12]. Previous oxydation studies of amorphous ZrNi alloys also have shown the formation of a ZrO_2 surface layer [7,8]. These authors observed a decrease of the Ni concentration at the surface as well.The structure of all samples annealed at 390°C is still amorphous as determined by X-ray diffraction. From RBS data we assume that the dominant reaction of the amorphous $Zr_{60}Ni_{25}Al_{15}$ alloy is the growth of an amorphous ZrAl-oxide surface layer. The heat of formation of ZrO_2 and Al_2O_3 is much more negative than for NiO[13].

In the temperature range of 310-410°C the square of the oxide thickness, x^2, versus the annealing time, t, is linear, as shown in Fig.4, hence $x^2=Kt$, where K is the effective diffusion coefficient. The Arrhenius plot of all effective diffusion coefficient is shown in Fig.4(d). The data follow quite well the Arrhenius equation over the temperature range 310°C < T < 410°C.

$$K=K_o\exp(-Q/kT),$$

where $K_o = 2.8\cdot10^{-5}m^2/s$ is the temperature-independent pre-exponential factor, the activation energy Q is about 1.7eV, k is the Boltzmann coefficient, and T is the oxidation temperature. It should be noted that this value is much too high to correspond to the activation energy for oxygen diffusion in Zirconia ($Q\approx0.9eV$) [14]. On the other hand, there is a reasonable agreement with activation energies measured for the diffusion of Ni in amorphous ZrNi by tracer diffusion ($Q\approx1.5eV$) [15] and during solid-state amorphization reaction ($Q\approx1.1eV$) [16]. We suggest that the oxydation process is controlled by the Ni diffusion back into the amorphous $Zr_{60}Ni_{25}Al_{15}$ alloy.

Bright-field cross-sectional transmission micrographs of the amorphous $Zr_{60}Ni_{25}Al_{15}$ microtomed samples annealed for 1h, 1.5h and 2h at 410°C in dry oxygen

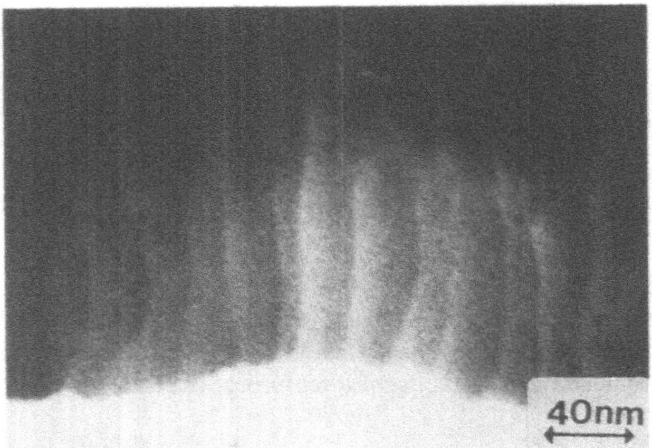

amorphous $Zr_{60}Ni_{25}Al_{15}$

← interface

amorphous ZrAl-oxide

Fig. 5(a) Cross-sectional transmission micrograph of an amorphous ZrNiAl/amorphous ZrAl-oxide interface formed after annealing in dry oxygen for 1h at 410°C.

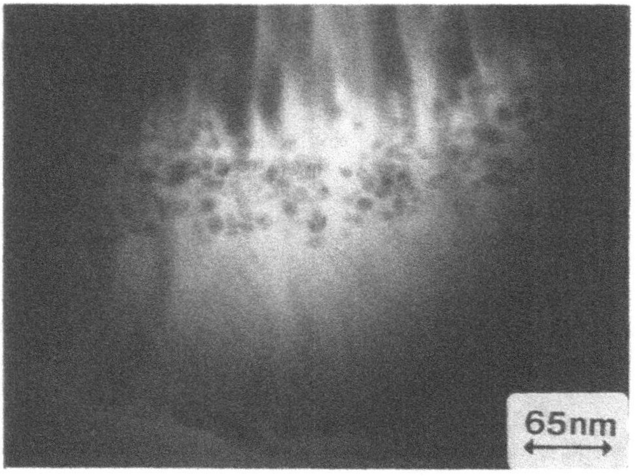

amorphous $Zr_{60}Ni_{25}Al_{15}$

nanocrystalline ZrO_2

amorphous ZrAl-oxide

Fig 5(b) Formation of the nanocrystalline ZrO_2 particles near the interface after annealing in dry oxygen for 1.5h at 410°C.

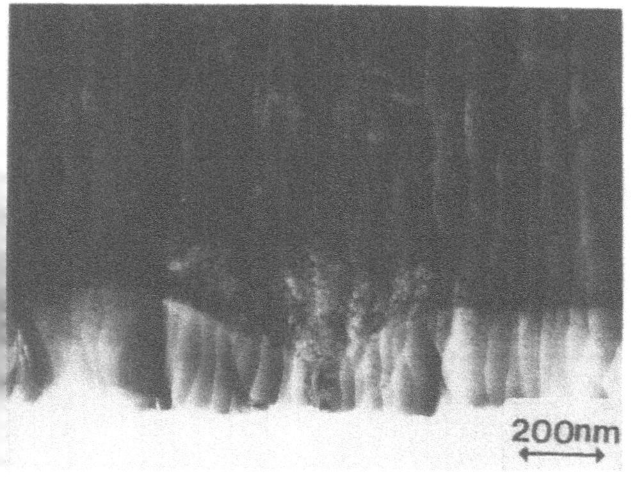

amorphous $Zr_{60}Ni_{25}Al_{15}$
+
intermetallic compounds

← interface

amorphous ZrAl-oxide

Fig. 5(c) Crystallisation of the amorphous $Zr_{60}Ni_{25}Al_{15}$ alloy after annealing in dry oxygen for 2h at 410°C.

are shown in Fig.5 a to c. Fine lines which are normal to the sample surface are found in all cross-sectional transmission micrographs micrographs. They are artefacts due to slip steps formed to relieve the shear stress during the cutting with the diamond knife [17]. Picture 5(c) shows also some knife marks parallel to the edge.

The first picture reveals clearly that a new amorphous oxide phase has formed at the surface. The uniform thickness of the amorphous oxide phase indicates a planar growth. The thickness of the oxide layer is about 110nm. After 1.5h at 410°C nanocrystalline particles appear near the amorphous ZrNiAl/amorphous ZrAl-oxide interface as shown in Fig.5(b). According to the X-ray diffraction and energy-dispersive X-ray spectroscopy investigation small ZrO_2 crystals have formed. The formation of these particles requires mobility of the Al atoms in addition to oxygen. For a temperature of 410°C we can estimate from the diameter of the ZrO_2 particles that the diffusivity of Al in the amorphous ZrAl-oxide is three times lower than the diffusivity of the Ni the amorphous $Zr_{60}Ni_{25}Al_{15}$ alloy. In this stage of the oxidation reaction we also see the first large crystals of the intermetallic compounds inside the amorphous matrix shown more clearly after 2h at 410°C in Fig 5(c). Most likely we see after crystallization a decomposition of the amorphous matrix into the two crystalline intermetallic phases $Zr_{50}Ni_{40}Al_{10}$ and $Zr_{66}Ni_{11}Al_{22}$.

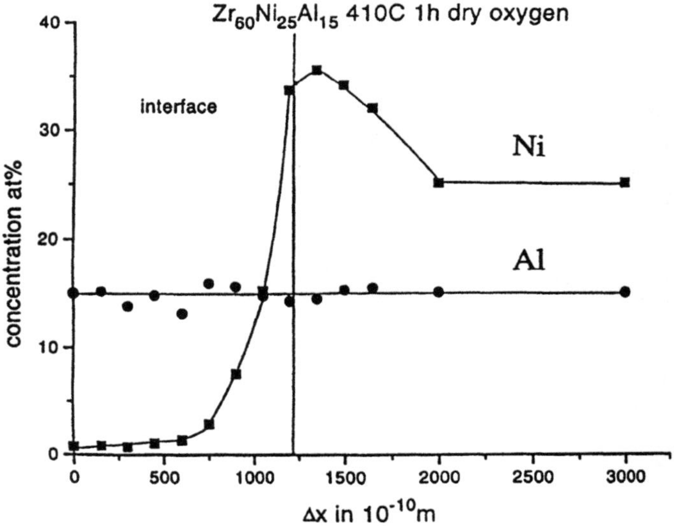

Fig.6 Concentration profile of the amorphous oxide surface layer formed after 1h at 410°C in dry oxygen taken by energy dispersive X-ray spectroscopy. The diameter of the electron probe is about 8nm.

For the interface region it seems reasonable to assume that the ZrO_2 particles cause heterogenous nucleation of a crystalline intermetallic compound. Preliminary investigations done by energy dispersive X-ray spectroscopy confirm that a Ni rich intermetallic compound has formed at the amorphous ZrNiAl/amorphous ZrAl-oxide interface. At this reaction stage the main part of the sample is still amorphous. For higher temperatures Köster and Meinhardt have found a formation of a nanocrystalline microstructure in the same $Zr_{60}Ni_{25}Al_{15}$ amorphous alloy [18].

It is evident that the Ni concentration in the amorphous oxide layer is less than 5 at %. The Ni concentration increases behind the interface up to 35 at%. Obviously, there is no gradient of the Al concentration after annealing in dry oxygen for 1h. This is in good agreement with the estimate of the relativly low diffusivity of the Al in this temperature range.

CONCLUSIONS

During dry oxidation at temperatures below the glass transistion it is possible to grow an amorphous oxide. The oxide is compositionally segregated and consists primarily of AlZrO. This is a thermodynamic effect that can express itself because the heat of formation for Ni oxide is much smaller than for formation of Zr and Al oxide and the diffusivity of Ni in the amorphous metallic alloy is high enough to allow its segregating there. The speed of the advancing oxidation front is obviously limited by the Ni diffusivity in the amorphous metallic alloy. The activation energy for the Ni diffusion in the amorphous alloy is about 1.7eV. The growth of the amorphous oxide layer is controlled by the Ni diffusion into the amorphous metallic alloy. For longer annealing durations above the glass transition temperature nanocrystalline ZrO_2 particles develope by preferential oxidation at the amorphous ZrNiAl/amorphous ZrAl-oxide interface. There is some indication that these particles cause heterogeneous nucleation of a Ni rich intermetallic compound.

ACKNOWLEDGMENTS

Financial support by the United States Department of Energy (No. DEFG-03-86ER45242 and by the U.S. Army Research Office) is gratefully acknowledged. Special thanks are due to C. Garland for the support with the Transmission electron microscope work, to R. Busch for the Differential scanning calorimetry measurements, and to M. Easterbrook for assistance with backscattering spectrometry. The authors thank R. Busch and K. Samwer for stimulating discussions.

REFERENCES

[1] A. Inoue, M. Kohinata, K. Otera, A.P. Tsai and T. Masumoto, Mater. Trans., JIM, **30** (1989) 378.

[2] K.H.J. Buschow, J. Phys.F.: Met. Phys.**14** (1984) 593.

[3] A. Inoue, T. Zhang and T. Masumoto, Mater. Trans., JIM, **31** (1990) 177.

[4] A. Inoue, K. Kita,T. Zhang and T.Masumoto, Mater. Trans., JIM, **30** (1989) 722.

[5] A. Peker and W.L. Johnson , Appl. Phys. Lett., **63** (1993) 2342.

[6] A.L. Greer, N. Karpe and J.Bøttiger, J. Alloy. Com.,**194** (1993) 199.

[7] J. C. Bertolini, J. Brissot, T. L. Moyne, H. Montes, Y. Calvayrac and J. Bigot, Appl. Surface Sci., **29** (1987) 29.

[8] D.L. Cocke, G. Liang, M. Owens. D.E. Halverson and D.G. Naugle, Mater. Sci. Eng., **99** (1988) 497.

[9] R. Busch and S. Schneider : unpublished research (1994)

[10] O. Kubaschewski, C.B. Alcock and P.J. Spencer, Materials Thermochemistry, Pergamon Press 1993

[11] P.Duwez and Odell, J. Am. Ceram. Soc., **33** (1950) 274

[12] K. Sugiyama, Y. Waseda and S. Kudo, ISIJ International, **31** (1991) 1362

[13] R. C. Weast, Handbook of Chemistry and Physics, ed. by R.C. Weast, 57th Edition, CRL Press, Cleveland (1976)

[14] K. Park and D. R. Olander, J. Electrochem. Soc., **138** (1991)1154

[15] K. Hoshino, R.S. Averback, H. Hahn and S. J. Rothman, J. Mater. Res. **3** (1988) 55

[16] E.J. Cotts, W.J. Meng, W.L. Johnson, Phys Rev. Lett. **57** (1986) 2295

[17] D.H. Kay, Techniques for Electron Microscopy, 2th edition, F.A. Davis Company, Philadelphia (1965)

[18] U. Köster and J. Meinhardt, Mat. Sci and Engg, **A178** (1994) 271

NON-CONTACT AC CALORIMETRY ON UNDERCOOLED ALLOYS

D. S. LEE, D. UFFELMAN† AND W. L. JOHNSON
Keck Laboratory of Engineering, California Institute of Technology
Pasadena, CA 91125, USA

ABSTRACT. A novel non-contact AC calorimetric technique based on inductive heating of a spherical metallic sample by a power-modulated radio frequency field under ultrahigh vacuum conditions has been developed. The sample's external relaxation response (due to radiative heat loss) and internal relaxation response (due to finite thermal conductivity) can be measured pyrometrically, allowing calculation of both the sample's heat capacity and thermal conductivity. The agreement between literature values (where available), predicted responses and measured data for (solid) Nb, $Ni_{60}Nb_{40}$, and $Zr_{76}Ni_{24}$ are excellent, demonstrating the feasibility and userfulness of this technique. This method will be applied to Zr, $Ni_{60}Nb_{40}$, and $Zr_{76}Ni_{24}$ in the undercooled regime during the upcoming Second International Microgravity laboratory (IML-2) space shuttle flight in July, 1994. On Earth, the power required to overcome the force of gravity is so large that undercooling in UHV conditions is not possible with inductive levitation. In contrast, the microgravity enviroment of space allows decoupling of the inductive positioner and heater power requirements, thus enabling a sample to be processed in the deeply undercooled region. From the data taken during the mission, entropies and free energies will be calculated and compared to various theories of glass formation and metastability.

1. Introduction

The measurement of the specific heat of undercooled metals and alloys provides important information regarding the thermodynamics of glass formation and metastability. The Gibbs free energy difference between the metastable liquid phase of a material and its stable solid phase can be determined experimentally by the following equation:

$$\Delta G(T) = (\Delta H_t) + \int_T^{T_t} \Delta C_p dT - \left(\Delta S_t + \int_T^{T_t} \frac{\Delta C_p}{T} dT \right), \quad (1)$$

M. A. Otooni (ed.), Science and Technology of Rapid Solidification and Processing, 327–337.

This quantity can be used to construct a metastable phase diagram and thus determine metastable equilibrium between the phases. The free energy difference is also used in classical nucleation theory, which predicts the nucleation rate of a stable solid solution from the undercooled liquid. Unfortunately, the specific heat of undercooled liquids are not generally available.

Traditionally, such measurements of glass forming alloys have been limited to alloys with a melting point below about 800K and for which a high degree of undercooling was obtainable by fluxing or dispersion techniques.[1, 2] At elevated temperatures, reactivity of any container with the melt made it difficult to prevent contamination of the experiment. In fact, the distinct lack of good thermophysical data, such as specific heat and emissivity, is directly due to the difficulties associated with selecting compatible crucible materials for containing the metal or alloy without contaminating the sample. In addition, container walls serve as heterogeneous nucleation sites for the molten metal or alloy and effectively limit the amount of undercooling attainable. To overcome these problems, previous investigators have used radio frequency (RF) electromagnetic levitation techniques to suspend molten specimens in a containerless manner. However, using such a technique in a 1g enviroment precluded measurements in the undercooled region because the power required to position (levitate) the sample could not be decoupled from the power needed to heat the sample; the electromagnetic force necessary for sample levitation resulted in the inductive heating of the sample to temperatures above its melting point. In the microgravity enviroment of Spacelab, positioner power is limited to holding the sample in a stable manner rather than generating a lifting force, and is thus much smaller than the power necessary to bring the specimen to elevated temperatures. As a result, the investigator can independently control the temperature of the sample by controlling the heating power input to the sample through a second set of RF coils. In TEMPUS, the ultrahigh vacuum electromagnetic positioning and heating facility onboard the Second International Microgravity Laboratory payload (IML-2), this technique is further refined by winding the positioner coil to be a lower coupling efficiency quadrupole field (100kHz). Inside of this coil sits a second coil which generates a much higher coupling efficiency dipole field (400kHz). Using this setup, with the quadrupole positioner coil powered to suppress dc (3×10^{-3} g) drift forces for our $Zr_{76}Ni_{24}$ sample, the resultant baseline temperature due to inductive heating of the specimen by the quadrupole field is about 875K, just over 60% of the minimum temperature we have achieved on the same sample in ground-based RF levitation systems.

The noncontact AC calorimetry technique[3] uses a modulated radio frequency field to inductively heat the sample. The specific heat and, under certain conditions, the thermal conductivity of the sample can be obtained from the pyrometrically measured temperature response of the sample to this field.

2. Experimental Setup

The experiments were performed using the laboratory model of TEMPUS, the

Spacelab facility for electromagnetic positioning and heating of metallic samples under ultrahigh vacuum conditions. Heating and positioning fields are controlled independently by two RF power supplies: one operating at 400kHz with its coil in a dipole field geometry for heating of the sample and one operating at 100kHz with its coil in a quadrupole field geometry for positioning of the sample within the dipole field. Other features of TEMPUS include self-oscillatory RF circuits allowing sample temperatures up to 2300K, UHV construction as well as inert gas enviroment processing, high speed video, optical pyrometers (100Hz) operating at visible (650nm) and infrared (1.0-2.5μm and 3.0-4.0μm) wavelengths and one high speed pyrometry channel for recalescence detection and analysis (1Mhz sampling speed). Further details have been presented elsewhere.[4] For the AC calorimetry setup, the RF power electronics of TEMPUS were modified through the addition of a function generator capable of generating sinusoidal power voltages with varying frequencies and amplitudes. A schematic of the experimental setup is shown in Figure 1. This function generator provided the modulation envelope on top of the

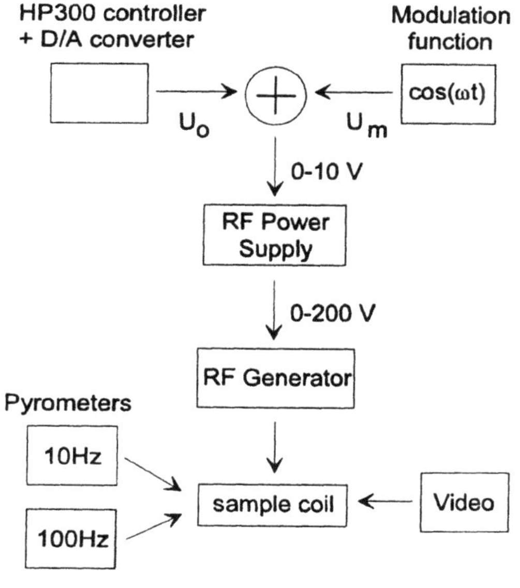

Figure 1. Schematic of the experiment setup.

oscillating circuit voltage to produce the modulated RF output power. Typical modulation frequencies used were in the range of 0.05Hz to 1.0Hz. A control signal in the range 0-10V defines the output voltage of the power supply of the RF generator. This control signal is composed of a D/A converted dc component from the TEMPUS facility controller and the AC signal from the function generator. This control voltage is applied t the power supply of the RF generator, generating an RF output signal with an amplitude i the range of 0-200V of the form:

$$U_{\omega} = U_o + U_m \cos(\omega_m t) \tag{2}$$

where U_o is the dc bias voltage, U_m is the modulation amplitude and ω_m is the modulation frequency, and are all user-controllable quantities. In the present set of experiments, the modulation amplitude was varied between 0.1 and 2.0V. Since $P_{\omega} \sim U_{\omega}^2$, the modulation of RF amplitude results in a modulation of the RF heating power proportional to:

$$P_{\omega} \sim U_o^2 + \tfrac{1}{2}U_m^2 + 2U_oU_m \cos(\omega_m t) + \tfrac{1}{2}U_m^2 \cos(2\omega_m t) \tag{3}$$

As we can see from the above equation, this type of power modulation will generate an increase of the average sample temperature superimposed over a periodic temperature modulation.

Accurate noncontact temperature measurement is essential to this experiment. TEMPUS is equipped with a DAPP Si pyrometer (650nm) capable of measuring temperatures in the range of 1300K to better than 2700K with an absolute accuracy of better than 1%. For lower temperatures, a Heimann infrared (InGaAs) pyrometer with tw channels (1.0-2.5μm and 3.0-4.0μm) is available. The relative resolution of both these pyrometers is better than 0.1K. The outputs of the pyrometers were collected at either 100Hz or 10Hz and were stored, together with the output voltage of the RF generator, oı a PC-compatible computer. Because the TEMPUS facility is not designed to levitate bulk samples on Earth, all samples were suspended in the center of the coil system by a 0.125mm diameter Pt-Rh wire. The samples were electrically discharge machined to have a 0.3mm hole through the diameter. The thin diameter of the Pt-Rh suspension wire allowed the conductive heat loss through the wire to be neglected relative to the radiative heat loss of the sample in the temperature range examined (800K to 1400K). All experiments were done under UHV conditions.

The samples were made by induction alloying Puratronic grade Zr, Ni and Nb in the proper ratios to make an 8m sphere of material on a water-cooled Silver boat in an Ar ga: atmosphere. The ingots were remelted several times to ensure thorough mixing and then transferred to a Buehler arc-melter. An 8mm hemispherical mold was machined into the baseplate of the arc-melter and the samples were remelted in it to produce a spheroidal shape.

3. Modulation Calorimetry

We can express the total power in the sample in Equation (3) as:

$$P_{tot} = P_o + \Delta P_{av} + P(\omega)\cos(\omega t) + P(2\omega)\cos(2\omega t) \tag{4}$$

where ΔPav is the increase in average dc power absorbed by the sample when the modulation is turned on, $P(\omega)$ is the power component at frequency ω, $P(2\omega)$ is the power component at frequency 2ω and Po is the constant power absorbed by the sample in the absence of any modulation. Po is related to the sample bias temperature by the Stefan-Boltzmann law:

$$P_o = A\varepsilon\sigma T_o^4 \tag{5}$$

with A the surface area, ε the total hemispherical emissivity of the sample and σ the Stefan-Boltzmann constant. For the modulation mode employed in this experiment, the Fourier coefficients of the power modulation are given by:

$$P(\omega) = 2\sqrt{2}\left[\Delta P_{av} P_o\right]^{1/2} \qquad P(2\omega) = \Delta P_{av} \tag{6}$$

For the analysis of heat transport in this problem, we consider only the radiative coupling of the sample to the heat bath and the finite thermal conductivity of the sample, represented by coupling constants $1/\tau_1$ and $1/\tau_2$, respectively. At the elevated temperatures in this experiment, heat conduction down the supporting wire can be neglected when compared to the radiative coupling of the sample to the heat bath. Analysis of the heat transport equations gives an expression for the amplitude of temperature modulation, ΔT_m, at modulation frequency ω, and the increase in average sample temperature, ΔT_{av}, due to the additional dc component of heat absorption from the ΔP_{av} term.[5,6] If the assumption is made that the temperature changes resulting from power modulation are much smaller than the sample's bias temperature T_o, i.e., that $\frac{\Delta T_m}{T_o}, \frac{\Delta T_{av}}{T_o} \ll 1$, then these terms are:

$$\Delta T_m = f(\omega_m, \tau_1, \tau_2) P(\omega_m) \frac{1}{\omega_m c_p} \tag{7}$$

$$\Delta T_{av} = \Delta P_{av} \frac{\tau_1}{c_p} = \frac{1}{4} \frac{\Delta P_{av}}{P_o} T_o \tag{8}$$

were $f(\omega_m, \tau_1, \tau_2)$ is a correction factor that accounts for radiative heat loss and the finite thermal conductivity of the sample. It is given by:

$$f(\omega_m, \tau_1, \tau_2) = \left\{1 + (\omega_m \tau_1)^{-2} + (\omega_m \tau_2)^2\right\}^{-1/2} \tag{9}$$

with relaxation time:

$$\tau_1 = \frac{c_p}{4A\varepsilon\sigma T_o^3} \tag{10}$$

$$\tau_2 = \frac{3c_p}{4\pi^3 R\kappa} \tag{11}$$

where R and κ are the sample radius and thermal conductivity and c_p is the sample heat capacity. Under the small temperature excursion limit, a change in the dc heating power will result in an exponential time dependence of the resulting temperature response of the sample with time constant τ_1.

For evaluation of c_p, the amplitude of the modulated power, $P(\omega)$, must be known. Note from Equations (5) through (8) that for the modulation mode used in the experiment ΔT_m be expressed in terms of ΔT_{av} and T_o. With known total hemispherical emissivity, the heat capacity of the sample can be obtained from measurements of ΔT_{av}, ΔT_m, and T_o:

$$\Delta T_m = f(\omega_m, \tau_1, \tau_2) 4\sqrt{2} P_o \left\{ \frac{\Delta T_{av}}{T_o} \right\}^{\frac{1}{2}} \frac{1}{\omega_m c_p} \tag{12}$$

P_o can be determined either by an independent bolometric measurement of the total radiant power from the sample or from the Stefan-Boltzmann law, Equation (5). For pure Nb and Zr, whose total hemispherical emissivities are known, the Stefan-Boltzmann law was used to calculate P_o. For other alloys, we are currently developing a total radiance bolometer which has a pyroelectric detector whose spectral response is flat from 0.6μm to 40.0μm. In the temperature range from about 700K to 2500K, this detector will measure better than 98% of the greybody spectrum from the radiating sample. Direct measurement of P_o has the advantage that the evaluation of c_p is much less susceptible to the accuracy of the T_o measurement.

Note also that this modulation calorimetry technique provides an intrinsic measurement of temperature which can be used to verify the accuracy of the pyrometry. Conbining Equations (5), (6), and (7), we obtain:

$$(\Delta T_m)^2 = f(\omega_m, \tau_1, \tau_2)^2 \, 2\sqrt{2} P_o \Delta T_{av} \frac{1}{\omega_m^{\ 2} \tau_1 c_p} \tag{13}$$

Using Equation (10) for τ_1 yields:

$$T_o = \frac{1}{2} f(\omega_m, \tau_1, \tau_2)^2 (\omega_m \tau_1)^2 \left\{ \frac{\Delta T_m^{\ 2}}{\Delta T_{av}} \right\} \tag{14}$$

so the absolute sample temperature corresponding to an input dc power of P_o can be measured just by measuring ΔT_m and ΔT_{av}.

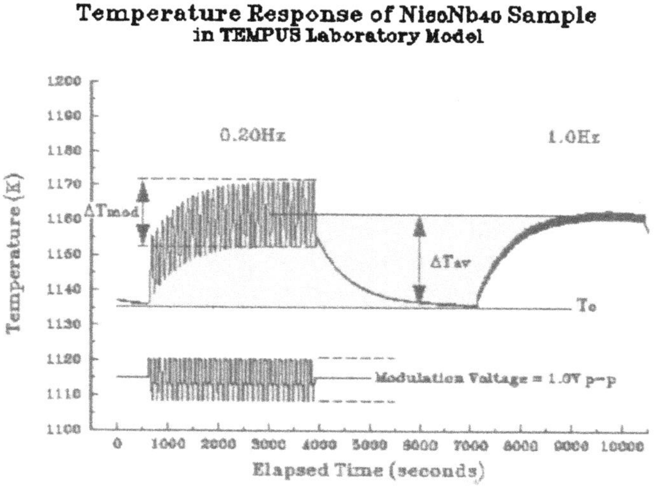

Figure 2. Temperature response of $Ni_{60}Nb_{40}$ sample to RF power modulation.

4. Results and Discussion

The dynamic temperature response of the sample upon application of heating power modulation is shown in Figure 2. The upper trace shows the RF control voltage modulation according to Equation (2) and the lower trace shows the temperature response of the sample. This temperature response can be written as follows:

$$T(t) = T_o + \Delta T_{av}\left[1 - \exp\left(-\frac{t}{\tau_1}\right)\right] + \Delta T_m \cos(\omega_m t) \tag{15}$$

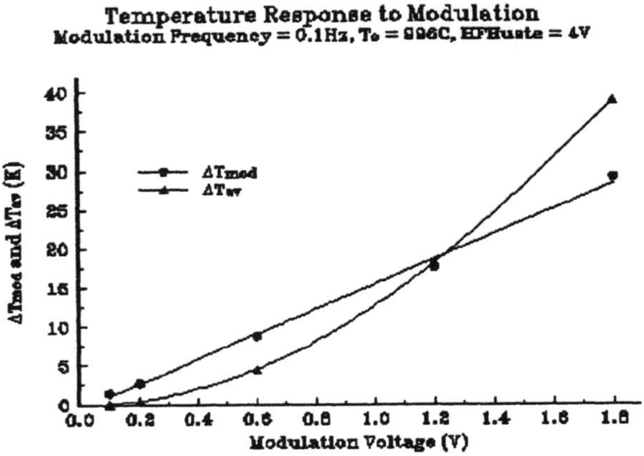

Figure 3. Dependence of ΔT_{mod} and ΔT_{av} on modulation voltage.

As can be seen from the figure, the increase in average temperature, ΔT_{av}, is independent of the modulation frequency, ω_m. No modulation component is seen at frequency (2ω) because the ratio of $P(\omega)/P(2\omega) = (4\sqrt{2})U_o/U_m$. For the control voltages used in this experiment, this ratio is ~50, and because the amplitude of temperature modulation is inversely proportional to the modulation frquency, the modulation term at 2ω contributes less than 1% of the observed temperature modulation. The dependence of ΔT_{av} and ΔT_m on modulation voltage, U_m, is shown in Figure 3. From Equation (2), a linear dependence for T_m and a quadratic dependence for T_{av} on U_m are predicted. This is in good agreement with experimental observations. In Figure 4, the frequency dependence of modulation is shown. Evaluation of c_p requires knowledge of the frequency dependent correction factor, $f(\omega, \tau_1, \tau_2)$, and the dc power absorbed by the sample, P_o. $f(\omega, \tau_1, \tau_2)$ has been calculated for many sets of parameters.[7] τ_1 is typically greater than τ_2 by a factor of 50 or so that the relaxation times are well separated in frequency. At lower frequencies, the function $f(\omega, \tau_1, \tau_2)$ is dominated by the $(\omega\tau_1)^{-2}$ term, allowing $f(\omega)$ to be determined solely from measurements of τ_1, either from the temperature decay of the sample in response to a small step change in dc power or from the time dependence of the increase in ΔT_{av} with heating power modulation. The time constants measured agree to better than 2% and show a purely exponential temperature dependence that is independent of modulation frequency. From these measurements of τ_1, we can determine τ_2 at higher modulation frequencies. Note also that because of the differences in the magnitudes of τ_1 and τ_2, there is always a range in modulation frequency for which the correction function $f(\omega)$ is approximately unity (to better than 1%). For this range of frequencies and slower, only τ_1 enters into the equations

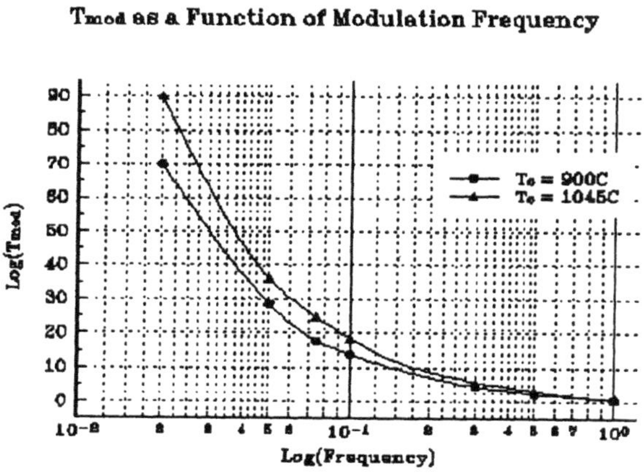

Figure 4. Dependence of the temperature modulation on modulation frequency.

and c_p can be determined with only measurements of ΔT_m, ΔT_{av} and T_o and a knowledge of ε. In fact, for the range in which $f(\omega)$ is approximately unity, c_p can be determined to better than 1% with no knowledge of τ_1. At higher modulation frequencies, a determination of τ_1 from lower frequencies allows us vary τ_2 as a free parameter and determine the thermal conductivity of the sample by fitting $f(\omega)$ to the theoretical model. Doing so for Nb gives $\tau_2 = 0.12$ sec., which yields $\kappa = 0.59$ W/cm-K, in qualitative agreement with experimentally measured values at lower temperatures. As such, the values of τ_1 and τ_2 differ by more than a factor of 100. In the liquid state, electromagnetic stirring forces may necessitate modeling before an estimate of the intrinsic thermal conductivity is possible. Nonetheless, the effects of both relaxation times are well-separated in frequency, allowing determination of c_p without knowledge of τ_2.

For evaluation of the heat capacity, c_p, ΔT_m, ΔT_{av} and τ_1 were measured for different modulation frequencies and samples over a wide temerature range. As a representative example, data for a 10mm Nb sphere has been evaluated at $T_o = 1840$K. The relaxation time τ_1 was measured to be 15.2 sec. With a total hemispherical emissivity of $\varepsilon = 0.23$,[8] we obtain a value of 31.95 J/K-mol, which agrees very well with literature data.[9] Results for $Zr_{76}Ni_{24}$ and $Ni_{60}Nb_{40}$ are also tabulated and compared with literature values where available.

Table I - Cp Values for Crystalline Nb (10mm diameter sample)[10]

T_o (K)	f_m (Hz)	C_p - measured (J/K-mol)	C_p - literature (J/K-mol)
1705	0.03	30.8	30.9
1840	0.03	32.0	31.72

Table II - C_p Values for Crystalline $Ni_{60}Nb_{40}$ (8mm diameter sample)

T_o (K)	f_m (Hz)	C_p - measured (J/K-mol)	C_p - drop cal. (J/K-mol)
1270	0.10	34*	32.7

*- assuming $\varepsilon_T = 0.3$

Table III - C_p Values for Crystalline $Zr_{76}Ni_{24}$ (8mm diameter sample)

T_o (K)	f_m (Hz)	C_p - measured (J/K-mol)	C_p - drop cal. (J/K-mol)
1108	0.10	33*	31.6

*- assuming $\varepsilon_T = 0.22$

5. Conclusions

The agreement between the theory, experiment and literature demonstrate the validity of this technique and the internal consistency of the approach. It also demonstrates that the total hemispherical emisivity, ε, is critical to proper analysis of the data. A method to measure the total hemispherical emissivity of molten samples using a pyroelectric radiometer has been developed and will soon be integrated onto a levitation system for testing and measurement.

6. Acknowledgements

This project is supported under NASA Grant No. NAG8-905. The authors would like to acknowledge the help of A. Diefenbach and R. Willnecker at DLR with use of the TEMPUS development and breadboard models, K. Ohsaka at JPL for drop calorimetry measurements, J. Li at Caltech for some sample preparation, S. Krishnan at Containerless Research Inc. for spectral emissivity measurements and many useful discussions with R. Wunderlich and H. Fecht at TU Berlin and W. Hofmeister at Vanderbilt University.

---

† DLR, Institute für Raumsimulation, W-5000 Köln 90, Germany.

1 H. J. Fecht, J. H. Perepezko, M. C. Lee and W. L. Johnson, J. Appl. Phys. **68**, 4494 (1990).

2 H. W. Kui and D. Turnbull, J. Non-Crystal. Solids **94**, 62 (1987).

[3] H. J. Fecht and W. L. Johnson, Rev. Sci. Instr. **62**, 1299 (1991).
[4] R. Knauf, W. Dreier and J. Pillar, Proc. VIIIth Eur. Symp. on Mat. and Fluid Sci. in Micrograv., **ESA SP-333** (1992).
[5] op. cit., Fecht and Johnson.
[6] P. F. Sullivan and G. Seidel, Phys. Rev. B **173**, 679 (1968).
[7] op. cit., Fecht and Johnson.
[8] W. Hofmeister, private communication.
[9] I. Barin, Thermochemical Data of Pure Substances, VCH Publishers (Weinheim, 1989), p.1010.
[10] Data taken by R. Wunderlich on TEMPUS breadboard..

DIFFRACTION STUDIES OF THE STRUCTURE OF AMORPHOUS AND NANOCRYSTALLINE Fe-W ALLOYS

C. N. J. WAGNER, E. YANG and M. S. BOLDRICK
Department of Materials Science and Engineering
University of California, Los Angeles
Los Angeles, CA 90024-1595
USA

ABSTRACT

Amorphous and nanocrystalline materials exhibit broad or diffuse peaks in their x-ray, neutron, or electron diffraction patterns. A detailed analysis of these patterns permits us to evaluate the atomic short-range order in amorphous materials, and the size of the coherently diffracting domains (particle size), the elastic strains within these domains, and the dynamic and static atomic displacements in nanocrystalline materials.

Amorphous metallic alloys have been prepared by rapid quenching from the vapor and liquid phases, by electron, proton or ion irradiation, by solid state reactions in multilayer films, and by mechanical alloying and mechanical working in high-energy ball mills. The term nanocrystalline metal has been introduced to describe compacted samples of metallic particles, clusters of particles, or grains of nanometer size, produced by quenching metallic vapor into an inert gas. Nanocrystalline powders can also be produced by precipitation from chemical solutions, and by mechanical grinding of micron-size particles.

As examples, the structures of amorphous and nanocrystalline Fe-W alloys, produced by mechanical grinding and alloying in a high-energy ball mill, will be described in terms of atomic distribution functions, particle sizes and microstrains.

1. Introduction

Amorphous metallic alloys have been prepared by rapid quenching from the vapor and liquid phases, by electron, proton or ion irradiation, and by solid state reactions in multilayer films [1], and their structures, i.e., the atomic short-range order, of metallic glasses have been deduced from their neutron and x-ray diffraction patterns. It has been observed that all metallic glasses exhibit a tendency towards chemical short-range order [2].

In recent years, it has been shown that amorphous metallic alloys can be prepared by mechanical alloying [1], which consists of intimate mixing and mechanical working of elemental powders in a high-energy ball mill [3,4]. Mechanical working or grinding of pure metals and intermetallic phases in a high-energy ball mill has also been employed to produce a nanocrystalline structure [5].

M. A. Otooni (ed.), Science and Technology of Rapid Solidification and Processing, 339–347.

Heavy plastic deformation, such as filing, introduces a high density of dislocations i metals and alloys which are responsible for the observed broadening of their powder pa tern peaks [6,7]. This broadening has been interpreted in terms of the reduction in size (the coherently diffracting domains and the appearance of microstrains within thes domains. Mechanical working or grinding in a high-energy ball mill is a form of heav deformation at high strain rates [8], which might produce nanocrystalline metals. There i considerable interest in these materials, since Gleiter and his co-workers [9,10] hav shown that nanocrystalline metallic powder compacts, produced by evaporation of pu metals in a He atmosphere, collecting the fine particles in a die and compacting them ir situ, exhibit interesting physical and mechanical properties.

In order to test the possibility that mechanical working of elemental powders or mechan cal alloying of powder mixtures can yield nanocrystalline samples [11-14], Wagner et a [15] have chosen W powder to be milled in a high-energy ball mill because W is elasticall isotropic and all (hkl) reflections can be employed in the Warren-Averbach analysis [6,7] (the diffraction profiles for the separation of particle size and strain broadening. They foun that the particle size decreased to 50 Å after milling for 10 h. However, an amorphous W Fe phase appeared after 20 h due to the Fe contamination from the steel vial and balls.

In this paper, we will describe the atomic structure of amorphous and nanocrystallin alloys of W-Fe, prepared by mechanical grinding and alloying of mixtures of Fe and \ powders in a high-energy ball mill. The structure of the amorphous and nanocrystallin materials will be deduced from the analysis of the entire diffraction pattern in terms of th interference function or structure factor and its Fourier transform, the atomic pair distribu tion function. For the nanocrystalline samples, we employ a Fourier analysis of the ind vidual powder pattern peaks and also apply the Warren-Averbach method to determine th particle or domain size and the internal microstrains [6,7].

2. Theoretical Background

2.1. DETERMINATION OF THE ATOMIC DISTRIBUTION FUNCTION

The atomic short-range order in binary amorphous alloys, consisting of elements of types and 2, is characterized by three atomic pair distribution functions $\rho_{ij}(r)$ describing the nun ber of j-type atoms per unit volume at the distance r about an i-type atom [16-18], i.e $\rho_{11}(r)$, $\rho_{22}(r)$ and $\rho_{12}(r)$, because $\rho_{21}(r)/c_1 = \rho_{12}(r)/c_2$, c_i being the atomic concentratic of element i.

We define the reduced intensity per atom $S_d(Q)$ by

$$S_d(Q) = I_a(Q) - [\langle f^2\rangle + \langle f\rangle^2 \rho_o V(Q)], \quad (1)$$

where Q is the length of the diffraction vector, i.e., $Q = 4\pi(\sin\theta)/\lambda$, $I_a(Q)$ is the scattere intensity per atom, $\langle f^2\rangle = c_1(f_1)^2 + c_2(f_2)^2$ is the mean square scattering factor , $\langle f\rangle = c_1f_1$ c_2f_2 is the mean scattering amplitude, and ρ_o is the average atomic density. The ter $\langle f\rangle^2\rho_oV(Q)$ represents the small-angle scattering, and V(Q) is the Fourier transform of tl size function v(r), i.e.,

$$v(r) = (1/V) \int s(u)s(u+r)\, du \quad (2)$$

where $s(r) = 1$ inside the volume V of the particle and zero outside [18,19].

The intensity per atom $S_d(Q)$ can be expressed as the weighted sum of the partial structure factors $I_{ij}(Q)$, i.e.,

$$S_d(Q) = (c_1f_1)^2[I_{11}(Q) - 1] + (c_2f_2)^2[I_{22}(Q) - 1] + 2c_1f_1c_2f_2[I_{12}(Q) - 1] \quad (3)$$

The Fourier transforms of $Q[I_{ij}(Q) - 1]$ yield the partial reduced atomic distribution functions $G_{ij}(r) = 4\pi r\{[\rho_{ij}(r)/c_j] - \rho_o\}$.

In general, the small-angle scattering $\langle f\rangle^2\rho_o V(Q)$ is not observable in amorphous metals, or is subtracted from the scattered intensity. For large values of Q, $I_a(Q) = \langle f^2\rangle$ and $S_d(Q) = 0$. Since $f_i(Q)$ is a slowly decreasing function with increasing Q for x-rays, it is customary to introduce the total structure factors or interference functions I(Q) or S(Q), defined as:

$$I(Q) = [S_d(Q)/\langle f\rangle^2] + 1 = \{I_a(Q) - [\langle f^2\rangle - \langle f\rangle^2]\}/\langle f\rangle^2 \quad (4)$$

$$S(Q) = [S_d(Q)/\langle f^2\rangle] + 1 = I_a(Q)/\langle f^2\rangle \quad (5)$$

The definition of S(Q) has the advantage that it is valid even when $\langle f\rangle^2 = 0$ as is the case for a "zero alloy", an alloy containing the appropriate amount of an element or isotope with negative scattering length for neutrons.

The total structure factors I(Q) or S(Q) permit us to evaluate the reduced atomic distribution functions $G_I(r)$ or $G_S(r)$, i.e.,

$$G_I(r) = (2/\pi) \int Q[I(Q) - 1]\, M(Q) \sin Qr \, dQ$$

$$= 4\pi r\, [\rho(r) - \rho_o]\, v(r) = \Sigma_i\Sigma_j\, [c_if_ic_jf_j/\langle f\rangle^2]\, G_{ij}(r)\, v(r) \quad (6)$$

$$G_S(r) = (2/\pi) \int Q[S(Q) - 1]\, M(Q) \sin Qr \, dQ$$

$$= 4\pi r\, [\rho(r) - \rho_o]\, (\langle f\rangle^2/\langle f^2\rangle)\, v(r) = \Sigma_i\Sigma_j\, [c_if_ic_jf_j/\langle f^2\rangle]\, G_{ij}(r)\, v(r) \quad (7)$$

where M(Q) is a modification function [6,18,20] and $\rho(r) = [c_if_if_j/\langle f\rangle^2]\, \rho_{ij}(r)$. The atomic distribution function G(r) becomes zero when $\rho(r) = \rho_o$ or $v(r) = 0$.

2.2. ANALYSIS OF THE PROFILES OF POWDER PATTERN PEAKS

The total interference function or structure factor I(Q) of a powder pattern can be expressed as the sum of the interference functions $I_{hkl}(Q)$ of the individual (hkl) reflections multiplied by the Debye-Waller factor $\exp(-\langle u^2\rangle Q^2)$, where $\langle u^2\rangle$ is the mean square displacements of the atoms from the equilibrium positions due to thermal vibration and static displacements, and the thermal diffuse scattering which modulates about the values of $[1 - \exp(-\langle u^2\rangle Q^2)]$:

$$I(Q) = \Sigma_{hkl}\, [I_{hkl}(Q) \exp(-\langle u^2\rangle Q^2)] + [1 - \exp(-\langle u^2\rangle Q^2)]\, W(Q) \quad (8)$$

where W(Q) is a modulation function characteristic of each Bravais lattice [6].

It is well known that each profile $I_{hkl}(Q)$ of a powder pattern peak can be expressed as a Fourier series whose coefficients are related to the effective particle size $\langle D_e\rangle$, due to the size of the coherently diffracting domains (grains or subgrains) and stacking faults in these grains, and to microstrains within the small domains in heavily cold-worked materials [6,7,19].

The function $I_{hkl}(Q)$ can be approximated as $I_{hkl}(Q) = I_{measured}(Q)/(\text{peak area})$ and can be written as Fourier series:

$$I_{hkl}(Q) = \Sigma_n |C(L)| \exp[iL(Q - Q_o)] \tag{9}$$

where $L = nd_{hkl}$ is a distance normal to the reflecting planes (hkl) with the interplanar spacing d_{hkl}, n is the harmonic number of the Fourier series, and Q_o is the Q-value of the peak maximum position of the (hkl) reflection.

The Fourier coefficients $|C(L)|$ can be expressed as the product of the particle size coefficients $A_p(L)$, due to the effective particle size $\langle D_e \rangle$, and the strain coefficients $|C_s(L)|$ produced by the strain fields of dislocations. For small values of L, one can approximate these coefficients by:

$$A_p(L) = 1 - L/\langle D_e \rangle = \exp(-L/\langle D_e \rangle) \tag{10}$$

$$|C_s(L)| = 1 - (2\pi^2 L^2)[\langle(\varepsilon_L)^2\rangle - \langle\varepsilon_L\rangle^2]/(d_{hkl})^2$$

$$= \exp\{-(2\pi^2 L^2)[\langle(\varepsilon_L)^2\rangle - \langle\varepsilon_L\rangle^2]/(d_{hkl})^2\} \tag{11}$$

and

$$\ln|C(L)| = \ln A_p(L) - (2\pi^2 L^2)[\langle(\varepsilon_L)^2\rangle - \langle\varepsilon_L\rangle^2]/(d_{hkl})^2$$

where the effective particle size $\langle D_e \rangle$ includes the size $\langle D \rangle$ of the coherently diffracting domains, and $[\langle(\varepsilon_L)^2\rangle - \langle\varepsilon_L\rangle^2]$ is a measure of the standard deviation of the strain distribution about the average strain $\langle\varepsilon_L\rangle = (\Delta L/L)_L$, which is small or zero in small powder particles.

Plotting $\ln|C(L)|$ as a function $1/(d_{hkl})^2$ [$= (h^2+k^2+l^2)/a^2$ in cubic materials with lattice parameter a] for constant values of L allows us to determine the particle size Fourier coefficients $A_p(L)$ and the mean square strains $[\langle(\varepsilon_L)^2\rangle - \langle\varepsilon_L\rangle^2]^{1/2}$ from the intercept of the straight line with the ordinate and its slope, respectively (Warren-Averbach method) [6]. These quantities can be correlated with the dislocation densities produced by the mechanical or cold working of the powders [21].

The sum of the Fourier coefficients $A_p(L)$ yields the integral breadth particle size D which is related to $\langle D_e \rangle$ by

$$D_I = \int A_p(L)\, dL = \langle(D_e)^2\rangle/\langle D_e \rangle \geq \langle D_e \rangle \tag{12}$$

3. Experimental Procedures

Alloys of $Fe_{100-x}W_x$ with x = 20, 33.5, 40, 50, and 75 at% were prepared by mechanical alloying in the high-energy Spex 8000 mixer/mill. In addition, pure Fe and W powders were mechanically worked in the Spex mill. 5 g charges of the powders with particle sizes of 6 - 9 μm for Fe and 12 μm for W were sealed under argon atmosphere in a hardened steel vial together with six 440C stainless steel balls (4 balls with 6.35 mm diameter and 2 balls with 12.7 mm diameter) and then milled for up to 50 h. The ball-to-powder ratio was thus 4:1. After a milling time of more than 20 h, the particle sizes of the powders were reduced to 1-2 μm, determined from scanning electron micrographs.

W is very hard and continued milling of elemental W powders will lead to Fe contamination [11,12,14,15]. Energy-dispersive x-ray fluorescence analysis revealed a combined Fe and Cr content of 24 ± 4 at% after 20 h and 50 ± 10 at% after 50 h of milling the originally pure W sample. The Fe and Cr uptake was small (< 5%) in the $Fe_{50}W_{50}$ alloy even after 100 h of milling.

The diffraction patterns of the Fe, W and Fe_xW_{100-x} powders were obtained with Co and Mo Kα radiation and a Ge solid state detector in the range of $Q = 4\pi(\sin\theta)/\lambda$ from 1 Å^{-1} to 15 Å^{-1}. The analysis of the total scattered intensities from the milled samples was carried out by the computer programs INTER and FOURIER written in FORTRAN for IBM compatible computers, based on an earlier program for main frame computers [22]. This program corrects the raw intensity for polarization, absorption, Compton and multiple scattering, and normalizes the data to scattered intensity per atom $I_a(Q)$ [18,23]. The Fourier analysis of the individual reflections of the nanocrystalline materials was performed with the PASCAL program LBA, based on an earlier FORTRAN program for main frame computers [24]. It allows the separation of the Kα profile of each peak into its $K\alpha_1$ and $K\alpha_2$ components, the correction for instrumental broadening using a well-annealed W powder as standard, and the separation of the particle size and strains using first and second order peak profiles [7].

4. Experimental Results and Discussion

The evolution of the nanocrystalline and amorphous phases is visible in the diffraction patterns shown in Figs. 1 and 2 for the pure W powder and the mixture of 50 at% Fe and W powders, respectively. It is readily seen that the broadening of the powder pattern peaks is already substantial even after one hour of milling. After about 20 h, a broad peak appears to develop at the position of the (110) reflection of Fe in pure W and the Fe-W mixture. It corresponds to the first peak in the diffraction pattern from an amorphous Fe-W alloy. Prolonged milling of the powders yielded a two-phase alloy, i.e., a nanocrystalline W solid solution and an amorphous $Fe_{50}W_{50}$ alloy. The appearance of the amorphous phase in the W sample is due to the incorporation of Fe and Cr from the steel vial and the stainless steel balls. The peaks in the diffraction patterns correspond to the body-centered cubic Bravais lattice and yield the lattice parameter a = 3.1525 ±0.0005 Å after 50 h of milling, indicating that a substantial amount of Fe and Cr is dissolved in the W lattice (a = 3.1653 Å). Assuming a linear dependence of the atomic volume of a substitutional W-Fe(Cr) alloy with concentration would yield an iron-chromium content of about 6 at% after 50 h milling.

The scattering patterns of Fe_xW_{100-x} are shown in Fig. 3 after 20 h of milling. The amorphous phase is only visible in W and the mixtures of W with 25 and 50 at% Fe.

The total scattering patterns of W(Fe) and $W_{50}Fe_{50}$ were converted into the structure factors or interference functions I(Q) [equation (4)], which are shown in Fig. 4 for milling times of 20, 50 and 100 h. They include the Bragg peaks and the thermal diffuse scattering (TDS) from the nanocrystalline phase and a contribution from the amorphous phase. The intensities of the Bragg reflections are reduced by the Debye-Waller factor $\exp(-\langle u^2\rangle Q^2)$, where $\langle u^2\rangle$ is the mean square displacement of the atoms due to thermal vibration and static

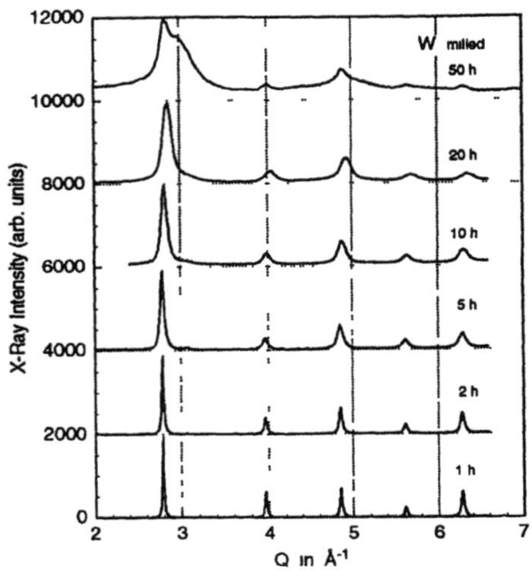

Fig. 1 X-ray diffraction patterns of W milled from 1 to 50 h in a high-energy ball mill.

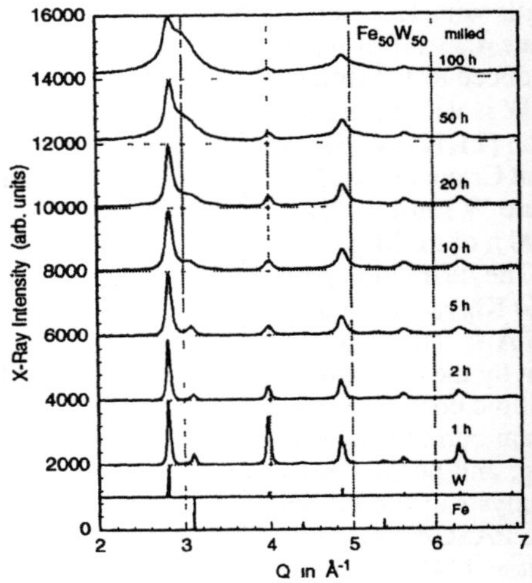

Fig. 2 X-ray diffraction patterns of a powder mixture of Fe with 50 at% W milled from 1 to 100 h.

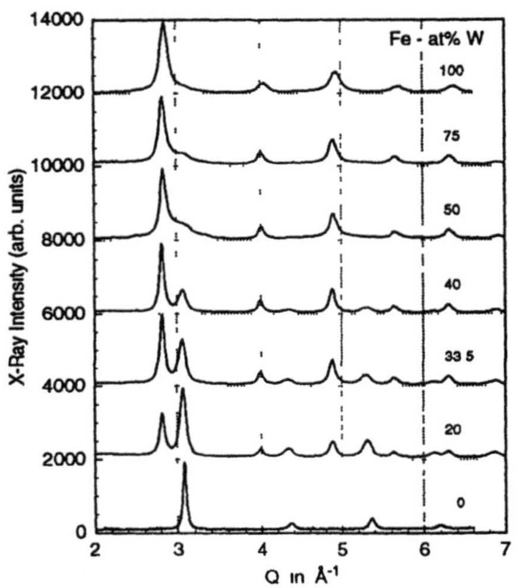

Fig. 3 X-ray diffraction patterns of powder mixtures of Fe and W milled for 20 h.

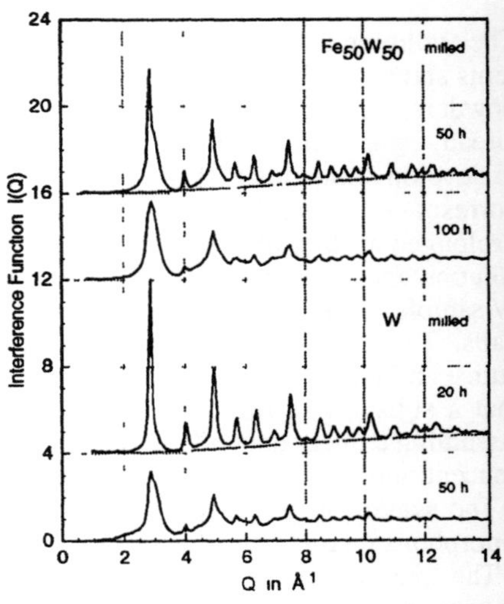

Fig. 4 Interference functions I(Q) of W and a powder mixture of Fe with 50 at% W milled for 20, 50 or 100 h.

displacements. The TDS modulates about the values $[1 - \exp(-\langle u^2\rangle Q^2)]$, which are also shown in Fig. 4 for the predominantly nanocrystalline samples. A value of $\langle u^2\rangle = 0.01$ was used in the TDS calculations. It is obvious from Fig. 4 that the TDS is largely responsible for the diffuse scattering in the nanocrystalline alloys [25,26].

The Fourier transforms of I(Q), i.e., the reduced atomic distribution functions $G_I(r)$ [equation (6)] are shown in Fig. 5. The patterns clearly show that W-Fe possess the body-centered cubic coordination which is similar in appearance to the face-centered cubic powder pattern. The size of the nanocrystals can be estimated from the $G_I(r)$ curves. The function G(r) becomes zero as soon as either $\rho(r) - \rho_o$ or v(r) becomes zero beyond a certain value of r [equation (6)]. Since the samples, except those milled for more than 50 h, are crystalline, v(r) must be the cause of the gradual damping of G(r). Taking the value of r beyond which G(r) fluctuates by no more than 5% of the maximum deflection about zero (this is estimated to be caused by the experimental errors in I(Q) due to the statistics of the photon counts and the corrections of the raw data), we deduce the values of the particle or coherently diffracting domain sizes D_G with an estimated error of ±10 Å, which are given in Table I.

When the peaks of the crystalline phase were removed from the structure factors I(Q) of in Fig. 4, the remaining pattern represented I(Q) of the amorphous W-Fe alloy. The distribution function $G_I(r)$ corresponding to this amorphous W-Fe phase in the samples, which contained about 50 at% Fe and Cr, is quite similar to that of the entire sample, except that the modulations in G(r) of the amorphous phase do not show much structure beyond 15 Å [26]. Since $G_I(r)$ is dominated by the W-W pairs because the weight factors $W_{ij} = c_i f_i c_j f_j / \langle f\rangle^2$ in equation (6) are 0.04, 0.30 and 0.66 for Fe-Fe, Fe-W and W-W, respectively, the position of the first peak at 2.70 Å corresponds predominantly to the W-W neighbor distance.

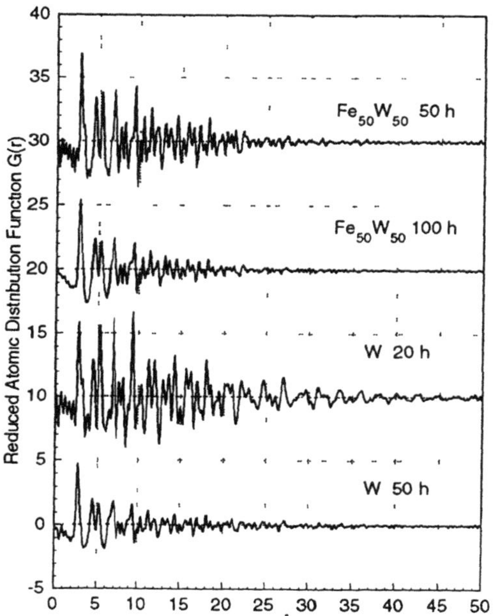

Fig. 5 Reduced atomic distribution function $G_I(r) = 4\pi r[\rho(r) - \rho_o]\, v(r)$ for W and a powder mixture of Fe with 50 at% W milled for 20, 50 or 100 h.

The reflections (110) to (310) of W in the W and $Fe_{50}W_{50}$ samples and the (110)-(220) pair of reflections of Fe were subjected to the Warren-Averbach analysis [6,7]. The effective particle size $\langle D_e\rangle$ was determined from the initial slope of the particle size Fourier coefficients $A_p(L)$ [equation (10)]. The sum of the Fourier coefficients $A_p(L)$ yielded the integral breadth particle size D_I [equation (12)]. The errors in $\langle D_e\rangle$ and D_I are estimated to be ±5 Å or ±10 Å. The values of $\langle D_e\rangle$ and D_I are given in Table I and are similar to the values D_G, determined from the atomic distribution function $G_I(r)$. The particle sizes of the W phase i the Fe-W alloys decreased more rapidly with milling time than those observed in the F powders. This is most likely due to the effect of the incorporation of Fe into each W parti cle. The particle sizes D_I are only slightly larger than $\langle D_e \rangle$ indicating that the particles hav a narrow size distribution.

The microstrains $\langle (\varepsilon_L)^2 \rangle^{1/2}$ at L = 50 Å [equation (11)], assuming that the average strain $\langle \varepsilon_L \rangle$ are zero or small in the individual powder alloy particles, are also given in Table] They increase initially in the W, Fe-W and Fe samples, but reach a plateau after prolonge milling, and are the consequence of the severe deformation during the mechanical workin in the high-energy ball mill.

Table I: Values of the particle size D_G determined from the reduced atomic distributio function G(r), the particle size $\langle D_e \rangle$, obtained from the initial slope of the particle siz Fourier coefficients $A_p(L)$, the particle size D_I, evaluated from the sum of the Fourie coefficients $A_p(L)$, and the root-mean square strains $\langle \varepsilon^2 \rangle^{1/2}$ at L = 50 Å for W $Fe_{50}W_{50}$ and Fe powders milled from 1 to 100 h.

Sample	Milling Time in h	D_G Å	$\langle D_e \rangle$ Å	D_I Å	$\langle \varepsilon^2 \rangle^{1/2}$ at L = 50Å %
W	1		260 ± 10	360 ± 10	0.27 ± 0.05
W	2		145 ± 10	220 ± 10	0.39 ± 0.05
W	5		90 ± 5	130 ± 5	0.49 ± 0.05
W	10		50 ± 5	70 ± 5	0.57 ± 0.05
W	20	50 ± 10	50 ± 5	70 ± 5	0.51 ± 0.05
W	50	35 ± 10			
Fe-W	1		170 ± 10	210 ± 10	0.29 ± 0.05
Fe-W	2		145 ± 10	195 ± 10	0.33 ± 0.05
Fe-W	5		95 ± 5	140 ± 5	0.45 ± 0.05
Fe-W	10		50 ± 5	75 ± 5	0.47 ± 0.05
Fe-W	20		50 ± 5	80 ± 5	0.41 ± 0.05
Fe-W	50	40 ± 10	50 ± 5	80 ± 5	0.40 ± 0.05
Fe-W	100	35 ± 10			
Fe	1		290 ± 10	430 ± 10	0.30 ± 0.05
Fe	2		190 ± 10	225 ± 10	0.40 ± 0.05
Fe	5		155 ± 5	190 ± 10	0.45 ± 0.05
Fe	10		115 ± 5	145 ± 5	0.55 ± 0.05
Fe	20		105 ± 5	120 ± 5	0.51 ± 0.05
Fe	50		90 ± 5	105 ± 5	0.54 ± 0.05
Fe	100		85 ± 5	100 ± 5	0.59 ± 0.05

5. Acknowledgment

This research was supported by the National Science Foundation under Grant DMR 89-10892.

6. References

1. W.L. Johnson,, Materials Science and Engineering, 97, (1988), 1.
2. C.N.J. Wagner, Amorphous Metals and Semiconductors, Haasen, P., Jaffee, R., eds., Acta-Scripta Metallurgica Proceedings Series, Vol. 3, Pergamon Press, Oxford, 1986, p. 54.
3. C.C. Koch, O.B. Cavin, C.G. McKamey and J.O. Scarbrough, Appl. Phys. Letters, 43, (1983), 1017.
4. A.W. Weeber and H. Bakker, Physica B, 153, (1988) 93.
5. R.B. Schwarz and C.C. Koch, Appl. Phys. Letters, 49, (1986), 146.
6. B.E. Warren, X-Ray Diffraction, Addison-Wesley, Reading, MA., 1968.
7. C.N.J. Wagner, Local Atomic Arrangements Studied by X-Ray Diffraction, J.B. Cohen and J.E. Hilliard, eds., Gordon and Breach, New York, 1966, p. 219.
8. A. Calka and A.P. Radlinski, Mat. Sci. Engr., A134 (1991) 1350.
9. R. Birringer and H. Gleiter, in Encyclopedia of Materials Science, R.W. Cahn, ed., Suppl. Vol. 1, Pergamon Press, Oxford, 1988, p. 339.
10. X. Zhu, R. Birringer, U. Herr and H. Gleiter, Phys. Rev. B35, (1987), 9085.
11. E. Hellstern, H. Fecht, Z. Fu and W.L. Johnson, J. Appl. Physics, 65 (1988) 305.
12. H. Fecht, E. Hellstern, Z. Fu and W.L. Johnson, Met. Trans., 21A (1990) 2333.
13. M. Oehring and R. Bormann, Mat. Sci. Engr., A134 (1991) 1330.
14. M. Trudeau and R. Schulz, Mat. Sci. Engr. A134 (1991) 1361.
15. C.N.J. Wagner, E. Yang and M.S. Boldrick, Advances in X-Ray Analysis, 35 (1992) 585.
16. D.T. Keating, J. Appl. Phys., 34, (1963), 923.
17. T.E. Faber and J.M. Ziman, Phil. Mag., 11, (1965), 153.
18. C.N.J. Wagner, Liquid Metals, Chemistry and Physics, S.Z. Beer, ed., Marcel Dekker, New York, 1972 , p. 257.
19. C.N.J. Wagner, Microscopic Methods in Metals, U. Gonser, ed., Topics in Current Physics, Vol. 40, Springer Verlag, Berlin, 1986, p. 117.
20. E.A. Lorch, J. Physics, C2 (1969) 229.
21. R.P.I. Adler, H.M. Otte and C.N.J. Wagner, Met. Trans., 1, (1970), 2791.
22. C.N.J. Wagner, Technical Report No. 2, National Science Foundation, Grant GP 3213, 1966.
23. C.N.J. Wagner, J. Non-Cryst. Solids, 31, (1978), 1.
24. C.N.J. Wagner, Technical Report No. 15, Office of Naval Research, Contract NONR 609(43), 1966.
25. C.N.J. Wagner and M.S. Boldrick, Mat. Sci. Engr., A133 (1991) 26.
26. C.N.J. Wagner, J. Non-Cryst. Solids, 150 (1992), 1.

STRUCTURAL CHARACTERISTICS OF THE NANOCRYSTALLINE MATERIALS SYNTHESIZED BY COMPLETELY CRYSTALLIZING AMORPHOUS ALLOYS

K. LU
National Key Laboratory for RSA
Institute of Metal Research
Chinese Academy of Sciences
Shenyang 110015, P.R. China

ABSTRACT. In this paper, some recent results were summarily reviewed on structural characteristics of the nanocrystalline materials synthesized by crystallizing amorphous alloys completely. It includes the morphologies, structures, defects, and the thermodynamic properties of the nanophases and interfaces in various nanocrystalline materials, which were investigated by means of the high resolution electron microscopy (HREM), x-ray diffraction techniques, positron annihilation spectroscopy, Mössbauer spectroscopy, and the differential scanning calorimetry (DSC), respectively. Emphases were given on the interfacial defects and the lattice structure of the nm-sized crystallites.

1. Introduction

Nanocrystalline (NC) materials are defined as polycrystals in which the grain size is in the nanometer region (normally less than 100 nm). The NC material is structurally characterized by ultrafine grains and a large fraction of atoms located on the internal interfaces, which enable many properties of the NC samples to be fundamentally different from those of the conventional polycrystals and the amorphous solids [1]. Therefore, this new class of materials has drawn increasing attentions in recent years.

Up to now, several synthesis methods for the NC materials have been developed following the classical method of the *in situ* consolidation of the ultrafine metal particles (UFP) [1, 2], for example, the mechanical alloying [3], electro-depositing [4], the complete crystallization method from the amorphous solids [5], and so on. Among these synthesis routes, the complete crystallization method possesses some unique advantages: (i) This method is very simple and convenient to be controlled in experimental procedures. Conventional annealing can realize the nanocrystallization in most alloy and element systems which can be formed in the amorphous states, and can produce a large quantity of the NC samples. Also, variable grain sizes can be easily obtained in the NC specimens by modifying the heat treatment conditions. (ii) The complete crystallization method is an

M. A. Otooni (ed.), Science and Technology of Rapid Solidification and Processing, 349–361.

efficient way to produce porosity-free NC samples [6]. Because no artificial consolidation process is involved and the nm-crystallites and their boundaries are formed naturally during the crystallization, the NC sample is dense and clean in the internal interfaces. (iii) The nanocrystallization itself provides an unique opportunity to study the interface formation process from the amorphous state. The nanocrystallization kinetics and thermodynamics of the amorphous solids are strongly affected by the present of plenty of interfaces in the crystallization products [7,8]. Consequently, it is possible to reveal some fundamental features of the interfaces in the NC materials from the transformation kinetics and thermodynamics[9].

Because of these features, the complete crystallization method has been intensively investigated in the past few years. Up to now, this method has been successfully applied not only in various alloy systems, such as in Fe-, Ni-, and Co-based alloys [10-14], but also in some elements (e.g., Se, Si) [15], in producing nanometer-sized structures. The NC materials synthesized this way have been studied on their formation processes, structures, properties, and stabilities [16-20]. In this paper, a brief review is made on the microstructural charateristics of the NC materials, with an emphasis on the interfacial defects and the lattice structure of the nm-sized crystallites.

2. Morphologies and Structures of the NC Materials

2.1. MORPHOLOGIES AND STRUCTURES OF THE CRYSTALLITES IN THE NC MATERIALS

Amorphous alloys crystallize by a nucleation and growth process, of which the driving force is the free energy difference between the glass and the appropriate crystalline phase(s). According to the composition of the amorphous alloys and the thermodynamic properties of the corresponding crystalline phase(s), three kinds of crystallization processes are always occuring in amorphous alloys:

(a) Polymorphous crystallization of one cystalline phase with the same chemical composition of the amorphous state.

(b) Eutectic crystallization of two crystalline phases simultaneously. In this case, the two phases grow in a coupled fashion analogously to the eutectic crystallization of liquids. The overall composition of the two crystalline phases is identical to that of the amorphous phase.

(c) Primary crystallization. During a primary crystallization, a primary crystalline phase nucleates and grows in the amorphous phase as the first step, then the residual amorphous phase crystallizes completely. The crysallization process of the primary phase is controlled by atomic diffusion, and there is a compositional pileup in front of the growing crystals.

Nanometer-sized grains or subgrains are available in all these three kinds of crystallization of amorphous alloys under proper heat treatment conditions. Figure 1 is a high resolution electrom microscopy (HREM) image of a NC Ni-P alloy which is a typical eutectic nanocrystallization product of a melt-spun amorphous $Ni_{80}P_{20}$(at.%) alloy [5]. There are two crystalline phases of a Ni(P) solid solution (with a fcc structure) and a Ni_3P compound (bct structure). It is clear that the two nm-sized crystallites are in the form of anisotropic-shaped blocks. The Ni_3P nanophases are separated by the Ni phases, and the

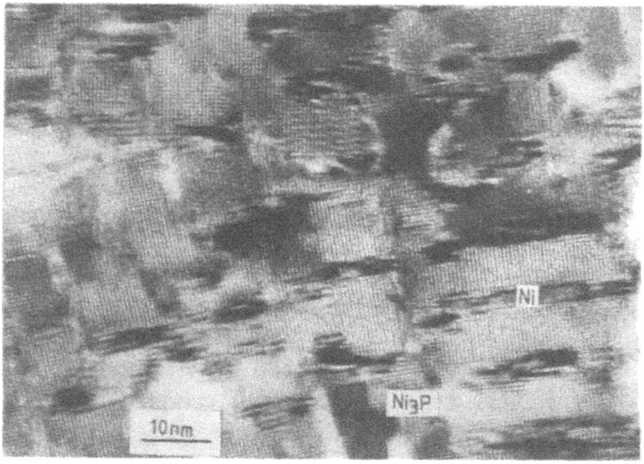

Figure 1. A HREM iamge of a NC Ni-P sample along [001] axis of Ni_3P.

orientations of the bct phase differ slightly from each other. The orientation relationship between Ni phase and the Ni_3P compound is: $<001>_{bct}//<110>_{fcc}$ and $<110>_{bct}//<111>_{fcc}$ [21]. Similar morphologies and structures have been observed in other eutectic nanocrystallization products, such as in the Fe-B binary alloy. And orientational relationships between the two crystalline phases are frequently observed as in the eutectic crystallization products of metallic glasses[22].

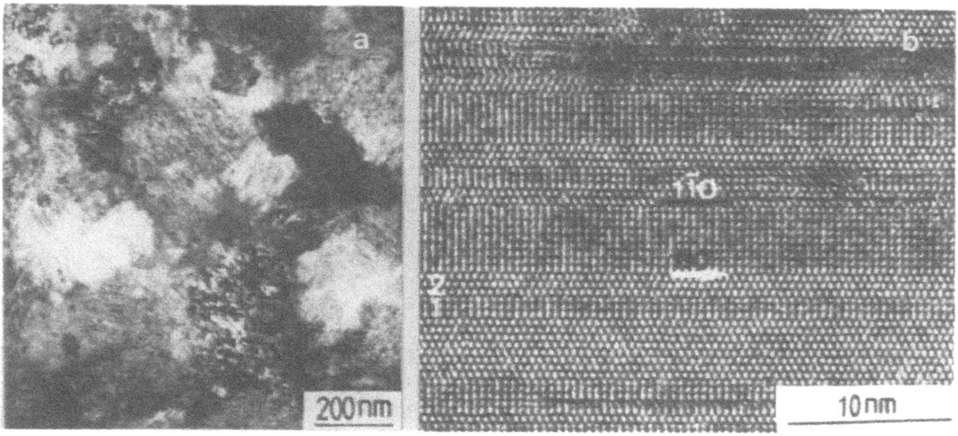

Figure 2. (a) A bright field TEM image of the NC $NiZr_2$ sample; (b) A HREM image of the find structure of the $NiZr_2$ nanophase.

Crystallization of an amorphous $Ni_{33}Zr_{67}$ (at.%) alloy is a typical polymorphous transformation, of which the products are of only one crystalline phase: a $NiZr_2$ intermetallic compound. Figure 2a shows a TEM photo of an as-crystallized $NiZr_2$ NC sample, which was made by annealing the glass at 753 K for about 15 min. There are irregular-shaped crystalline regions within which nanostructured phases or domains are visible. Figure 2b shows a HREM image inside a crystalline region in Fig. 2a. It can be seen that each crystalline region (of about 200-500 nm across) is composed of lamellar domains of which the thickness is about a few nanometers and the length is below 100nm. Between the neighbouring lamellae there is a coherent twin boundary, and the orientation relationship between the two grains is: $[001]_1//[111]_2$ and $[110]_1//[110]_2$ [23].

Figure 3 shows a TEM image of a NC Fe-Cu-Si-B alloy which is a typical primary nanocrystallization product from an amorphous $(Fe_{99}Cu_1)_{78}Si_9B_{13}$ alloy [24]. It contains two crystalline phases of a α-Fe(Si) and a Fe_2B compound. X-ray diffraction measurements (XRD) and the differential scanning calorimetry (DSC) analyses of the nanocrystallization process indicated that the α-Fe(Si) crystallizes as a primary phase, followed by a complete crystallization of the residual amorphous phase. The shapes of these two nanophases are approximately isotropical and their grain sizes are uniform with a narrow size distribution. The selected area diffraction pattern shows orientations of the two nanophases are random and no orientational relationship between the two phases is expected.

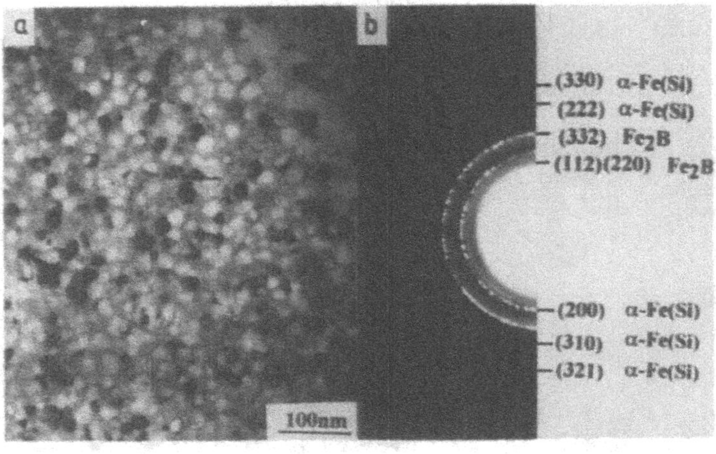

Figure 3. A bright field TEM image (a) and the corresponding diffraction pattern (b) of the NC Fe-Cu-Si-B alloy.

2.2. INTERFACIAL STRUCTURE IN THE NC MATERIALS

For the polycrystalline materials with nanometer-sized grains, the grain boundaries or internal interfaces constitute a large volume fraction. Therefore, the structure of these interfaces would play an important role in the properties of NC materials. The interfacial structures in the NC samples synthesized by the complete crystallization method have been investigated in several systems. Three samples will be discussed here to illustrate the interfacial structural characteristics in these NC materials.

As discussed above, polymorphous nanocrystallization leads to formation of one crystalline phase and grain boundaries between the crystallites, as in the $NiZr_2$ nanophase samples. In the case of $NiZr_2$, most grain boundaries are straight and well-fitted, as in Fig. 2b. They are coherent twin boundaries with a specific structure. It can be expected such kind of boundaries are in a low energetic configuration with very small excess volumes.

In the primary nanocrystallization products of the Fe-based alloys, incoherent grain boundaries and interphase boundaries are always observed. Figure 4a and 4b show HREM images of the α-Fe(Si) grain boundary and the α-Fe(Si)/Fe_2B interface in a NC $(Fe_{99}Mo_1)_{78}Si_9B_{13}$ alloy [24]. The α-Fe(Si) grain boundary is very flat and seems to be a normal large-angle grain boundary. And the interphase boundary in Fig. 4b exhibits flat facets separated by steps. But in both interfaces, no extended contrast effects caused by the "disordered boundaries" can be detected. As the orientations of the crystallites are random and there is no orientation relationship between the neighbouring grains in this sample, most interfaces are expected to be incoherent and with relative larger excess volume than the coherent or semicoherent ones as in the polymorphous crystalllization products.

In the Ni-P eutectic nanocrystallization products, most internal interfaces are the $Ni(P)/Ni_3P$ interphase boundaries. As there is an orientation relationship between the two phases, the interfacial structure should be specified. Calorimetry measurement during the grain growth yielded that the interfacial energy of the Ni/Ni_3P is about only 0.155 J/m^2 (and is about 0.111 J/m^2 for the Ni/Ni boundaries) [25], which is close to the value of a small-angle grain boundary. This means the interfaces in the NC samples are in low energetic configurations.

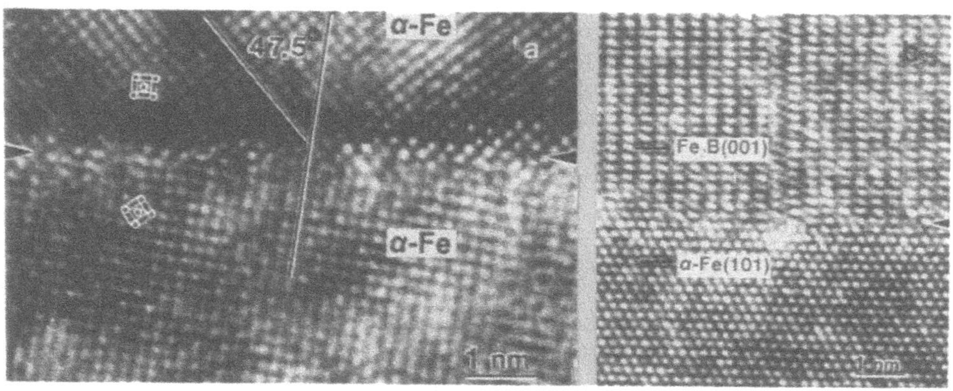

Figure 4. HREM images of the α-Fe grain boundary (a) and the interface between α-Fe/Fe_2B in the NC Fe-Mo-Si-B alloy.

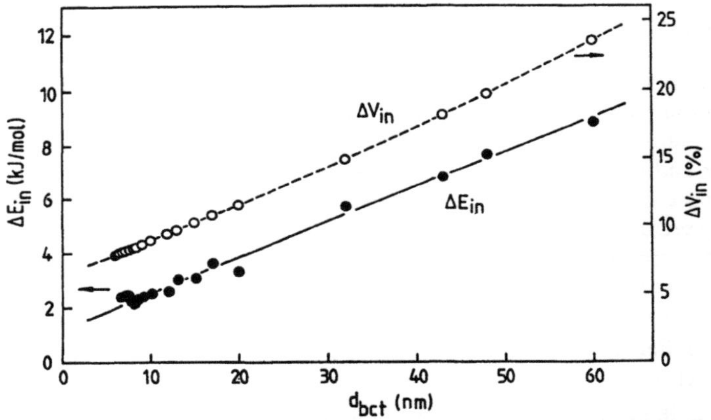

Figure 5. Variations of the interfacial excess energy and the interfacial excess volume with the average grain size in the Ni-P nanophase samples.

By means of the thermodynamics of a transformation from the amorphous to the NC state, the interfacial excess energies in the NC Ni-P materials with different grain sizes were experimentally measured by using a microcalorimeter [9]. It was found that the average interfacial excess energy decreases significantly with a reduction of grain size in an approximate linear relation, as shown in Fig. 5. In the sample with a grain size of a few nanometers, the interfacial energy is rather small, close to the energy of a small-angle grain boundary. The variation of the interfacial excess energy with the grain size is found to be correlated with that of the interfacial excess volume (which is the most significant parameter designating the grain boundary energy) deduced from the experimental results [26], see Fig. 5. The interfacial excess volume in the NC Ni-P samples is decreasing evidently with the reduction of grain size. These results suggest that the interfacial structure in the NC materials might be changable and dependent on the grain size.

It can be seen the interfacial structures are different in different alloy systems with different crystallization processes. For the polymorphous and the eutectic nanocrystallization products, comparatively, the interfaces seem to be in lower energetic configurations than those in the primary case. This feature can also be found from the positron annihilation spectroscopy results as presented in the follwing section.

3. Defects in the NC Materials

3.1. TWO KINDS OF DEFECTS IN THE NC MATERIALS

The positron annihilation spectroscopy (PAS) technique has been proved to be a powerful method in surveying the characteristics and distributions of microdefects in solids and has been successfully applied in the study of the defects in the NC materials [27]. Systematic PAS investigations on the defects in the NC samples synthesized by using the complete crystallization method were carried out in the Ni-P [6], Fe-Si-B [28], and Fe-Mo-Si-B [29] alloys. Analyzing the positron lifetime spectra for these samples, we found some common features in the NC materials.

Table 1 lists the positron lifetime results of these NC samples. It is clear that all positrons injected into the NC samples are annihilated with lifetimes clearly longer than those in the delocalized free states in bulk crystals of Ni and Fe. The short lifetime component τ_1 may be thought of as resulting from the positron localized and annihilated at free volumes with a size smaller than a monovacancy. And the intermediate lifetime component τ_2 is corrresponding to the positrons annihilated at the intersections of two or three crystallites interfaces and/or grain boundaries, or we called as microvoids, of which the size is about 10-15 vacancies. The long lifetime component, τ_3, with a very small intensity of about 1-3%, may be attributed to the formation of orthopositronium (*o*-Ps) at the internal surfaces of large voids. In the stack samples of the small piece of ribbons, it is reasonable to believe the *o*-Ps would be formed at the surfaces between the source and the sample, as well as the internal surfaces among the ribbon pieces. No large voids were detected in microscopy observations of the Ni- and Fe-based NC samples. This feature is fundamentally different from that in the NC samples prepared by consolidation of UFPs.

For comparison, the positron lifetime results of a NC Fe sample synthesized by the consolidation method [27] was also listed in Table 1. It is clear there are two long lifetime components, τ_3 and τ_4, of which the intensity is about 10-30%, besides the short and the intermediate ones. These long lifetime components were thought of as resulting from the large voids inside the NC sample which were formed during the compaction process of the powders. Existance of the sample porosities inside the NC sample was proved by the microscopic observations that there are pores with diameters as large as about 1 μm in the Fe NC specimen.

Therefore, it can be concluded that there are only two kinds of defects in the NC materials crystallized from the amorphous state: type-I of free volumes of which the size is smaller than a monovacancy, and type-II of microvoids. The complete crystallization method from the amorphous alloys can produce porosity-free NC samples which contain denser interfaces and less defects than those prepared by the consolidation method.

Table 1. Positron lifetime results for three NC samples synthesized by means of the complete crystallization method, in comparison with that for a NC Fe sample made by consolidation of UFPs.

Sample	τ_1 (ps)	τ_2 (ps)	τ_3 (ps)	τ_4 (ps)	I_1 (%)	I_1/I_2	I_3+I_4 (%)
Ni-P	152.3±1.3	350±20	1470±96	--	94.4±1.3	21.0	1.1±0.1
Fe-Si-B	145.5±1.4	376±23	1845±41	--	89.5±0.9	11.3	2.5±0.8
Fe-Mo-Si-B	139.4±1.5	240±12	1582±42	--	90.3±1.0	10.7	1.2±0.1
NC Fe	180±15	360±30	1200±200	4000±500	--	0.5~1.0	10~30

3.2. VARIATION OF THE DEFECTS WITH GRAIN SIZE

Distributions of the microdefects in the NC samples prepared by crystallization of the amorphous alloys were found to be influenced by the grain size. Figures 6a-6d show the lifetimes τ_1, τ_2, the intensity ratio I_1/I_2, and the mean positron lifetime versus the grain size in the Ni-P [6] and the Fe-Si-B [28] NC samples, respectively. It is evident that for both NC samples, the lifetimes τ_1 and τ_2 increase with a reduction of the mean grain size. A small increase in the values of τ_1 indicates volumes of the type-I defects in the samples increase slightly with a reduction of grain size. Similarly, the intermediate lifetime corresponding to the type-II defects decreases with an increasing grain size, revealing that type-II defects shrink in size or the microvoids are agglomerated by less vacancies when the grain size is larger.

It can be seen from Fig. 6c that the intensity ratio I_1/I_2 clearly decreases with an increment of the grain size for the both alloys. The mean positron lifetime $\overline{\tau} = (\tau_1 \cdot I_1 + \tau_1 \cdot I_2)/(I_1 + I_2)$ keeps a constant around 160 ps for the Ni-P NC samples, but decreases with an increasing grain size for the Fe-Si-B alloy, as shown in Fig. 6d.

According to the trapping model, concentrations (C) of the two kinds of defects in the interfaces may be correlated to the intensities by

$$I_i = \sigma_i C_i/(1/\tau_0 - 1/\tau_i) \qquad (1)$$

where τ_0 is the residence time of positrons in the delocalized lattice state and σ_i denotes the specific trapping rates of various types, i, of traps. The ratio

$$\sigma_1 C_1/\sigma_2 C_2 \propto I_1/I_2 \qquad (2)$$

can be derived. Assuming equal trapping rates per unit area for interfaces with the two kinds of defect, the interface-area ratio of the two kinds of defect may be estimated by means of the experimental data of I_1/I_2.

For the Ni-P and the Fe-Si-B NC alloys, the ratio I_1/I_2 is increasing when the grain size is reduced. Then it is reasonable to believe that the area ratio of interfaces with type-I defects to those with type-II defects is increased with a decreasing grain size. This means in a unit area of interfaces there are more type-I defects and fewer type-II in the samples with smaller grains. In other words, the interfacial density increases with a decrease in the grain size in the both alloys, which agrees with the measurement results of the interfacial excess volume in the NC Ni-P samples (Fig. 5).

From Fig. 6c one may also notice that values of the ratio I_1/I_2 for the Ni-P NC sample with an eutectic nanocrystallization product is much larger than those for the primary nanocrystallized Fe-Si-B alloy with the same grain size. Larger values of I_1/I_2 imply there are more type-I defects and fewer type-II in a unit area of interfaces. It suggests that the interfaces in the Ni-P NC samples are denser and containing less excess volumes in a unit interfacial area than the Fe-Si-B samples. Similar feature is valid for the primary nanocrystallization of Fe-Mo-Si-B samples, as listed in Table 1. This evidence indicates that the eutectic nanocrystallization may creat much denser interfaces than the primary nanocrystallizations, which is in a satisfactory agreement with the structure analysis results as presented in the last section.

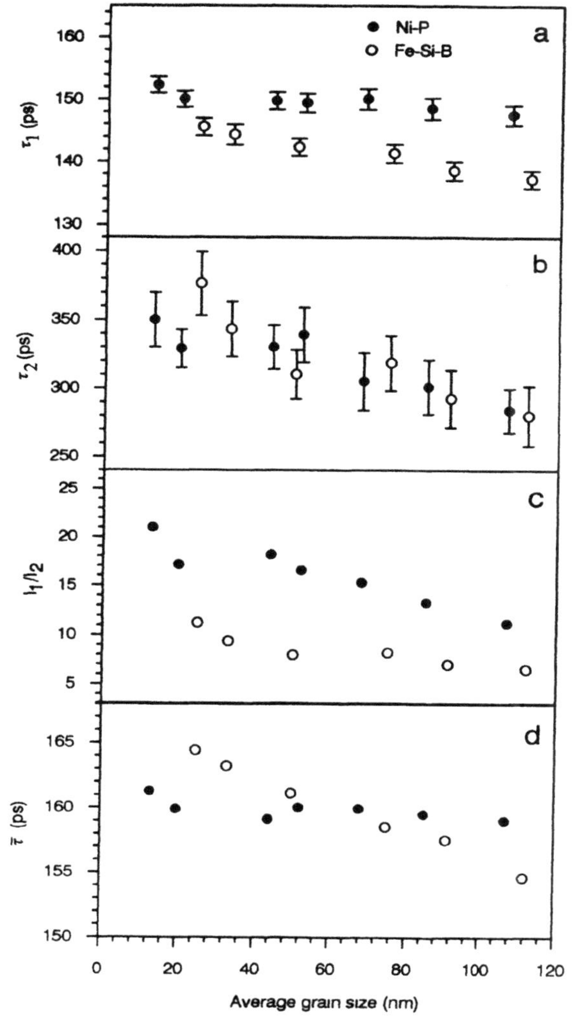

Figure 6. Plots of the positron lifetime results of NC Ni-P and Fe-Si-B alloys versus the average grain size.

4. The Lattice Structure of the nm-size Crystallites

The structure of nm-sized crystallizes in NC materials is always thought of as the same as that in the perfect crystal lattice and hence has drawn seldom attention. However, an important question is actually open to whether or not the lattice structure is changed relative to the perfect crystal when the crystallites are refined down to a few nanometers.

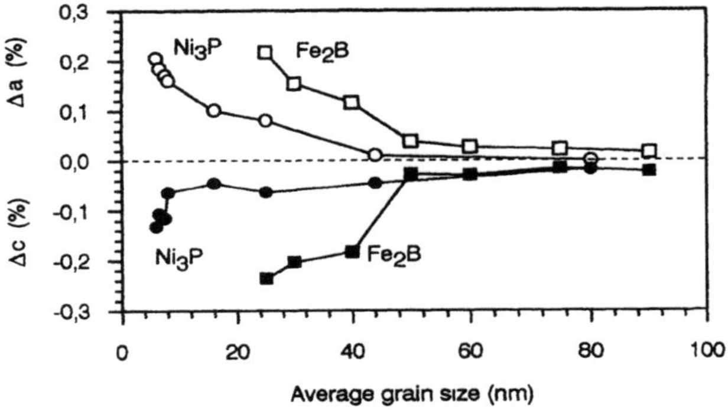

Figure 7. Variations of Δa and Δc with the average grain size fo the Ni_3P and Fe_2B nanophases.

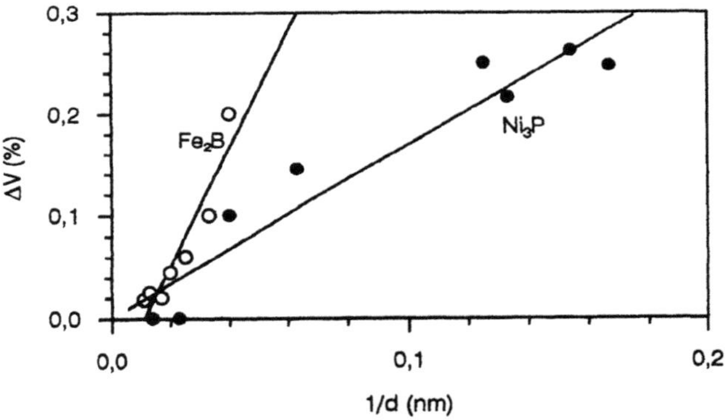

Figure 8. Plots of ΔV versus $1/d$ for the Ni_3P and Fe_2B nanophases.

The lattice structures of the nanometer-sized crystallites in various NC samples crystallized from the amorphous materials were studied by means of the XRD and the Mössbauer spectroscopy. Qantitative XRD measurement of the lattice parameters of the nm-sized Ni_3P (bct) compound in the NC Ni-P alloy [17], and of the Fe_2B (bct) nanophase in the Fe-Mo-Si-B alloy [18], revealed a fact that the lattice parameters of these two nm-sized stoichiometic compounds evidently deviated from the equilibrium values of the perfect crystal lattice. The value of a is always larger than the standard value of a_0 while c is always smaller than c_0 for both bct nanophase compounds. Meanwhile, the values of a and c are changing with the grain size: with a reduction of the average grain size, the value of a increases and c decreases in both cases. The relative Fe_2B deviations of the two parameters reach as much as Δa=0.22% and Δc= -0.24% for the with a grain size of about 25 nm, and Δa=0.21% and Δc= -0.13% for the 6 nm-Ni_3P. Figure 7 shows the variation of lattice parameters with grain size in the two nanophases. It is clear that the deviation of lattice parameters in the Fe_2B nanophase is greater than that in the Ni_3P with the same grain size, which might be attributed to the difference in the interfacial structures in the two NC samples.

The unit cell volume of each compound, $V=a^2c$, was found to be enlarged with respect to the corresponding equilibrium value. That is the lattice of the nm-sized crystallite is dilated relative to the perfect crystal structure. With decreasing grain size, the lattice expansion increases for the both nanophase, as shown in Fig. 8. The variation of the relative unit cell volume change ($\Delta V=V/V_0$-1, $V_0=a_0{}^2c_0$) with $1/d$ may be approximated by a straight line with a positive slope for both nanophases, as in Fig. 8. The slope for Fe_2B is larger than that for the Ni_3P.

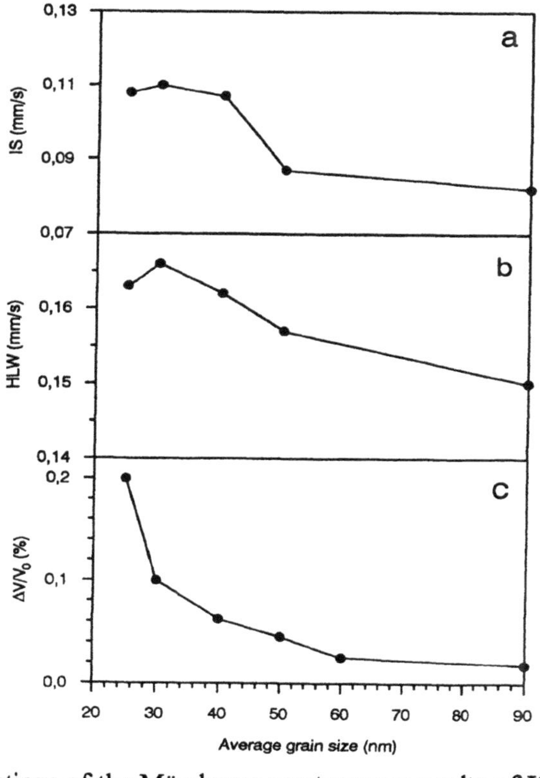

Figure 9. Variations of the Mössbauer spectroscopy results of IS (a) and HLW (b), and the unit cell volume change ΔV (from the XRD results) with the average grain size for the Fe_2B nanophase.

Mössbauer spectroscopy investigations on the microstructure of the Fe_2B nanophases with different grain sizes revealed that the isomer shift (IS) and the half-linewidth (HLW) for the Fe_2B subspectrum are increasing with a decrease of grain size, as shown in Fig. 9a and 9b. An increase of IS means the density of s-electron of Fe atoms in the Fe_2B phase decreases, which reflects a dilated lattice structure. The XRD measurement results of the unit cell volume change was plotted versus the average grain size in Fig. 9c. While the increasing HLW value indicates the distribution of the atomic coordination distances in the Fe_2B phase becomes wider and the degree of disordering is enhanced. This result is in agreement with the XRD measurements that the distortion in the nanophase lattice is increasing with a reducing grain size.

The intrinsic reason for the lattice distortion and expansion in the nanophase stoichiometric compounds, which have no solubitiy of other elements in the equilibrium state, is still absent. According to the classical thermodynamics theory, it is known that the free energy state of a crystallite with a finit size will be influenced by the existance of grain boundaries. The enhancement in the free energy of the crystallite relative to the perfect crystal lattice will increase when the grain becomes smaller. Therefore, for the nm-sized crystallites, there is a large increase in the free energy which will consequently enhance the solubility of other elements and vacancies. Following the Gibbs-Thomson equation the behavior of lattice distortion and expansion was analyzed, which might be reasonably attributed to a supersaturation of either elements (e.g., P or Ni in the Ni_3P) or vacancies [30]. Further investigations are still in progress.

5. Summary and Conclusions

NC materials synthesized by the complete crystallization from amorphous alloys possess some specific structural characteristics which might differ from those of the conventional NC materials made by consolidation of the ultrafine particles. However, some of their structural features might be intrinsic and representive for the nanostructured materials, which lead to the crystallized NC samples exhibit similar properties as other NC materials do (such as the specific heat, thermal expansion, the abnormal Hall-Petch relation, etc.). For instance, decreasing interfacial energy with a reduction of grain size in NC state has been recently reported in a consolidated TiO_2 NC materials [31]. Lattice distortions in various NC samples have been observed [32].

The porosity-free NC samples are very useful for a deep investigation on the microstructures and properties of the internal interfaces. Therefore, the complete crystallization of amorphous alloys might be a promising route to creat model materials for researches on the interfacial structures in NC materials and on the structure-property relationship in solids. More solid and systemetic studies on the microstructures of the crystallized NC materials should be done in the future.

Acknowledgements: This work is financially supported by the Chinese Academy of Sciences and the National Science Foundation of China. The author thanks his colleagues, Drs. X.D. Liu, M.L. Sui, and Mr. H.Y. Tong for their contributions to this work, and to U. Köster, A.L. Greer, and R. Lück for helpful discussions.

References

1. H. Gleiter, Prog. Mater. Sci., **33**, 223 (1989).
2. R. Birringer, U. Herr, H. Gleiter, Trans. Jpn. Inst. Met. Suppl., **27**, 43 (1986)
3. C.C. Koch, Nanostructured Mater., **2**, 109 (1993).
4. U. Erb, A.M. El-Sherik, G. Palumbo, K.T. Aust, Nanostr. Mater., **2**, 383 (1993).
5. K. Lu, J.T. Wang, and W.D. Wei, J. Appl. Phys., **69**, 522 (1991); Scripta Metall. Mater., **24**, 2319 (1990).
6. M.L. Sui, K. Lu, W. Deng, L.Y. Xiong, S. Patu, and Y.Z. He, Phys. Rev. B, **44**, 6466 (1991); J. Appl. Phys., **69**, 4451 (1991).
7. R. Lück, K. Lu, and W. Frantz, Scripta Metall. Mater., **28**, 1071 (1993)
8. K. Lu, R. Lück, and B. Predel, J. Alloy Comp., (1993), in print.
9. K. Lu, R. Lück, and B. Predel, Scripta Metall. Mater., **28**, 1387 (1993).
10. M.M. Nicolaus, H.-R. Sinning, F. Haessner, Mater. Sci. Eng., **A150**, 101 (1992).
11. T. Spassov and U. Köster, J. Mater. Sci., **28**, 2789 (1993).
12. H.Y. Tong, J.T. Wang, B.Z. Ding, H.G. Jiang, K. Lu, J. Non-cryst. Solids, **150**, 444 (1992)
13. H.Q. Guo, T. Reininger, H. Kronmüller, M. Rapp, V.Kh. Skumrev, Phys. Stat. Sol., **A127**, 519 (1991).
14. X.D. Liu, J.T. Wang, B.Z. Ding, Scripta Metall. Mater., **28**, 59 (1993).
15. H.Y. Zhang, K. Lu, Z.Q. Hu, Acta Phys. Sinica, (1994), to be published.
16. K. Lu, Acta Metall. Sinica, **30**, B1 (1994).
17. M.L. Sui, and K. Lu, Mater. Sci. Eng. A, (1994), in print.
18. X.D. Liu, K. Lu, Z.Q. Hu, B.Z. Ding, Nanostr. Mater., **2**, (1993).
19. K. Lu, J.T. Wang, W.D. Wei, J. Phys. D: Appl. Phys, **25**, 808 (1992).
20. K. Lu, Scripta Metall. Mater., **25**, 2047 (1991); Nanostr. Mater., **2**, (1993).
21. M.L. Sui, K. Lu, Y.Z. He, Phil. Mag., **B63**, 933 (1991).
22. M.G. Scott, in: Amorphous Metallic Alloys, ed. F.E. Luborsky, Butterworth, London, 144 (1983).
23. C. Beeli, H.-U. Nissen, Q. Jiang, R. Lück, Mater. Sci. Eng., **A133**, 346 (1991).
24. X.D. Liu, Ph.D. Thesis, Institute of Metal Research Academia Sinica, (1994).
25. K. Lu, W.D. Wei, J.T. Wang, J. Appl. Phys., **69**, 7345 (1991).
26. M.L. Sui, and K. Lu, Acta Metall. Sinica, **30**, (1994), in print.
27. H.E. Schaefer and R. Wurschun, Phys. Lett., **119A**, 370 (1987).
28. H.Y. Tong, B.Z. Ding, J.T. WAng, K. Lu, J. Jiang, J. Zhu, J. Appl. Phys., **72**, 5124 (1992).
29. X.D. Liu, B.Z. Ding, J.T. Wang, J. Mater. Sci. Lett., **12**, 1826 (1993).
30. K. Lu, and M.L. Sui, J. Mater. Sci. Tech., **9**, 419 (1993).
31. C.D. Terwilliger, Y.M. Chiang, MRS Symp. Proc., **289**, 15 (1993).
32. K. Lu, M.L. Sui, and X.D. Liu, Proc. ICAM'94 (Tokyo), (1993), in print.

THEORY OF THE CRYSTALLIZATION OF PZT THIN FILMS VIA RAPID THERMAL ANNEALING

E.K.F. DANG and R.J. GOODING
Department of Physics, Queen's University,
Kingston, Ontario K7L 3N6, Canada

ABSTRACT. The crystallization of lead zirconate titanate thin films from an amorphous state is known to be greatly aided by rapid thermal annealing (RTA). We have examined this effect within a simple phenomenological model which incorporates the presence of an experimentally observed secondary crystalline phase, and have studied the approach to steady state using Langevin simulations.

1. Introduction

In recent years, ferroelectric thin films have been a subject of considerable interest because of their important potential applications, especially in non-volatile memory devices or dynamic random access memory (DRAM). Among the most studied of these materials is the ceramic lead zirconate titanate $Pb(Zr_xTi_{1-x})O_3$ (PZT), with x typically about $0.52 - 0.58$ [1-4]. One important aspect of the research effort is the development of reliable fabrication techniques of these films. A variety of deposition processes have been employed, including metal–organic chemical–vapour deposition, dc and rf magnetic sputtering, and sol–gel processing. However, in all of these processes, the as–deposited films are amorphous and non-ferroelectric. Post–deposition annealing treatments are required to enable the films to crystallize into the ferroelectric perovskite structure. Conventional furnace annealing (CFA) typically involves heating rates of order $1 - 10°$C/s, with final temperatures of about 600°C, and annealing times of the order of an hour. However, since many of the technological applications of ferroelectric ceramics films demand successful integration of the films into semiconductor devices, it is crucial to reduce either the annealing temperature or annealing time, or both, in order to avoid strains and other defects at the film–substrate interface [5]. Thus, alternative annealing processes have been attempted, e.g. laser, electron beam, and radiation annealing. Among these, rapid thermal annealing (RTA), as characterized by an extremely large heating rate (e.g. 100°C/s), has been shown to be particularly promising [2-4]. RTA was initially developed for semiconductor processing [6]. When applied to PZT thin films, it not only significantly reduces the annealing temperature–time product, but also yields improved ferroelectric properties, as well as a smoother surface morphology compared to films processed by CFA [7].

A comprehensive theoretical understanding of the differences between CFA and RTA, or more precisely, of the role of heating rates in the crystallization of PZT, will involve

M. A. Otooni (ed.), Science and Technology of Rapid Solidification and Processing, 363–371.

a detailed study of the chemical and structural transformation path that begins with the amorphous film and ends with the final perovskite structure. However, it is also useful, in the first instance, to investigate the process in terms of a simple phenomenological model with the aim of identifying the essential physics involved. In this report, we present such an investigation using a Landau model and Langevin dynamical simulations. The Landau potential is an effective free energy functional determined by a small number of order parameters characterizing the transformation, such that it represents the competition between locally stable phases. From our point of view, it allows one to study the paths that a system may follow when transforming from one phase to another. Our model involves a single scalar field and incorporates the possibility that the transformation always proceeds via an intermediate defect–pyrochlore state, such as was observed in CFA experiments [1, 3, 4, 8, 9]. We have found that a necessary feature of a model which successfully accounts for the observed experimental behaviour involves long–ranged elastic interactions. We have chosen a simple potential to represent this effect. This interaction is motivated by the work of Littlewood and Chandra [10], who showed that long–ranged elastic forces produce a well-defined delay time in nucleation and growth processes, in good agreement with experimental results [11] for ferroelectric barium titanate ($BaTiO_3$).

2. Experimental results on PZT Crystallization

First, let us clearly review the experimental data on the subject. In experiments the composition of the PZT films is routinely determined by X-ray diffraction. In addition, since the dielectric constants of both the amorphous and defect–pyrochlore phases are about 60, while that of the perovskite is almost 1400, dielectric measurements offer a good indication of the fraction of the film that has transformed to the ferroelectric perovskite phase [4]. It is well known that in CFA the composition and structure of the ceramics PZT film depend strongly on both the annealing temperature and annealing time [1, 2, 4, 8, 9]. It is found that at intermediate annealing temperatures, in the region of 450°C – 600°C, the resulting film predominantly consists of the defect–pyrochlore ($Pb_2(Zr_xTi_{1-x})_2O_6$) phase. Above 600°C, the system starts to transform to the perovskite ($Pb(Zr_xTi_{1-x})O_3$) phase. Even at such high temperatures, the transformation is still not entirely complete after soaking for one hour.

Martin [9] suggested that a reason why the defect–pyrochlore structure tends to be formed as an intermediate phase is that the perovskite system is a very dense structure in comparison to the more open, less dense defect–pyrochlore phase. One might expect the less dense material to be more readily formed from the amorphous phase. Apart from this, we believe that there is a second reason for the presence of an intermediate defect-pyrochlore state. It is known [3, 12] that one must have a certain abundance of lead to allow for the formation of perovskite to proceed. We believe that this may be attributed to the formation of lead oxide from the perfect pyrochlore state, the former being found in abundance after the transition has been completed [12]. Schematically, this corresponds to

$$\text{excess } Pb + Pb_2(Zr_xTi_{1-x})_2O_7 \rightarrow \text{excess } PbO + Pb_2(Zr_xTi_{1-x})_2O_6 \quad (1$$

Then, having obtained an intermediate (defect–pyrochlore) state which has an *identic* stoichiometry to that of the desired cubic perovskite, the final transition can be completed without any diffusion [13].

The crystallization kinetics are drastically changed with RTA processing, which utilizes a heating rate of about 100°C/s. The essential differences between CFA and RTA, as observed, e.g., by Hu *et al.* [4], are twofold:

1. In RTA, the defect–pyrochlore phase starts to form at a lower temperature, about 340°C, and the perovskite phase forms above 415°C.
2. With RTA, for all annealing temperatures above 450°C, the transformation to perovskite is very rapid, achieving complete transition with a soaking time of *only* one second.

Wu *et al.* [2] also investigated the RTA process as a function of heating rates. Their results supported the notion that an increase in heating rate reduces both the annealing temperature and time required for transformation to the perovskite.

Based on their observations, Hu *et al.* [4] suggested that in CFA, the complete transformation of the PZT film is a three-stage process. First, the amorphous film undergoes some type of relaxation process, which is strongly exothermic. The relaxed amorphous film then transforms to the metastable defect–pyrochlore phase. Finally, the film transforms to the stable perovskite phase. Because the heating is slow, the heat released in the relaxation of the amorphous film is slowly dispersed, and plays no role in the crystallization process. In RTA, however, the rapid increase in temperature means that defect–pyrochlore formation may be *bypassed.* Furthermore, the heat released in the relaxation of the amorphous film is not dispersed, but in fact aids in further nucleation and crystallization of the perovskite phase. The process is therefore self-promoting and this accounts for the fact that the films produced via RTA crystallize both faster and also at a lower temperature than in CFA. A schematic of their model is shown in Fig. 8 of their paper [4].

We now present an alternative model that is significantly different from the above picture. Specifically, we do not attempt to exclude the defect–pyrochlore phase from the transition path. Rather, we rely on enhanced *fluctuations*, viz. we drive the system towards equilibrium with rapidly elevated temperatures as in RTA.

3. The Model

The model we propose is described by a Landau–type free energy that is a functional of a single scalar field, $\psi = \{\psi_i\}$, where the subscript i denotes discrete sites on a $L \times L$ lattice, and $-\infty < \psi_i < \infty$:

$$\mathcal{F}[\psi] = \sum_i \{V(\psi_i) + \tfrac{1}{2}K(\nabla\psi_i)^2 + \tfrac{1}{2}\ell(\psi_i - \overline{\psi})^2\}. \quad (2)$$

The first term, $V(\psi_i)$, in the summation in Eq. (2), is a one-dimensional potential. Its precise form is discussed below. The second term is the discrete version of the squared gradient. Via the third term, a long–ranged interaction is also introduced, intended to mimic the elastic forces that arise from the presence of deformed inhomogeneities [14] in a mean-field manner [10]. Here, $\overline{\psi} = 1/L^2 \sum_j \psi_j$, is the spatial average of the order parameter, and ℓ is the strength of this interaction.

The potential $V(\psi_i)$ that we have considered has a three-well structure. For simplicity

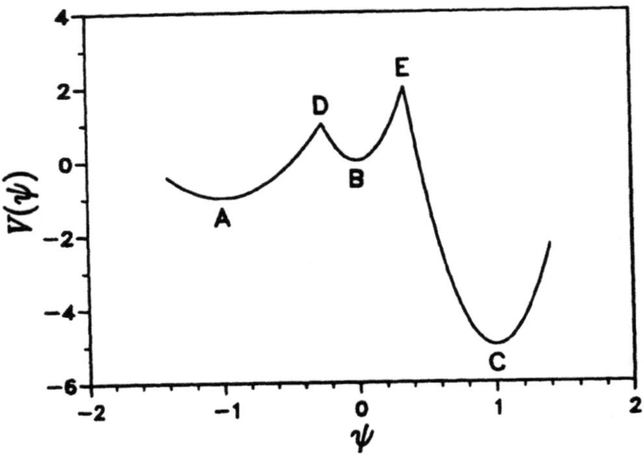

Fig. 1. A piecewise quadratic potential with three energy minima. The B well represents an amorphous phase, and all our simulations begin with the entire system localized in this well. The A well is taken to represent the defect–pyrochlore state, while the C well represents the cubic perovskite phase.

of implementation (especially in adjusting the well widths and barrier heights) we consider a piecewise quadratic form (Fig. 1). In particular, we are interested in the following conditions:

$$V_D < V_E \tag{3}$$

and

$$V_C < V_A < V_B. \tag{4}$$

The rationale for choosing this particular form of the potential is as follows. Equation (3) means that the energy barrier relevant to the transition from the phase B to the phase A is lower than that for the transition from phase B to phase C. The second condition, Eq. (4), defines phase C as the ground state, while phases A and B are metastable local energy minima. If, at some initial time, the system was in well B, this model represents two competing phases (A and C) into which an initial high energy phase (B) may transform. Because of the lower energy barrier, $V_D - V_B$ compared with $V_E - V_B$, initially more of the system will transform to A. However, the lower free energy of C means that eventually this will be the dominant phase. Physically, here we associate the wells at A, B, and C with the defect–pyrochlore, amorphous and perovskite phases, respectively. Then, Eq.(3) is consistent with the experimentally–observed [4] formation of the defect-pyrochlore in the initial stages of the annealing process, even though the perovskite structure is the more stable phase.

While the free energy $\mathcal{F}[\psi]$ determines the equilibrium state of the system, the dynamical evolution from an arbitrary initial configuration towards equilibrium can be generated b a Langevin equation:

$$\frac{d\psi_i}{dt} = -\Gamma\frac{\delta\mathcal{F}[\psi]}{\delta\psi_i} + \eta_i(t). \tag{5}$$

Here Γ is a kinetic coefficient and η_i is a Gaussian thermal noise with zero mean satisfying the ensemble average

$$\langle\eta_i(t)\,\eta_j(t')\rangle = 2\Gamma k_B T\delta_{ij}\delta(t-t'). \tag{6}$$

In Eq. (6), k_BT is the usual product of the Boltzmann constant and temperature. Note that in the present study, temperature is introduced into the model *solely* via the stochastic thermal noise — the Landau free energy (2) is temperature independent.

The simplest numerical procedure to solve Eq. (5) is given by the Euler method [15], which has been employed in many Langevin simulations. In this algorithm, time is discretized into steps of the size δt. After substituting the free energy $\mathcal{F}[\psi]$, as given by Eq. (2), into the Langevin equation Eq.(5), and rescaling the variables appropriately, we obtain

$$\psi_i(t+\delta t) = \psi_i(t) + \delta t\left\{-\frac{d\tilde{V}(\psi_i)}{d\psi_i} + \tilde{K}\sum_{\mathbf{a}}(\psi_{i+\mathbf{a}} - \psi_i) - \tilde{\ell}(\psi_i - \overline{\psi})\right\} + \sqrt{2\delta t}\xi_i(t), \tag{7}$$

where $\mathbf{a}$ is a nearest neighbour lattice vector, and $\xi_i(t)$ satisfies

$$\langle\xi_i(t)\,\xi_j(t')\rangle = \epsilon\delta_{ij}\delta_{tt'}. \tag{8}$$

Here, ϵ is the scaled temperature in dimensionless units. All other quantities, including the time t, the potential $\tilde{V}$ and the coupling constants $\tilde{K}$ and $\tilde{\ell}$ are also dimensionless. From this equation we see that the long–ranged interaction manifests itself as a field, dependent upon the average state of the entire system [10].

In order to mimic the experimental situation in our simulations, we must include the heating rate in our dynamical model. To this end, we let the scaled temperature ϵ increase linearly from zero to the annealing temperature, which we denote by ϵ_a, and then fix the temperature at this value for the remainder of the simulation. While this should lead to equilibrium expectation values determined solely by the annealing temperature, our simulations demonstrate that the competing intermediate metastable phases actually play an important role.

4. Results

We have carried out simulations on a variety of lattice sizes with periodic boundary conditions, as well as run many different thermal fluctuation force sequences, and have obtained converged values of the averaged order parameter $\langle\psi\rangle$ at each instant in time. Some typical results are described below — a more detailed account of our simulation data will be presented elsewhere.

When no long–ranged interaction is included ($\tilde{\ell} = 0$), evolutions similar to those shown in Fig. 2 are obtained. For this system one finds that although the very short time dynamics are a function of the heating rate, the equilibrium average is not, and thus the

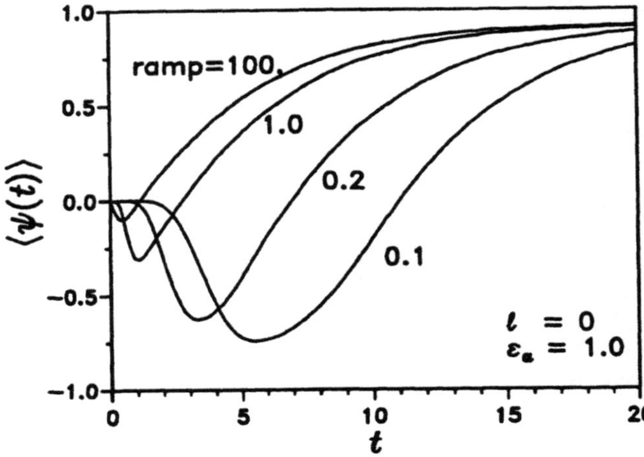

Fig. 2. The time evolution of the order parameter for $\ell = 0$ and $\epsilon_a = 1.0$, plotted for the various ramp rates shown. Evolutions of 40 different random force sequences have been used in determining the average value of the order parameter at each instant in time. We have chosen $(V_A, V_B, V_C, V_D, V_E) = (-1, 0, -5, 1, 2)$, and $(\psi_A, \psi_B, \psi_C) = (-1, 0, +1)$. Further, $\psi_B - \psi_D = 0.2$. The harmonic coupling between sites corresponds to $\tilde{K} = 0.1$. The inverse of the ramp rate shown is equal to the number of time units before the annealing temperature $\epsilon_a = 1$ is reached.

true equilibrium states are easily accessed. In no way are the experiments similar to this behaviour.

When long–ranged interactions are included ($\tilde{\ell} \neq 0$), we have determined the long–time averages of the order parameter as a function of both the annealing temperature and heating rate. This allows us to make a direct comparison to the measurements of the normalized X-rays intensities shown in Fig. 4 of Ref. [2], and to the dielectric constant vs. time data shown in Fig. 6 of Ref. [4]. We find that, as shown in our Fig. 3, for a change of heating rate of *only* a factor of two, there exists a range of ϵ_a over which the faster heating rate leads to a completed transition, whereas the slower heating rate leads to incomplete formation of the cubic perovskite phase. The larger the difference in heating rates, the more pronounced this difference becomes, and one can reproduce the numerical values of Ref. [4] with an appropriate choice of $\tilde{\ell}$. Clearly, these trends are similar to the experimental results of Refs. [2] and [4], indicating that our model is a good starting point for studies of the crystallization kinetic of PZT thin films via RTA.

5. Discussion

We now propose a simple explanation of our simulation results. Motivated by the elastic misfit problem, our model has included a long–ranged interaction. Its significance is that

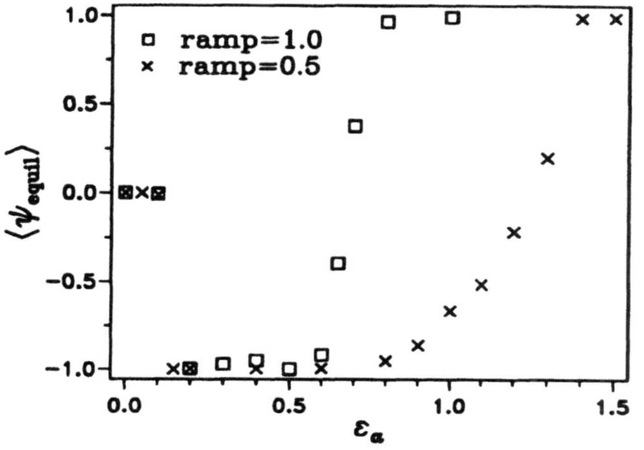

Fig. 3. The long–time asymptotic limit of the average order parameter $\langle \psi_{\mathrm{equil}} \rangle$ as a function of the annealing temperature, plotted for two ramp rates. The potential parameters are the same as those of Fig. 2, except that now $\tilde{\ell} = 5$, and averages are performed over 400 different random force sequences.

it imposes an effective bias–field that is proportional to the average order parameter $\langle \psi \rangle$. Consider first the results of Fig. 2 for which long–ranged interaction is absent ($\tilde{\ell} = 0$). Each curve in the figure demonstrates the competition between the metastable phase $\psi_A = -1$ and the true equilibrium state $\psi_C = +1$. As the ramp rate is reduced, it takes a longer time for thermal fluctuations to reach large enough amplitudes to overcome the large energy barrier to the ground state C. Therefore, initially, a larger fraction of the system is transformed to phase A, resulting in a more negative value of the average order parameter $\langle \psi \rangle$, as shown in the figure.

When the long–ranged interaction is included, a larger negative $\langle \psi \rangle$ produces a stronger negative bias, so that the system is much more likely to be trapped in the metastable state. This provides a clear explanation of why a higher annealing temperature is required for transformation to the stable state (Fig. 3). Physically, when either the defect–pyrochlore or perovskite phase nucleates in the material, elastic forces favour the transformation of the rest of the system to the same phase. In the present mean-field description of the long–ranged forces, the overall bias is towards the competing phase that initially exists in a larger fraction. Although the defect–pyrochlore is thermodynamically less favourable compared with the perovskite, kinetically it is more accessible from the amorphous phase. With a low heating rate, therefore, the relative ease of defect–pyrochlore formation results in the defect–pyrochlore being the dominant product phase up to a higher annealing temperature.

6. Conclusion

In an attempt to understand the underlying physics of RTA, we have performed Langevin simulations of the crystallization of PZT thin films based on a simple phenomenological model, and have obtained qualitative agreement with experiments, viz. a reduction in the required annealing temperature when a large heating rate is employed. Our results suggest that this drastic influence of heating rates on PZT crystallization is a consequence of the competition between the lowest energy crystalline state and a secondary metastable state, the competition being enhanced by long-ranged elastic strains. Here, the competing phases are the stoichiometrically identical perovskite and defect–pyrochlore structures. We expect that similar behaviours will also be observed in other systems in which competing crystalline phases exist.

In order to substantiate the validity of the above theoretical picture, new experimental data are required. An ideal experimental comparison would involve an in-situ determination of all structural units in the PZT thin film during the entire annealing process. In particular, one would like to establish whether or not the secondary defect–pyrochlore phase is present even under large heating rates, rather than being bypassed as suggested by Hu *et al.* [4]. However, such an analysis may be difficult to perform in view of the extremely short time-scales present in RTA processing. On the other hand, the model presented here may be refuted if a system can be found in which a secondary crystalline phase is absent but nevertheless follows the same drastic influence of the heating rates.

Lastly, it is clear that one should extend our use of the 1D Landau potential to something more relevant to the compounds employed in the formation of PZT. A possible extension is one based on an expansion of the free energy in terms of density–waves representing the two competing phases [16]. This will enable a closer comparison between theory and existing experimental data, and, in general, serve as a method by which the the usefulness of RTA with regard to a particular material's processing history may be determined theoretically.

Acknowledgments

We are grateful to Mike Sayer, Ellen Griswold, and David Barrow for helpful comments on the experimental details of this problem. This work was supported by the OCMR, an Ontario Centre for Excellence.

References

[1] C.V.R. Vasant Kumar, M. Sayer, R. Pascual, D.T. Amm, Z. Wu and D.M. Swanston, Appl. Phys. Lett. **58**, 1161 (1991).

[2] Z. Wu, R. Pascual, C.V.R. Vasant Kumar, D. Amm and M. Sayer, Mat. Res. Soc. Symp. Proc. **224**, 305 (1991).

[3] C.V.R. Vasant Kumar, R. Pascual and M. Sayer, J. Appl. Phys. **71**, 864 (1992).

[4] H. Hu, C.J. Peng and S.B. Krupanidhi, Thin Solid Films **223**, 327 (1993).

[5] For example, see K. Sreenivas *et al.* , Ferroelectric Thin Films, Vol.200, MRS Symp. Proc. 1990 (Ed. E.R. Myers and A.I. Kingon, Material Research Society 1990), p.255; Z. Wu and M. Sayer, Proceedings of the 8th International Symposium on Applications of Ferroelectrics (unpublished).

[6] See, e.g., R. Singh, J.Appl. Phys. **63**, R59 (1988); Rapid Thermal and Integrated Processing II, MRS Symp. Proc. 303 (1993).

[7] J. Chen, K.R. Udayakumar, K.G. Brooks and L.E. Cross, J. Appl. Phys. **71**, 4465, 1992.

[8] A. Okada, J. Appl. Phys. **49**, 4495 (1978).

[9] F.W. Martin, Physics and Chemistry of Glasses **6**, 143 (1965).

[10] P.B. Littlewood and P. Chandra, Phys. Rev. Lett. **57**, 2415 (1986); P. Chandra, Phys. Rev. A **39**, 3672 (1989).

[11] D.B. McWhan, G. Aeppli, J.P. Remeika and S. Nelson, J. Phys. C **18**, 1307 (1985); D.A. Neumann, D.B. McWhan, P.B. Littlewood, G. Aeppli, J.P. Remeika and R.G. Maines, Phys. Rev. B **32**, 1866 (1985).

[12] M. Klee, R. Eusemann, R. Waser, W. Brand and H. Vanhal, J. Appl. Phys. **72**, 1566 (1992).

[13] The details of the continuous crystallographic transformation path connecting the two crystalline phases will be discussed elsewhere.

[14] J.D. Eshelby, Proc. R. Soc. London Ser. A **241**, 376 (1957); A.C.E. Reid and R.J. Gooding, Phys. Rev. B **46**, 6045 (1992).

[15] O.T. Valls and G.F. Mazenko, Phys. Rev. B **34**,7941 (1986); *ibid*, **42**, 6614 (1990).

[16] S. Alexander and J. McTague, Phys. Rev. Lett. **41**, 702 (1978).

INDEX

www.ingramcontent.com/pod-product-compliance
Ingram Content Group UK Ltd.
Pitfield, Milton Keynes, MK11 3LW, UK
UKHW020132290926